A Blueprint for Greenhouse Gas Reduction

This comprehensive book details greenhouse gas (GHG) emissions across a range of sources, sectors, and countries, with particular emphasis on the United States and other developed nations. It presents an in-depth analysis, covering the roles of industries, policymakers, and regulatory frameworks in contributing to and mitigating GHG emissions. The authors discuss emission measurement and monitoring methodologies, supported by real-life case studies illustrating how corporations, policymakers, and industry leaders can implement responsible decarbonization strategies. Additionally, it covers the critical verification aspect of enforcing climate policies and regulations.

Features

- Provides a detailed sector-wise breakdown of GHG emissions across energy, transportation, industry, and agriculture.
- Explains the scientific basis of emissions and their regulatory frameworks governing them under global agreements such as the Paris Agreement and the U.S. EPA policies.
- Presents recommendations for energy-efficient strategies and key actions that industries and policymakers can adopt to reduce GHG emissions.
- Highlights AI, big data, and IoT as tools for carbon monitoring and emission tracking.
- Offers guidance for professionals and scholars on estimating emissions and implementing reduction methods at both organizational and policy levels.

This book is essential for environmental and energy professionals, policymakers, corporate sustainability officers, researchers, and industry leaders engaged in emissions reduction, regulatory compliance, and climate policy. It is also a valuable resource for EHS personnel, facility administrators, students, and faculty studying and teaching the environmental impacts of GHG emissions.

A Blueprint for Greenhouse Gas Reduction

Confronting the Climate Crisis

Rengasamy Kasinathan and Arun Kanchan

CRC Press
Taylor & Francis Group
Boca Raton London New York

CRC Press is an imprint of the
Taylor & Francis Group, an **informa** business

Designed cover image: Shutterstock

First edition published 2026
by CRC Press
2385 NW Executive Center Drive, Suite 320, Boca Raton FL 33431

and by CRC Press
4 Park Square, Milton Park, Abingdon, Oxon, OX14 4RN

CRC Press is an imprint of Taylor & Francis Group, LLC

ISBN: 978-1-032-76692-8 (hbk)
ISBN: 978-1-032-76700-0 (pbk)
ISBN: 978-1-003-47966-6 (ebk)

DOI: 10.1201/9781003479666

Typeset in Times
by codeMantra

To my mother, Mrs. Thengani Ammal, father Mr. Rengasamy, mother-in-law Mrs. Maragathavalli, father-in-law Mr. Deivasigamani who always reminded me that words have the power to change the world. My beloved wife Dr. Sumathi Kasinathan, MD., has been my most profound inspiration. My daughter Dr. Sushma Kasinathan, MD., never fails to fill my heart with joy, and the light of my life is my son Shriram Kasinathan. They all never stopped believing in me!

Dr. Rengasamy Kasinathan

For my parents, who shaped my path, and my family, who walked it with me.

Dr. Arun Kanchan

Contents

Foreword

State University of New York
COLLEGE OF ENVIRONMENTAL SCIENCE AND FORESTRY
Department of Chemical Engineering

A BLUEPRINT FOR GREENHOUSE GAS REDUCTION: CONFRONTING THE CLIMATE CRISIS

Greenhouse gas emissions have risen drastically during the industrial era, and modern scientific studies have identified them as the most likely primary driver for global climate change in this century. The consequences of such projected climate changes could be catastrophic unless mitigating actions are taken.

This book, *A Blueprint for Greenhouse Gas Reduction*, is a critical contribution to our understanding of the seriousness of the challenge posed by climate change, the roles of different institutions, industry, and regulation in addressing this, and the importance of incorporating sustainability and stewardship of natural resources into our future.

Besides its solid contribution in summarizing academic literature on the carbon footprint and climate change, it also includes a call to action. This includes a succinct and clear approach to the urgent need for decarbonization and provides maps for achieving net-zero emissions.

A notable feature of this book is its clear exposition of carbon footprints, illustrating the complex interplay between greenhouse gas emissions and global warming. It discusses how natural processes and human activities impact the climate. Nuanced discussions of climate change processes follow. In successive chapters, the authors explore several input factors contributing to greenhouse gas emissions, analyzing industries sector by sector. Agricultural methods and the contributions of deforestation are also included. The authors recommend a holistic response to the challenge—one that takes into account the multifarious facets of climate change.

Climate change is projected to have far-reaching social and economic impacts. Environmental degradation affects public health and food security in real time. This book discusses how ecosystems, economies, and communities are intertwined, urging policymakers, business leaders, and citizens alike to adopt sustainable practices.

One of the standout features of this work is its practical approach to quantifying greenhouse gas emissions. Chapters dedicated to understanding Scope 1, 2, and 3 emissions provide readers with the tools to measure, report, and ultimately reduce carbon outputs. This is valuable for environmental experts, corporate decision-makers, and those involved in enforcing or creating of compliance strategies. Guidance is offered on robust data collection, and the methodologies for accurate measurement further reinforce the book's commitment to actionable, science-based solutions.

The role played by regulatory bodies, such as the U.S. Environmental Protection Agency, in shaping climate policy is explored. Strategic governance can facilitate the transition to a low-carbon economy, and in a global context, comparing emission patterns across nations underscores that climate change has a much broader impact and requires international cooperation.

Technology and innovation are addressed in later chapters, providing some of the solutions which are being proposed for carbon management. From carbon capture and storage to renewable energy innovations, artificial intelligence, and big data analytics, the book provides a forward-looking perspective on how technological advancements can drive down emissions while stimulating economic growth. The discussion on decarbonization is particularly useful, introducing the reader to the technical aspects of reducing emissions but also to the socioeconomic opportunities that arise from green investments.

A special feature of this book is the deep and insightful case studies of carbon reduction strategies. These real-world examples serve as both inspiration and a blueprint for industries and policymakers striving to balance operational efficiency with environmental stewardship.

A Blueprint for Greenhouse Gas Reduction offers more than an analysis of climate change—it provides a strategic vision for an inclusive, sustainable global economy. It challenges us to rethink our energy management practices, to adopt innovative technologies, and to commit to policies that safeguard both the environment and our shared future.

I invite you to engage deeply with the insights presented in this book. If you are involved in EHS, facilities administration, or are a student interested in the environmental challenges of our time, this book provides you with thought-provoking and insightful analyses and much data to guide your thinking and inspire action.

Bandaru V. Ramarao, PhD Fellow AIChE
Professor & Department Chair
Syracuse, New York
February 24, 2025

Preface

This work provides a comprehensive analysis of carbon footprints (CFPs) and underscores the vital requirement for decarbonization to tackle the issues of climate change. It starts by presenting the concept of CFP, underlining its importance in relation to global warming and elaborating on the carbon cycle, along with the disruptive effects of greenhouse gas (GHG) emissions (Chapter 1). The text explores the multiple contributors to greenhouse gas emissions, highlighting human behaviors such as the use of fossil fuels, land clearing, and agricultural methods, alongside natural phenomena. A thorough analysis of GHG emissions across different sectors is provided, stressing the need for a comprehensive approach to mitigation (Chapter 2).

Chapter 3 examines the important human and ecological impacts of climate change, pointing out the crucial requirement to adopt a sustainable economic approach. It outlines the severe environmental repercussions, such as ecosystem disturbances, risks to food security, and the rise in natural disasters. The human impact is also discussed, focusing on public health emergencies, economic inequalities, and social injustices exacerbated by climate change. Chapter 4 offers a thorough scientific and technical analysis of climate change, featuring a detailed look at the Earth's climate system, its influential factors, and the alterations that have been observed. It traces the development of climate science, highlighting key milestones in recognizing human influence, from early theories to modern studies.

A substantial part of the book is devoted to comprehending and quantifying GHG emissions, specifically Scope 1, 2, and 3 emissions as detailed in Chapter 5. It offers comprehensive guidance on how to calculate and report emissions for each scope, highlighting the significance of precise data collection and strong methodologies. Chapter 6 investigates the role of the U.S. Environmental Protection Agency (EPA) in climate initiatives, outlining its regulatory strategies, voluntary programs, collaborations, and educational efforts aimed at reducing emissions and fostering sustainability.

The book transitions in Chapter 7 to the worldwide context of carbon emissions, analyzing the ranking of countries from the highest to the lowest emitters. It highlights emission patterns, sector contributions, and the factors driving emissions in leading countries, emphasizing the necessity for a unified global approach. Furthermore, Chapter 8 investigates the intricacies of climate change regulations across various levels, from local to global. It analyzes the participation of diverse groups, including local authorities, international bodies, and non-profits, in crafting and enforcing climate policies.

In this chapter (Chapter 9), the book examines the latest in technology and methods for minimizing carbon emissions, concentrating on novel solutions including carbon capture, utilization, and storage (CCUS), strides in renewable energy, and enhancements in energy efficiency. It also addresses the significance of carbon offsetting, artificial intelligence, electrification, green hydrogen, and big data in mitigating carbon emissions. The text further examines waste valorization, biofuels, bio-based polymers, and emission monitoring as essential elements of a holistic carbon reduction strategy.

It is imperative to recognize the pressing need to redefine energy management to tackle the environmental and economic challenges posed by traditional energy sources. Chapter 10 delves into the drawbacks associated with fossil fuels and nuclear power, highlighting their roles in climate change, resource exhaustion, and environmental contamination.

Chapter 11 outlines the complex journey of decarbonization, viewed as a vital strategy for securing net-zero emissions and tackling climate change. It analyzes the carbon cycle and the repercussions of human endeavours on greenhouse gas emissions, emphasizing the urgent requirement for systemic transformations across energy production, transportation, industrial practices, and land management. The approaches suggested involve techniques for offsetting carbon emissions, efforts towards carbon sequestration, the employment of renewable energy innovations, and the rollout of advanced solutions like carbon capture and storage. Additionally, it discusses the economic prospects linked to decarbonization, such as job creation in green sectors and advancements in renewable energy systems, while addressing challenges like financial implications and the need for workforce reskilling.

Chapter 12 spotlights the newest Intergovernmental Panel on Climate Change (IPCC) findings, emphasizing the crucial demand for timely and significant responses. This part delves into the critical role of global teamwork, information exchange, and climate negotiations in addressing the worldwide climate emergency, along with frameworks like the Conference of Parties (COP) and their impact on developing climate pacts, such as the Kyoto Protocol and the Paris Agreement.

Even though maintaining regulatory compliance can demand additional personnel and occasionally create difficulties for facility operators, industries, or countries collectively, the rewards are considerable, encompassing the fulfilment of ESG targets, CSR pledges, and mitigation of global warming. Chapter 13 delves into the rewards of compliance and climate initiatives, emphasizing their economic, environmental, and social ramifications. It investigates economic motivators such as carbon taxes, cap-and-trade mechanisms, and emission reduction credits as catalysts for promoting sustainable practices and realizing net-zero emissions. This chapter highlights the significance of environmental assessments, compliance monitoring systems, and innovations like real-time tracking in guaranteeing adherence to regulatory standards.

Finally, case studies presented in Chapter 14 demonstrate how key greenhouse gas–emitting industries have started to adopt carbon footprint reduction technologies while simultaneously optimizing their operational costs.

Each chapter concludes with a series of review questions to help readers quickly grasp the essential points. Thus, this work is a vital reference for environmental experts, academic analysts, facility directors, ESG leaders, and compliance bodies, supplying frameworks and strategies to understand and attain sustainability goals.

About the Authors

Dr. Rengasamy Kasinathan is a graduate of Annamalai University (one of the oldest engineering schools in India), holding a Bachelor of Science Degree in Chemical Engineering (1980). In 1987, he earned a Master of Science degree in Engineering/Atmospheric Science from New York University, followed by 1.5 years of PhD work-studies. In 2017, Dr. Kasinathan received his Ph.D. in Bioprocess Engineering from the State University of New York – College of Environmental Science and Forestry, and in 2018, he received his MBA from Fayetteville State University of NC. He is a licensed PE in four states (NY, NC, FL, and CA), and a Fellow of Institutions of Environmental Sciences (FIEnvSc).

From 1980 to 1985, he worked for BHEL (Bharat Heavy Electricals, Ltd), a boiler, heavy machinery, valve and compressor manufacturing company in India. While there, Dr. Kasinathan researched coal combustion and gas clean-up technologies. Impressed by the emerging clean gas/scrubber technologies, from 1985 to 1989, at NYU's research lab funded by USDOE, he handled a flue gas analysis system where he conducted research to study the influence of (Ka+Na) alkali in the boiler flue gas stream. As one of the key members of the NYU research team, under the auspices of the DOE grant, they evaluated the performance of various fluidized bed combustion systems and tested several air pollution control (scrubber) technologies, such as wet and dry scrubbers, baghouses, etc., to improve boiler efficiencies in an environmentally acceptable manner, i.e., low (NOx and SOx). Due to limited research funds, DOE shut down coal combustion research, thus disabling several PhD works at NYU, which led Dr. Kasinathan to pursue other APC-related endeavors. This is where he came across medical waste, beach wash-ups, and other related environmental issues rampant in the late 80s.

Being well trained in combustion with the main focus on the environment and air pollution and the fundamentals of the 3 R's (Reduce, Reuse, Recycle), Dr. Kasinathan decided to dedicate his career to the environmental engineering aspects of medical waste-related incineration air, waste, and water. Learning the respective Code of Federal Rules and State and local Environmental Conservation laws, he joined a firm that specialized in this area and became an expert in the design and development of medical waste incinerators, air dispersion modeling, multipath health risk analysis, and optimum stack designs, which were the key components to get any projects permitted in the United States. Further, he mastered the specifics of the impact of PCDD/PCDF (Dioxins and Furans) from the medical waste incineration (MWI) systems, which must be thoroughly studied to demonstrate that no adverse impact occurs. To this end, many states promulgated MWI rules. This involved a very complex and systematic permutation and combination of multiple parameters of interactive calculations. Every single project, from conception to completion, was a challenge. Nevertheless, with the fundamental research knowledge of combustion and environmental impact assessment, Dr. Kasinathan got many plants permitted throughout the nation.

Being in charge of air quality/regulatory compliance, he was able to successfully execute projects in more than 20 states, serving 400+ hospitals. Almost all the projects involve emissions (criteria/hazardous) inventory, waste management, reduction methods, air quality dispersion modeling, and environmental impact analysis. Some supplemental environmental projects included emissions offset and voluntary emissions mitigation efforts. This knowledge, along with progressive learning of the rules and adopting them in the field with state-of-the-art technologies, saved thousands of dollars for clients and ultimately created a cleaner environment around the facilities (hospitals) and surrounding receptors (neighbors).

Encouraged by the overwhelming client acceptance, Dr. Kasinathan founded EESPC in 1998, now serving more than 150 clients in the New York/New Jersey Metropolitan area. Air pollution control, emissions assessment, mitigation, monitoring, recording, and reporting are key to mitigating pollution. Unless the Environmental/Regulatory community enforces their rules, compliance will not be fully achieved. To this end, both communities need to be participating in effective communication. Short of these, violations, litigations, and hefty penalties will cause adverse economic impacts. Based on his more than 30 years of experience, Dr. Kasinathan states that the core topics of environmental ethics, policy, and governance, enforcing authority's roles and accountability of polluters, resource management, public participation, mass media, and human behavior are the contributing entities for a cleaner and stable environment. Dr. Kasinathan is currently providing consulting services to many institutions and industries concerning GHG emission estimation, reporting, and compliance status, including mitigation measures for decarbonization to accomplish the required CFP mandate.

Dr. Kasinathan has written NCEE P.E. Board Exam Question, he is a Certified USEPA Lead-based Paint Risk Assessor, Diplomate of the American Academy of Environmental Engineers (DEE), Certified Energy Manager (CEM) by the American Association of Energy Engineers, and LEED AP BD+C. He holds memberships in the American Institute of Chemical Engineers (AIChE), Air and Waste Management Association (A&WMA), Association of Professional Industrial Hygienists (APIH), and American Academy of Environmental Engineers (AAEE). He is also a Fellow of the Institution of Environmental Sciences (FIEnvSc), UK.

His published works include *Design and Development of Rotary Kiln Gasifier, A Feasibility Study,* 1984; *Particulate and Alkali Capture from PFBC Flue Gas Utilizing Granular Bed Filter (GBF), 10th International Conference of PFBC,* 1989; *Environmental and Health Risk Analysis of Medical Waste Incinerators Employing State of the Art Emission Controls,* June 1991; and *Effect of Waste Stream Modification and Other Factors on Ground Level Concentrations Resulting from Medical Waste Incineration Compared with Acceptable Air Quality*, June 1992.

Dr. Arun Kanchan, Ph.D., FIEMA, has over 35 years of experience in environmental and climate change consulting. His work spans greenhouse gas accounting, social and ecological impact assessments, sustainability strategies, and financial analysis across various sectors, including oil and gas, petrochemicals, cement, electric utilities, and manufacturing.

He holds a Ph.D. in Urban Innovation and Sustainability from the Asian Institute of Technology, where his research focuses on the co-benefits of climate change mitigation, air quality improvement, and public health in rapidly growing industrial corridors, including the Delhi-Mumbai Industrial Corridor. Dr. Kanchan also holds an MBA from Northwestern University's Kellogg School of Management, an M.S. in Engineering Management from NJIT, and a B.Eng. in Mechanical Engineering from the Manipal Institute of Technology.

Dr. Kanchan has worked across the United States, Saudi Arabia, Qatar, Kuwait, and Asia, advising governments and industries on air quality management, emissions reduction, and sustainable infrastructure. His consulting practice has contributed to national-scale planning, regulatory compliance, and stakeholder engagement efforts in diverse environmental and economic contexts.

In addition to his consulting work, Dr. Kanchan is active in academic teaching and capacity-building programs focused on urban sustainability, climate-health linkages, and environmental governance. He has delivered training workshops for government agencies and technical audiences across the United States, Kuwait, Qatar, Saudi Arabia, and the United Arab Emirates. His peer-reviewed publications appearing in journals such as *Land* and conference presentations at international forums, including the Society of Petroleum Engineers and the Royal Commission of Jubail, reflect a commitment to applying rigorous, interdisciplinary research to real-world environmental challenges.

A Fellow of the Institute of Environmental Management and Assessment (FIEMA, UK) and the Asia Pacific Initiative for Rural Impact (AIRI), Dr. Kanchan also supports community-focused climate resilience and nature-based solutions across the ASEAN region. His work contributes a practical, cross-regional perspective to this volume, grounded in science, policy, and implementation.

Acknowledgments

We sincerely thank Dr. Nishanta Sharma, Michelle Hanks, and Priyanka Dave for their time, patience, and invaluable insights. Their contributions were crucial in shaping this book and ensuring its accuracy.

1 Introduction to Carbon Footprint

This chapter provides a comprehensive introduction to the concept of carbon footprint (CFP), emphasizing its significance in the context of global warming and climate change. It delves into the intricacies of the carbon cycle, elucidating how greenhouse gas (GHG) emissions disrupt this natural process and contribute to adverse environmental effects. This chapter explores the reasons behind the growing concern over CFP, highlighting its impact on climate change, ecosystems, economies, and human well-being. It also discusses the role of various sectors, including energy, agriculture, and industry, in contributing to GHG emissions. Additionally, this chapter examines global efforts to mitigate climate change, referencing international agreements like the Kyoto Protocol and the Paris Agreement, and emphasizing the importance of reducing CFP to achieve a sustainable future.

1.1 WHY SHOULD WE BE CONCERNED ABOUT CFP?

CFP measures the amount of carbon dioxide and other carbon compounds, including GHG, emitted due to natural or anthropogenic factors, and is also called a GHG footprint. The Kyoto Protocol to the United Nations Framework Convention on Climate Change, 1997, listed carbon dioxide (CO_2), methane (CH_4), nitrous oxide (N_2O), hydrofluorocarbons (HFCs), perfluorocarbons (PFCs), and sulfur hexafluoride (SF_6) as GHG inventory. During the 2011 Durban negotiations, nitrogen trifluoride (NF_3) was added to the GHG. The Third Assessment Report of the Intergovernmental Panel on Climate Change (IPCC) introduced the concept of CO_2 equivalent (CO_2-eq) (UNFCCC, 1997).

1.1.1 The Carbon Cycle and GHG Emissions

The carbon cycle is a fundamental Earth system process involving the exchange of carbon among the atmosphere, biosphere, geosphere, and oceans. This cycle maintains a balance of carbon essential for sustaining life and climate stability. In the atmosphere, carbon primarily exists as CO_2, which plants absorb during photosynthesis to create organic matter while releasing oxygen. In the biosphere, plants and marine organisms use CO_2 for photosynthesis, providing food for animals and, through respiration, releasing CO_2 back into the atmosphere.

When plants and animals die, decomposers break down their organic matter, releasing CO_2 and CH_4 into the atmosphere or soil. In the geosphere, carbon is stored in fossil fuels and carbonate rocks for millions of years and can be released into the atmosphere through volcanic eruptions. The oceans absorb CO_2 from the atmosphere, where it dissolves in seawater and forms bicarbonate and carbonate ions. Marine organisms use CO_2 for photosynthesis, and their remains can sink to the ocean floor, becoming part of the sediment.

GHGs disrupt this natural carbon cycle significantly. Burning fossil fuels for energy releases large amounts of CO_2 stored in the geosphere over millions of years, rapidly increasing atmospheric CO_2 levels. Deforestation for agriculture, urban development, and logging reduces the number of trees available to absorb CO_2 and releases stored carbon from trees and soil into the atmosphere. Agricultural practices, urbanization, and other land-use changes disturb the soil, releasing stored carbon as CO_2 and CH_4, with practices like rice cultivation and livestock farming further contributing to methane emissions. Industrial processes, such as cement production, release CO_2 and other GHGs, contributing significantly to atmospheric carbon levels. Additionally, increased atmospheric CO_2 levels lead to

DOI: 10.1201/9781003479666-1

higher CO_2 absorption by oceans, resulting in ocean acidification that negatively impacts marine life, particularly organisms that rely on calcium carbonate for their shells and skeletons.

These disruptions cause several adverse effects, including global warming, ecosystem imbalance, and ocean acidification. Elevated levels of atmospheric CO_2 and other GHGs enhance the greenhouse effect, leading to global temperature increases and climate change. Altered carbon dynamics affect ecosystems and biodiversity, potentially causing the loss of species and habitats. Ocean acidification impacts marine life, especially organisms crucial to oceanic food webs and carbon sequestration processes.

The carbon footprint is a critical measure for addressing climate change, sustainable development, and environmental impact at various levels, from individual activities to international trade (Ya-Min & Yong-Sheng, 2014; Caro, 2019). The concern over CFP has become increasingly paramount in the contemporary world due to its pivotal role in contributing to global warming and climate change. CFP contributes to climate change, which poses significant risks to ecosystems, economies, and human well-being. The effects of climate change, exacerbated by carbon emissions, are already being felt globally, with direct experiences influencing individuals' behavioral intentions toward sustainable consumption (Ding, 2017). CFP causes inefficiencies and adverse environmental impacts across various sectors, such as agriculture, energy, transportation, and waste management (Font Vivanco et al., 2024; Huang et al., 2024; Huo et al., 2024).

1.1.2 Greenhouse Gases: Heating Our Planet

The greenhouse effect is a natural process where Earth's atmosphere traps some of the Sun's energy, warming the planet, as shown in Figure 1.1. While a third of the Sun's energy is reflected into space, the remaining two-thirds is absorbed by Earth's surface and atmosphere. Earth then radiates energy

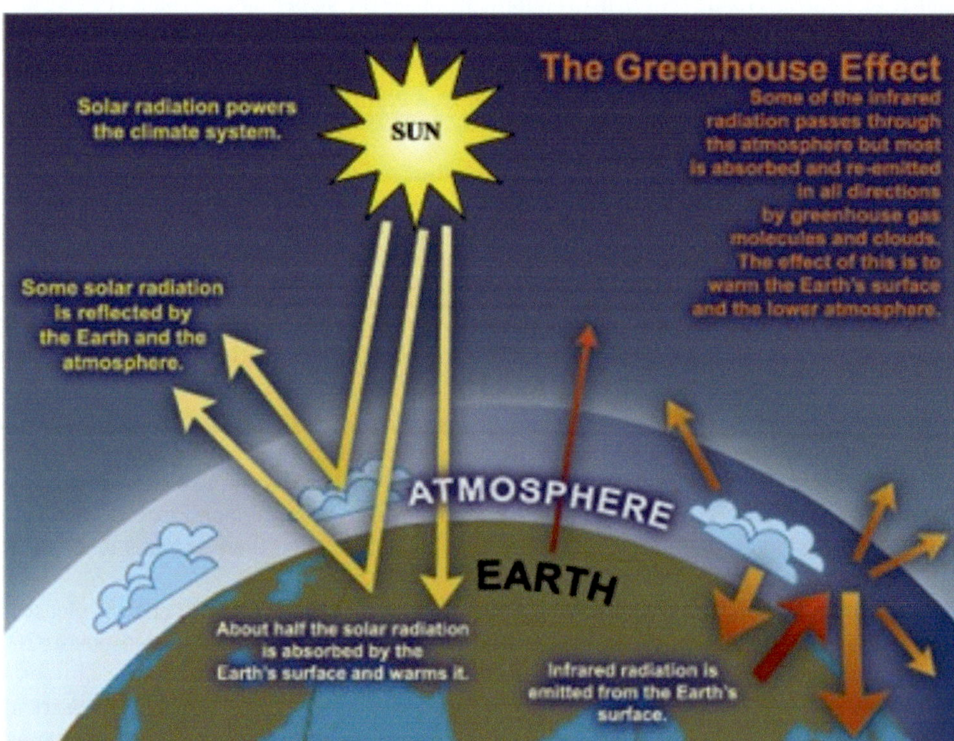

FIGURE 1.1 An idealized model of the natural greenhouse effect. (Source: IPCC, 2007 https://www.global-climate-change.org.uk/6-1.php.)

back in the infrared spectrum, which is absorbed and re-radiated by atmospheric gases, warming the planet. This effect, essential for life, is intensified by human activities like burning fossil fuels and deforestation, leading to global warming. Key greenhouse gases include water vapor, carbon dioxide, methane, and nitrous oxide. Feedback mechanisms, such as increased water vapor from warming, further amplify this effect. Clouds also play a significant role by both absorbing infrared radiation and reflecting solar energy, with their overall impact on climate being an area of active research.

The Climate Change 2023 Synthesis Report (IPCC, 2023) illustrates the causal chain from emissions to atmospheric heating effects since 1850 AD. There has been a rapid increase in GHG emissions, which include CO_2 from fossil fuel combustion and industrial processes (CO_2-FFI); net CO_2 from land use, land-use change, and forestry (CO_2-LULUCF); CH_4; N_2O; and fluorinated gases (HFCs, PFCs, SF_6, and NF_3). This rise in emissions has led to heightened atmospheric concentrations of these gases, contributing to a global surface temperature increase of approximately 1.1°C since 1850–1900.

Comparisons to natural warm periods, such as those occurring 6,500 and 125,000 years ago (The Last Interglacial), driven by slow orbital variations, underscore the significance of current warming. Formal studies employing detection and attribution methods indicate that all observed warming from 1850–1900 to 2010–2019 is human-caused, based on a breakdown of factors including GHG concentrations, aerosols, ozone, land-use changes, solar and volcanic influences, and internal climate variability.

Humans produce GHG emissions from burning gasoline when driving, burning oil or gas for home heating, or using electricity from coal, natural gas, and oil. It varies among individuals depending on their location, habits, and personal choices. Globally, CFP is highly concentrated in a small number of dense, high-income cities and affluent suburbs; 100 cities drive 18% of global emissions, which means concerted action by a small number of local mayors and governments has the potential to reduce national total carbon footprints significantly (Moran et al., 2018).

A significant increase in the CFP associated with global material production is likely due to economic growth and infrastructure development. The CFP of global material production increased by 120% from 5 billion metric tons CO_2-equivalent ($GtCO_2e$) in 1995 to 11 Gt in 2015, primarily driven by a substantial rise in investments, particularly in emerging economies like China, resulting in a shift toward capital-intensive development projects and infrastructure (Hertwich, 2021).

Coal is the dominant source of CO_2 emissions related to electricity generation. One pound (454 g) of CO_2 gas occupies 0.2426 m^3 (8.566 ft^3, 64 US gallons, 243 L). It would fill a cube 62.4 cm high (24.6 inches) or a sphere 77.4 cm across (30.5 inches). CO_2 emitted per gallon of gas is about 8,887 g. Every gallon of gasoline burned creates about 8,887 g of CO_2, whereas one (1) gallon of diesel creates about 10,154 g of CO_2. Diesel cars tend to have lower volumetric fuel consumption figures than comparable gasoline vehicles. However, the benefit of CO_2 emissions is significantly lower, as the combustion of 1 L of diesel fuel releases approximately 13% more CO_2 than the same amount of gasoline fuel.

CO_2 cannot be used as a fuel because it is fully oxidized and cannot be burned (oxidized) further. However, CO_2 is used in fire extinguishers precisely because it does not burn but can put out fires by displacing air from the fire.

Fossil fuel is the primary source of CO_2. Further, CO_2 is also emitted from direct human-induced impacts on forestry and other land use, such as deforestation, land clearing for agriculture, and soil degradation. Factory farming produces GHGs throughout the 'supply chain.' For example, forest clearance to grow crops and rear the animals reduces vital carbon 'sinks' and releases gases previously stored in the soil and vegetation. Factory farming also requires large amounts of energy to function. It isn't just to rear the animals but also to grow the vast feed they need. According to a study by The Royal Society, feed is the dominant energy user, taking around 75% of the total energy required (Woods et al., 2010). Gases, including CH_4 and N_2O are also produced in significant quantities and are released through various sources, including animal waste and fertilizer use.

Livestock farming produces 37% and 65% of global CH_4 and N_2O emissions, respectively (Steinfeld et al., 2006). Both gases are much more potent than CO_2.

Climate change is already harming food production, and these impacts are projected to increase over time, with potentially devastating effects (Jones, 2011). Higher temperatures, for example, could place further stress on water-scarce regions and make it harder to rear animals and grow food crops. According to the Convention on Biological Diversity, climate change may affect plant growth and production by promoting the spread of pests and diseases, increasing exposure to heat stress, and encouraging soil erosion due to stronger winds. The more CO_2 we release into the atmosphere, the more the Earth's temperature rises ("Summary for Policymakers," 2023), as illustrated in Figure 1.2. Studies show a near-linear relationship between these two factors. This means limiting the total CO_2 emitted (cumulative emissions) is crucial for controlling global warming. Reaching net-zero emissions, where we remove as much CO_2 as we emit, stabilizes temperatures.

Facts of GHG in the United States

1. The average CFP for a person in the USA is 16 tons, one of the highest rates in the world.
2. Burning fossil fuels for electricity, heat, and transportation is the biggest cause of carbon footprint in the USA. The EPA tracks total U.S. emissions by publishing the Inventory of U.S. Greenhouse Gas Emissions and Sinks.
3. Texas is responsible for the most carbon emissions in the USA, at 706.5 million metric tons.
4. Provo, Utah, had the largest CFP, averaging 10.55 metric tons of CO_2 per household per year.

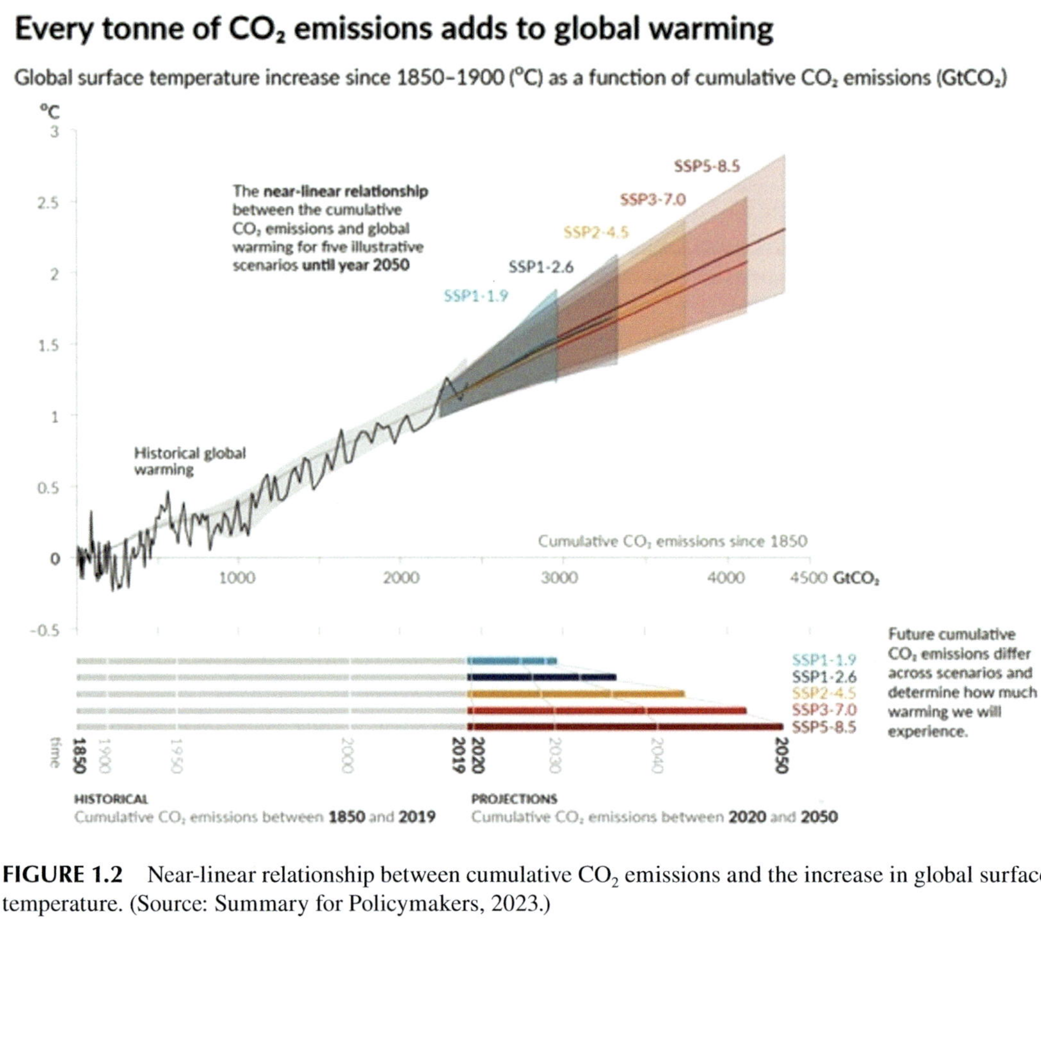

FIGURE 1.2 Near-linear relationship between cumulative CO_2 emissions and the increase in global surface temperature. (Source: Summary for Policymakers, 2023.)

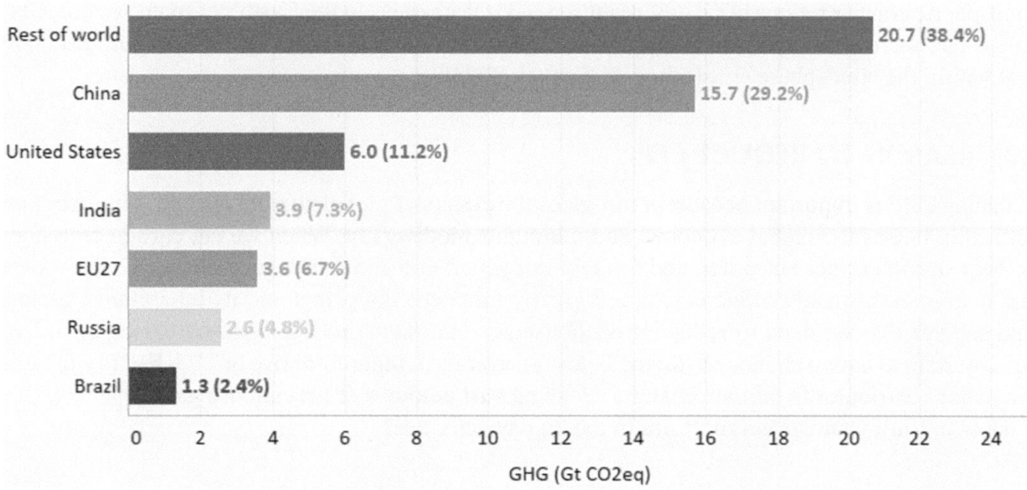

FIGURE 1.3 GHG emission highlight. (Source: Crippa et al., 2023.)

Data from the Joint Research Centre (JRC), the European Commission's science and knowledge service, shows a worrying trend in GHG emissions (Crippa et al., 2023). Despite a brief dip in 2020, emissions in 2022 surpassed 2019 levels by a significant 2.3%. Fossil fuels remain the primary culprit, responsible for over 70% of these emissions. Several countries stand out as major emitters, with China, the United States, India, the European Union (EU27), Russia, and Brazil collectively contributing over 60% of global GHG emissions. Interestingly, these same countries account for more than half of the world's population (50.1%), economic output (61.2% of GDP), and fossil fuel usage (63.4%). While the EU27 has reduced emissions by 27% compared to 1990, emissions from China, the US, and India climbed in 2022. India's increase was the most significant, estimated at around 5%, as shown in Figure 1.3.

Most of the GHG pollution comes from just a few countries: China, for example, generates around 30% of all global emissions, while the United States is responsible for almost 14%. In the ranking below, you can find the ten countries that produce the most emissions, measured in millions of tons of CO_2 in 2019 (ClimateTrade, 2021):

1. China, with more than 10,065 million tons of CO_2 released
2. United States, with 5,416 million tons of CO_2
3. India, with 2,654 million tons of CO_2
4. Russia, with 1,711 million tons of CO_2
5. Japan, 1,162 million tons of CO_2
6. Germany, 759 million tons of CO_2
7. Iran, 720 million tons of CO_2
8. South Korea, 659 million tons of CO_2
9. Saudi Arabia, 621 million tons of CO_2
10. Indonesia, 615 million tons of CO_2

1.1.3 Climate Feedback Loops

GHG emissions and climate feedback loops are central to understanding and addressing climate change. The relationship between GHG emissions and climate change forms a vital feedback loop, where climate change influences GHG emissions and vice versa (Wang et al., 2021; Bhatti et al., 2024). Human activities, such as industrialization and deforestation, have led to increasing atmospheric concentrations of GHGs, resulting in large changes to the Earth's climate system (Cloy, 2017). The increased concentrations of GHGs contribute to climate change by trapping excessive heat within the atmosphere (Easterling & Kunkel, 2023).

1.2 REASON TO REDUCE CFP

Reducing CFP is important because it mitigates the effects of global climate change, improves public health, boosts the global economy, and maintains biodiversity. When we cut carbon emissions, we help ensure cleaner air, water, and food for our generation and future generations. We now know that man-made climate change is real and greatly threatens the planet and its inhabitants. Current data suggest that we need to reduce greenhouse gas emissions in developed countries by at least 80% by 2050 to have a chance of staying below an average temperature rise of 2°C. Factory farming is a major contributor to climate change, releasing vast amounts of greenhouse gases.

How to limit your carbon footprint in day to day activities?

- Consume local and seasonal products (forget strawberries in winter)
- Limit meat consumption, especially beef
- Select fish from sustainable fishing
- Bring reusable shopping bags and avoid products with excessive plastic packaging
- Make sure to buy only what you need, to avoid waste.

1.2.1 Facts on Carbon Reduction

1. What plant takes in the most carbon dioxide?

   Bamboo absorbs five times more greenhouse gases and produces 35% more oxygen than an equivalent volume of trees. It has a very important CO_2 retention capacity since one hectare of bamboo grove can capture up to 60 tons of CO_2 each year.
2. What house plant removes the most CO_2?

   The most effective plant for absorbing carbon dioxide (CO_2) and reducing its carbon footprint is a Prayer plant, a member of the Calathea family. The prayer plant took in 14.4% of the total CO_2 in the chamber, which is about 0.0003 kg.
3. What is the best absorber of CO_2?

   Lithium hydroxide is the best chemical absorbent for CO_2.

According to the IPCC glossary, "carbon neutrality" and "net zero CO_2 emissions" are closely related concepts. Carbon neutrality balances the amount of CO_2 released into the atmosphere and the amount removed. It can be applied globally or at smaller scales, such as countries, regions, or even individual companies. Net-zero CO_2 emissions specifically focus on balancing CO_2 emissions under an entity's direct control (e.g., a nation or company) and the CO_2 removed from the atmosphere.

Achieving carbon neutrality is significant, but it focuses on a company's direct operations (Scope 1) and purchased energy (Scope 2). It allows for offsetting emissions elsewhere, which doesn't directly reduce the company's footprint. Hilti Corporation of Liechtenstein has taken an extra step by including business travel emissions (part of Scope 3) in carbon neutrality goals as outlined in Figure 1.4. The company tackles its entire impact for true net-zero carbon emissions, including indirect emissions across its value chain (Scope 3). This means actively removing carbon dioxide from the atmosphere through natural solutions like planting trees or technological advancements like carbon capture.

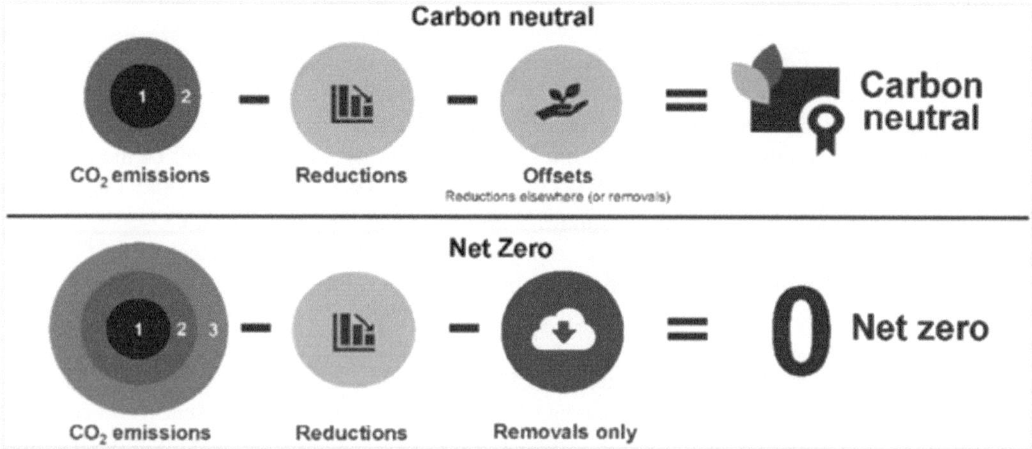

FIGURE 1.4 Net Zero and CO_2-Neutrality at Hilti Corporation. (Source: Zharkova, 2023.)

Both carbon neutrality and net-zero emissions aim to achieve a state where human activity doesn't result in a net increase of CO_2 in the atmosphere. While they share the goal of balancing emissions with removals, key differences exist. Carbon neutrality is a policy of not increasing carbon emissions and achieving carbon reduction through offsets. While net-zero carbon means making changes to reduce carbon emissions to the lowest amount – and offsetting as a last resort. As part of the Paris Agreement, countries agreed to pursue efforts to limit global warming to well below 2°C, preferably to 1.5°C, compared to pre-industrial levels. To achieve this, countries must reduce GHG emissions to 'net zero' by around 2050.

Net ZERO-Aligned Offsetting Principles

1. Cut emissions, use high-quality offsets, and regularly revise the offsetting strategy as best practice evolves
2. Shift offsetting toward carbon removal
3. Shift offsetting toward long-lived storage
4. Support the development of net zero-aligned offsetting.

Countries won't be able to reach carbon neutrality only by reducing their domestic emissions. They must offset much of their carbon footprint on international carbon markets. The latest climate science is clear: Limiting global warming to 1.5°C (2.7°F) is still possible. However, to avoid the worst climate impacts, global GHG emissions will need to drop by nearly half by 2030 and ultimately reach net zero.

Which companies are net zero?

Here are seven companies that set particularly noteworthy net-zero commitments in 2021:

- Coca-Cola HBC. Sector: Food and Beverage
- Sasol. Sector: Oil and gas
- China Petroleum & Chemical Corporation (Sinopec) Sector: Oil and gas
- General Motors (GM) Sector: Automobiles
- Engie. Sector: Electric utilities
- Nippon Steel.

1.2.2 Carbon Dioxide Equivalent (CO_2e)

The concept of CO_2 equivalent (CO_2e) is a critical tool in understanding and managing GHG emissions, as it allows for the aggregation of different gases into a single metric based on their global warming potential (GWP) over a specified time horizon, typically 100 years. This standardization facilitates reporting and policymaking by providing a common unit of measurement for diverse emissions, such as CH_4 and N_2O, which have GWPs of 25 and 298, respectively, compared to CO_2's GWP of 1. For any quantity and type of greenhouse gas, CO_2e signifies the amount of CO_2 that would have the equivalent global warming impact. A quantity of GHG can be expressed as CO_2e by multiplying the amount of the GHG by its GWP. For example, if 1 kg of methane is emitted, this can be expressed as 29.8 kg of CO_2e (1 kg CH_4 * 29.8 = 29.8 kg CO_2e). A million metric tons equals about 2.2 billion pounds or 1 trillion grams. For comparison, a small car will likely weigh more than one metric ton. Thus, a million metric tons are roughly the same mass as 1 million small cars. The U.S. Inventory uses metric units for consistency and comparability with other countries. For reference, a metric ton is slightly more (approximately 10%) than a U.S. "short" ton.

"CO_2e" is a very useful term for several reasons: it allows "bundles" of greenhouse gases to be expressed as a single number, and it allows different bundles of GHGs to be easily compared (in terms of their total global warming impact). However, one caution when comparing CO_2e totals is that it is important to know that the same GHGs are included in the totals being compared to be sure that like-for-like comparisons can be made. It is also worth noting that "CO_2e" is also sometimes written as "CO_2eq," "CO_2equivalent," or even "CDE," and these terms can be used interchangeably.

However, the use of CO_2e is not without complexities. It is a construct shaped by international law, which frames carbon as a fungible object, enabling market-based climate governance. The conventional CO_2e metric, while useful, does not fully capture the distinct behaviors of long-lived climate pollutants (LLCPs) and short-lived climate pollutants (SLCPs), leading to potential inaccuracies in assessing their impact on global warming. To address this, alternative metrics like CO_2-forcing-equivalent (CO_2-fe) emissions have been proposed, which maintain a linear relationship with warming and are more suitable for multi-gas scenarios.

Additionally, the GWP approach offers a refined method for calculating warming-equivalent emissions, particularly for SLCPs, by equating their emission rate changes to a one-off CO_2 pulse, thus aligning better with long-term temperature goals. Despite its simplicity and familiarity, the CO_2e metric requires transparency and justification in its application, as different bases for equivalence can significantly alter the interpretation and implications of reported emissions.

1.2.3 Global Warming Potentials

Global Warming Potentials are used to estimate, compare, and aggregate the relative climate effects of various GHGs. They measure the relative radiative impact of a given substance compared to another, integrated over a chosen time horizon (UNFCCC, 2004). By multiplying one gas's emissions in tons and GWP, we can obtain the CO_2-eq of this gas. By such a method, we can standardize the warming effect of different GHGs.

GHGs are present in the Earth's atmosphere in varying concentrations, measured in parts per million, billion, or trillion. These gases have different lifetimes in the atmosphere, ranging from a few years to thousands of years, but all are well mixed globally. GWP is a metric for comparing the heat-trapping ability of different gases to CO_2 over a specific period, typically 100 years. A higher GWP indicates a greater warming impact. GWP considers two factors:

a. **Radiative Efficiency**: This refers to how effectively a GHG traps atmospheric heat
b. **Atmospheric Lifetime**: This refers to how long a gas stays in the atmosphere before it breaks down or is removed through natural processes

GWPs of given substances are calculated with the following equation (IPCC Third Assessment Report, 2001):

$$\mathrm{GWP}(x) = \frac{\mathrm{AGWP}\ (x)}{\mathrm{AGWP}\ (r)} = \frac{\int_0^{\mathrm{TH}} a_x \cdot [x(t)]\,dt}{\int_0^{\mathrm{TH}} a_r \cdot [r(t)]\,dt} \tag{1.1}$$

where

x=Substance in question (usually a GHG)
r=Reference gas (usually CO_2)
GWP=Global Warming Potential
AGWP=Absolute Global Warming Potential
TH=Time horizon
a=Radiative efficiency (W/m/kg)
$[x(t)]$=Time-dependent decay in abundance of the instantaneous release of substance x
$[r(t)]$=As above, but for the reference gas r.

If the GWP is to be calculated for a GHG, and the reference gas and the time horizon are chosen to be CO_2 and 100 years, respectively, the equation to calculate the GWP is a specific case of Equation 1.1, and reads as follows:

$$\mathrm{GWP}(x) = \frac{\mathrm{AGWP}\ (\mathrm{GHG})}{\mathrm{AGWP}\ (\mathrm{CO_2})} = \frac{\int_0^{100\ \text{years}} a_{\mathrm{GHG}} \cdot [\mathrm{GHG}(t)]\,dt}{\int_0^{100\ \text{years}} a_{\mathrm{CO_2}} \cdot [\mathrm{CO_2}(t)]\,dt} \tag{1.2}$$

Facts on GWP (USEPA, n.d.):

- CO_2, by definition, has a GWP of 1 regardless of the time used, because it is the gas being used as the reference. CO_2 remains in the climate system for a very long time: CO_2 emissions cause increases in atmospheric concentrations of CO_2 that will last thousands of years.
- Methane (CH_4) is estimated to have a GWP of 27–30 over 100 years. CH_4 emitted today lasts about a decade on average, much less time than CO_2. However, CH_4 also absorbs much more energy than CO_2. The net effect of the shorter lifetime and higher energy absorption is reflected in the GWP. The CH_4 GWP also accounts for some indirect effects, such as the fact that CH_4 is a precursor to ozone, and ozone is a GHG.
- Nitrous Oxide (N_2O) has a GWP 273 times that of CO_2 for a 100-year time scale. N_2O emitted today remains in the atmosphere for more than 100 years, on average.
- Chlorofluorocarbons (CFCs), hydrofluorocarbons (HFCs), hydrochlorofluorocarbons (HCFCs), perfluorocarbons (PFCs), and sulfur hexafluoride (SF_6) are sometimes called high-GWP gases because, for a given amount of mass, they trap substantially more heat than CO_2.
- The GWP values are adjusted occasionally, based on new science or changes in atmospheric gas concentrations that affect their heat-trapping abilities.

Table 1.1 below presents GWP for various GHG's.

TABLE 1.1
Kyoto Gases (IPCC 2021 - 6th Assessment Report Values)

Greenhouse Gas	Global Warming Potential (GWP)
Carbon dioxide (CO_2)	1
Methane (CH_4)	29.8
Nitrous oxide (N_2O)	273
Hydrofluorocarbons (HFCs)	5–14,600
Perfluorocarbons (PFCs)	78–12,400
Sulfur hexafluoride (SF_6)	25,200

Source: Brander (2021).

1.3 GLOBAL CONCERN

The global effort to mitigate climate change has evolved significantly over the years. Beginning with the establishment of the IPCC and the UNFCCC in the late 20th century, milestones include the adoption of the Kyoto Protocol in 1997 and the Paris Agreement in 2015, which set ambitious targets to limit global warming. Additional measures, like the Kigali Amendment to phase out HFCs and youth-led climate strikes, underscore growing awareness and activism. Despite setbacks, such as the COVID-19 pandemic, renewed commitments—like the Glasgow Climate Pact in 2022 and the EU's "Fit for 55" package in 2023—demonstrate ongoing efforts to address the urgent challenge of climate change on a global scale.

The Conference of Parties (COP) summits address climate change and biodiversity. Established after the 1992 Rio Summit, these annual meetings bring together 197 countries to discuss climate action. Since 2015, all participating nations have acknowledged climate change and aim to limit greenhouse gas emissions while achieving net zero by the end of the century.

A significant turning point came at COP21 in Paris (2015) when all parties agreed on the human-caused nature of climate change. The resulting Paris Agreement aims to limit global warming to below 2°C, ideally 1.5°C. Despite ongoing efforts, recent COPs, like COP27 (Carlin, 2022), highlight challenges. Fossil fuel reliance remains high, and progress on mitigation has been slow. There has been some progress on financing climate change's negative impacts on vulnerable countries (Corigliano, 2023).

1.3.1 Paris Agreement 2015

The Paris Agreement is a historic international treaty aimed at tackling climate change. Adopted in 2015, it became legally binding in 2016 and boasts near-universal participation (196 countries). The core objective is curbing global warming. Ideally, this means keeping the average temperature rise well below 2°C compared to pre-industrial levels, with even stronger efforts to limit it to a maximum of 1.5°C. Experts warn that exceeding this threshold could trigger much harsher climate consequences, including more frequent and intense droughts, heatwaves, and heavy rainfall (Figure 1.5).

The Paris Agreement operates on a five-year cycle to enhance global climate action, with each phase aiming for greater ambition, as shown in Figure 1.6. Countries submit their nationally determined contributions (NDCs), outlining their plans to mitigate greenhouse gas emissions and adapt to climate impacts. These NDCs are revised every five years to reflect more ambitious targets in line with the agreement's goal of limiting global warming. Additionally, countries are encouraged to develop long-term low-GHG emission development strategies (LT-LEDS) to provide a broader context and direction for their climate efforts. These strategies complement NDCs by offering a vision for sustainable development beyond short-term targets.

FIGURE 1.5 The Rio Earth Summit established the UNFCCC and began the COP process. (Source: Gamma-Rapho Via Getty Images; Forbes.)

Wealthier nations are lending a hand to developing countries by providing financial aid, better technology, and training. This helps developing countries reduce emissions and become more resilient to climate change's effects. The Paris Agreement set this framework in motion. It created a system where developed countries provide money for climate action in developing countries. This money goes toward reducing emissions (mitigation) and adapting to climate change (adaptation). Technology is another area of support. The agreement established a system to share technology that helps reduce emissions and build resilience. This could include things like new solar panels or ways to grow crops more resistant to drought. Finally, the agreement recognizes that some countries need extra help building the skills and knowledge to tackle climate change. So, developed countries provide training and other forms of capacity-building to developing countries.

Tracking progress on climate change is performed through a system established by the Paris Agreement called the Enhanced Transparency Framework (ETF). This system requires countries to report on their actions to reduce emissions (mitigation), adapt to the effects of climate change (adaptation), and the support they are providing or receiving for these efforts. These reports are reviewed internationally to assess how well we are doing collectively. So far, the Paris Agreement has spurred the development of new low-carbon technologies and markets. More and more countries, regions, cities, and companies are setting goals to achieve carbon neutrality. Zero-carbon solutions are becoming cost-competitive with traditional options in some sectors, like energy and transportation. This is creating new business opportunities, and by 2030, these solutions could be competitive in over 70% of the global emissions sectors (UNFCCC, n.d.).

1.3.2 COP-28 Dubai

At COP28, held in Dubai in December 2023, an agreement was reached, signaling the "beginning of the end" for the fossil fuel era, emphasizing a swift, just, and equitable transition toward

FIGURE 1.6 Key elements of the Paris Agreement. (Source: United Nations, n.d.)

renewable energy. This agreement, part of the world's first 'global stocktake,' aims to strengthen climate action to keep global temperature rise within 1.5°C. The stocktake recognizes the need to cut global greenhouse gas emissions by 43% by 2030 compared to 2019, urging nations to triple renewable energy capacity and double energy efficiency improvements by 2030. Additionally, it calls for a phase-down of unabated coal power and the phasing out of inefficient fossil fuel subsidies. Parties must submit ambitious emission reduction targets aligned with the 1.5°C limit in their next climate action plans by early 2025 (UNFCCC, n.d.).

The COP28 set the four pillars for COP28, aiming to achieve key outcomes in global climate action. The first pillar is fast-tracking a just, orderly, and equitable energy transition. It focuses on fast-tracking a just and equitable energy transition, which is evident through initiatives like the Global Renewables and Energy Efficiency Pledge and collaborative alliances such as the Powering Past Coal Alliance. The second pillar is fixing climate finance, with endorsements such as the UAE Leaders' Declaration and collaborative endeavors like Assets to Flows II. The third pillar focuses on people, lives, and livelihoods, as demonstrated by initiatives like the Sharm el-Sheikh Adaptation Agenda and commitments to enhance resilience in various sectors. Furthermore, the fourth pillar underpins everything with full inclusivity, as seen through the appointment of a Youth Climate Champion and initiatives like the Podong Indigenous Peoples Initiative to leverage technology and cultural sectors for inclusive climate action (UNFCCC COP28, 2023).

1.3.3 COP-29 Baku

At the last COP29 held in Baku, Azerbaijan (UNFCCC COP 29, 2024), key outcomes included a pledge from developed nations to mobilize $300 billion annually by 2035 to support developing countries in addressing climate impacts. However, this commitment fell short of expert recommendations for $1.3 trillion annually. Delegates finalized rules for international carbon credit trading under Article 6 of the Paris Agreement, aiming to enhance transparency and standardization in global carbon markets. Additionally, the Baku Communiqué, endorsed by Small Island Developing States, highlighted vulnerable regions' urgent climate challenges. However, progress on phasing out fossil fuels stalled, and critiques over insufficient funding and human rights concerns in Azerbaijan underscored lingering challenges in achieving equitable and inclusive climate action.

1.4 CONCLUSION

Understanding the carbon footprint is fundamental to grasping the scale and urgency of the climate challenge. This chapter has traced the journey from the scientific underpinnings of the carbon cycle and greenhouse gas emissions to the global frameworks and individual actions shaping our response. By examining the interplay between natural processes and human activities, we have seen how cumulative emissions drive global warming, disrupt ecosystems, and threaten economic and social stability. The evidence is clear: meaningful progress hinges on both systemic change and individual responsibility, underpinned by international cooperation and innovation.

As we move forward, it is essential to recognize that the carbon footprint is not an abstract metric but a direct result of identifiable sources and activities. The next chapter builds on this conceptual groundwork by providing a detailed analysis of the origins of greenhouse gas emissions, dissecting the roles of fossil fuel combustion, deforestation, industrial processes, agriculture, and transportation, and highlighting their interconnectedness and cumulative impact on climate change. This transition from foundational concepts to a sector-by-sector exploration ensures that our understanding progresses from the "why" to the "how," equipping readers with the knowledge needed to engage with practical solutions and policy responses in the chapters ahead.

By bridging the scientific, societal, and policy dimensions of the carbon footprint, this chapter prepares us to confront the complexities of emission sources and the integrated strategies required for meaningful climate action. The foundation laid here will support a deeper engagement with mitigation pathways, technological innovations, and the urgent need for collective global action to ensure a sustainable and habitable planet for future generations.

REVIEW QUESTIONS

1. What is a carbon footprint, and why is it important in understanding climate change?
2. How do the carbon cycle and greenhouse gas (GHG) emissions impact global warming?
3. Why are climate feedback loops significant in the study of climate change?
4. How does the Paris Agreement aim to address carbon emissions on a global scale?
5. What actions can individuals take to reduce their carbon footprints?
6. How do different greenhouse gases contribute to global warming potentials?
7. In what ways does reducing CFP impact human health and ecosystems?
8. How does international collaboration influence carbon reduction efforts?
9. What role do natural vs. human-made emissions play in the global CFP?
10. How is CFP measurement and reporting essential in environmental policies?

REFERENCES

Bhatti, U.A., Bhatti, M.A., Tang, H., Syam, M.S., Awwad, E.M., Sharaf, M., Ghadi, Y.Y. (2024). Global production patterns: Understanding the relationship between greenhouse gas emissions, agriculture greening and climate variability. *Environmental Research*, 245. https://doi.org/10.1016/j.envres.2023.118049

Brander, M. (2021). Greenhouse Gases, CO_2, CO_2e, and Carbon: What Do All These Terms Mean?, https://bluemangrove.fund/wp-content/uploads/2021/03/Glossary-on-different-CO2-terms.pdf

Carlin, D. COP 27: How the UN's Global Climate Conference First Started, https://www.forbes.com/sites/davidcarlin/2022/11/05/cop-27-how-the-uns-global-climate-conference-first-started/?sh=43cd2e284799

Caro, D. (2019). Carbon footprint. In S. Elias (Ed.), *Encyclopedia of Ecology* (pp. 252–257). Elsevier.

Climate Trade. (n.d.). World's Countries Biggest Carbon Polluters | ClimateTrade, https://climatetrade.com/which-countries-are-the-worlds-biggest-carbon-polluters/

Cloy, J.M., Smith, K.A. (2017). Greenhouse gas sources and sinks. In D.A. DellaSala, M.I. Goldstein (Eds.), *Encyclopedia of the Anthropocene* (pp. 391–400). Elsevier.

Corigliano, G.G. (2023). An Overview of the Latest COPs on the Road to Dubai 2023 | Laudato Si' Movement, https://laudatosimovement.org/news/an-overview-of-the-latest-cops-on-the-road-to-dubai-2023/

Crippa, M., Guizzardi, D., Pagani, F., Banja, M., Muntean, M., Schaaf, E., Becker, W., Monforti-Ferrario, F., Quadrelli, R., Risquez Martin, A., Taghavi-Moharamli, P., Koykka, J., Grassi, G., Rossi, S., Brandao De Melo, J., Oom, D., Branco, A., San-Miguel, J., Vignati, E. (2023). GHG Emissions of All World Countries, https://edgar.jrc.ec.europa.eu/booklet/GHG_emissions_of_all_world_countries_booklet_2023report.pdf

Ding, G.K.C., Banihashemi, S. (2017). Ecological and carbon footprints: The future for city sustainability. In M.A. Abraham (Ed.), *Encyclopedia of Sustainable Technologies* (pp. 43–51). Elsevier.

Easterling, D.R., Kunkel, K.E. (2023). Climate change in the Earth system. In M.J. Kennish, H.W. Paerl, J.R. Crosswell (Eds.), *Climate Change and Estuaries* (pp. 21–40). CRC Press.

Font Vivanco, D., Bassa Echaurren, M., Aulinas Masó, M., Giavini, M., Doubell, M., Bagatella, E., Venturi, R. (2024). Carbon footprint and life cycle economic costs of pilot Know-as-you-Throw schemes in Italy and Spain. *Waste Management*, 177, 46–55. https://doi.org/10.1016/J.WASMAN.2024.01.027

Hertwich, E.G. (2021). Increased carbon footprint of materials production driven by rise in investments. *Nature Geoscience*, 14, 151–155. https://doi.org/10.1038/s41561-021-00690-8

Huang, L., Yoshida, Y., Li, Y., Cheng, N., Xue, J., Long, Y. (2024). Sustainable lifestyle: Quantification and determining factors analysis of household carbon footprints in Japan. *Energy Policy*, 186, 114016. https://doi.org/10.1016/J.ENPOL.2024.114016

Huo, Y., Mi, G., Zhu, M., Chen, S., Li, J., Hao, Z., Cai, D., Zhang, F. (2024). Carbon footprint of farming practices in farmland ecosystems on the North and Northeast China plains. *Journal of Environmental Management*, 354, 120378. https://doi.org/10.1016/J.JENVMAN.2024.120378

IPCC. (2007). FAQ 1.3 – AR4 WGI Chapter 1: Historical Overview of Climate Change Science, https://archive.ipcc.ch/publications_and_data/ar4/wg1/en/faq-1-3.html

IPCC. (2023). *Climate Change 2023: Synthesis Report. Contribution of Working Groups I, II and III to the Sixth Assessment Report of the Intergovernmental Panel on Climate Change*. IPCC, Geneva, Switzerland.

IPCC Third Assessment Report. (2001). Climate Change 2001: The Scientific Basis, https://www.ipcc.ch/site/assets/uploads/2018/03/WGI_TAR_full_report.pdf

Jones, N. (2011). Climate change curbs crops. *Nature*. https://doi.org/10.1038/news.2011.268

Laksmawati, W.K., Hsieh, C.M., Yang, S.H. (2024). Social influence and climate change issues affecting consumer behavioral intention toward carbon footprint label: A study of Taiwanese consumers. *Journal of Cleaner Production*, 444, 141092. https://doi.org/10.1016/J.JCLEPRO.2024.141092

Moran, D., Kanemoto, K., Jiborn, M., Wood, R., Többen, J., Seto, K.C. (2018). Carbon footprints of 13 000 cities. *Environmental Research Letters*, 13, 064041. https://doi.org/10.1088/1748-9326/aac72a

Steinfeld, H., Gerber, P., Wassenaar, T., Castel, V., Rosales, M., de Haan, C. (2006). *Livestock's Long Shadow*. FAO.

Summary for Policymakers. (2023). *Climate Change 2021 – The Physical Science Basis*, pp. 3–32. Cambridge University Press.

UNFCC. (2004). *United Nations Information on Global Warming Potentials*. UNFCC.

UNFCC COP28. (2023). *Summary of Global Climate Action at COP 28*. UNFCC

UNFCC COP29. (n.d.). Summary of Global Climate Action at COP 29 UNFCC: COP 28: What Was Achieved and What Happens Next? | UNFCCC, https://unfccc.int/cop28/5-key-takeaways#end-of-fossil-fuels

UNFCCC. (n.d.). The Paris Agreement | UNFCCC, https://unfccc.int/process-and-meetings/the-paris-agreement

UNFCCC. (1997). *Kyoto Protocol to the United Nations Framework Convention on Climate Change*.

United Nations. (n.d.). *The Paris Agreement*. United Nations.

USEPA. (n.d.). Understanding Global Warming Potentials | US EPA, https://www.epa.gov/ghgemissions/understanding-global-warming-potentials

Wang, J., Quan, Q., Chen, W., Tian, D., Ciais, P., Crowther, T.W., Mack, M.C., Poulter, B., Tian, H., Luo, Y., Yu, G., Niu, S. (2021). Increased CO_2 emissions surpass reductions of non-CO_2 emissions more under higher experimental warming in an alpine meadow. *Science of the Total Environment*, 769. https://doi.org/10.1016/j.scitotenv.2020.144559

Woods, J., Williams, A., Hughes, J.K., Black, M., Murphy, R. (2010). Energy and the food system. *Philosophical Transactions of the Royal Society B: Biological Sciences*, 365, 2991–3006. https://doi.org/10.1098/rstb.2010.0172

Ya-Min, Z., Yong-Sheng, F. (2014). The strategy and technology selection for non-CO_2 greenhouse gas emission control. *Advances in Climate Change Research*, 5, 28–33. https://doi.org/10.3724/SP.J.1248.2014.028

Zharkova, O. (2023). Net zero and CO_2-neutrality at Hilti | Hilti Corporation, https://www.hilti.group/content/hilti/CP/XX/en/company/media-relations/hilti-stories/net_zero_co2_neutrality_sbti_explainer.html#nav/close

2 Sources of Greenhouse Gas Emissions

In today's world, human actions and natural processes significantly affect our environment, shaping its delicate equilibrium. Global warming, mainly caused by human-generated greenhouse gas (GHG) emissions, is a central concern within this interconnected system. It is crucial to comprehend where these emissions come from to address the complex challenges of climate change and work toward a sustainable future.

Burning fossil fuels, such as coal, oil, and gas, is crucial in increasing GHG levels in the atmosphere. This continuous combustion process, essential for industrial development, releases substantial amounts of carbon dioxide and nitrous oxide, disrupting the balance of gases in our atmosphere (Nunes, 2023; Bhardwaj, 2024; Hassan, 2024). At the same time, deforestation is a significant factor driving climate change. As human activities expand into natural habitats, cutting down trees disrupts an important process where forests store carbon. This disruption releases stored carbon dioxide, further intensifying the greenhouse effect (Nunes, 2023; Bhardwaj, 2024; Hassan, 2024).

Livestock farming emits significant amounts of GHG, particularly in agriculture. Cows and sheep, crucial for food production, release substantial quantities of methane as part of their digestive processes (Nunes, 2023; Seftiani et al., 2024). This methane emission contributes significantly to the planet's overall warming. Nitrogen-based fertilizers represent another significant source of emissions below the surface of agricultural areas. While these fertilizers are vital for improving crop productivity, they release nitrous oxide into the atmosphere (Nunes, 2023; Chowdhuri & Gupta, 2023). Additionally, modern technology's widespread use of fluorinated gases threatens our climate. These gases have a warming potential thousands of times greater than carbon dioxide, making their emissions a potent catalyst in the progression of global warming (Chowdhuri & Gupta, 2023).

The recent analysis of GHG emissions in the United States offers a comprehensive understanding of the intricate network of human activities influencing climate change. As outlined in the EPA's *Inventory of U.S. Greenhouse Gas Emissions and Sinks*, key contributors include electricity production, transportation, industry, commercial and residential sectors, and agriculture. By acknowledging the interconnectedness of these emission sources and their impact on our environment, we empower ourselves to enact meaningful mitigating measures to address climate change and foster a sustainable future for future generations.

2.1 ELECTRIC POWER

The electricity generation sector is a major contributor to greenhouse gas (GHG) emissions, primarily due to the reliance on fossil fuels such as coal, which, despite accounting for only 23% of electricity generation in the United States in 2021, was responsible for 59% of emissions from electricity production (see Figure 2.1). This disproportionate impact underscores the need for a transition to lower-emission energy sources. Efforts to reduce emissions have led to a 15% decrease since 1990, driven by shifts toward natural gas, nuclear power, and renewables (Ghosh et al., 2022; Babatunde et al., 2019). Integrating renewable energy sources, such as solar and wind, along with advancements in smart grid technologies, is pivotal in this transition, offering sustainable and environmentally friendly alternatives to fossil fuels (Kabeyi & Olanrewaju, 2020; Sahoo, 2019). Moreover, strategies like carbon capture and storage (CCS), fuel balancing, and the decommissioning of aging fossil fuel plants are essential for further decarbonization (Arif et al., 2020; Babatunde et al., 2019). Adopting these technologies, alongside stringent low-carbon policies, is crucial to maintaining a

DOI: 10.1201/9781003479666-2

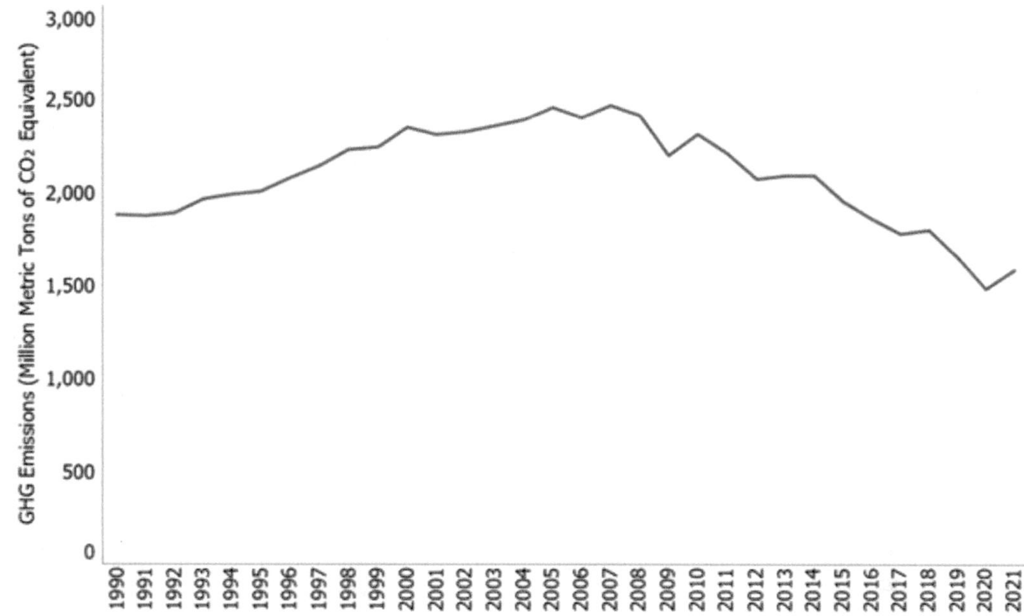

FIGURE 2.1 Greenhouse gas emissions from electric power, 1990–2021. All emission estimates are from the *Inventory of U.S. Greenhouse Gas Emissions and Sinks: 1990–2021.*

rise in global temperature below 2°C, as outlined in the Paris Agreement (Babatunde et al., 2019). However, the variability in emission factors across different regions, influenced by factors such as plant age and fuel quality, highlights the complexity of achieving uniform reductions in emissions (Ghosh et al., 2022). An integrated approach that combines various decarbonization technologies and policies is necessary to enhance low-carbon electricity generation and ensure a sustainable energy future (Babatunde et al., 2019).

Electricity is used by other end-use sectors—in homes, businesses, and factories—and the GHG emissions from electricity generation can be allocated to the sectors that use the electricity. Looking at GHG emissions by end-use sector can help us understand energy demand across sectors and changes in energy use over time. When emissions from electric power generation are allocated to the industrial end-use sector, industrial activities account for a much larger share of U.S. GHG emissions, as illustrated in Figure 2.2. GHG emissions from commercial and residential buildings also increase substantially when emissions from electricity end-use are included, due to the relatively large share of electricity use (e.g., heating, ventilation, and air conditioning, lighting, and appliances) in these sectors.

Total emissions in 2021 were 6,340 million metric tons of CO_2 equivalent. Percentages may not add up to 100% due to independent rounding. Land use, land-use change, and forestry in the United States are net sinks that offset 12% of these GHG emissions. This net sink is not shown in the above diagram. Details of how electric power sectors reduce their GHG emissions are illustrated in Table 2.1.

All emission estimates are from the *U.S. Greenhouse Gas Emissions and Sinks: 1990–2021.*

2.2 INDUSTRY

Industrial activities are significant contributors to GHG emissions, both directly and indirectly, through processes such as combustion, chemical reactions, and electricity consumption. In the U.S., industrial operations accounted for 23% of direct GHG emissions in 2021, with this figure rising to

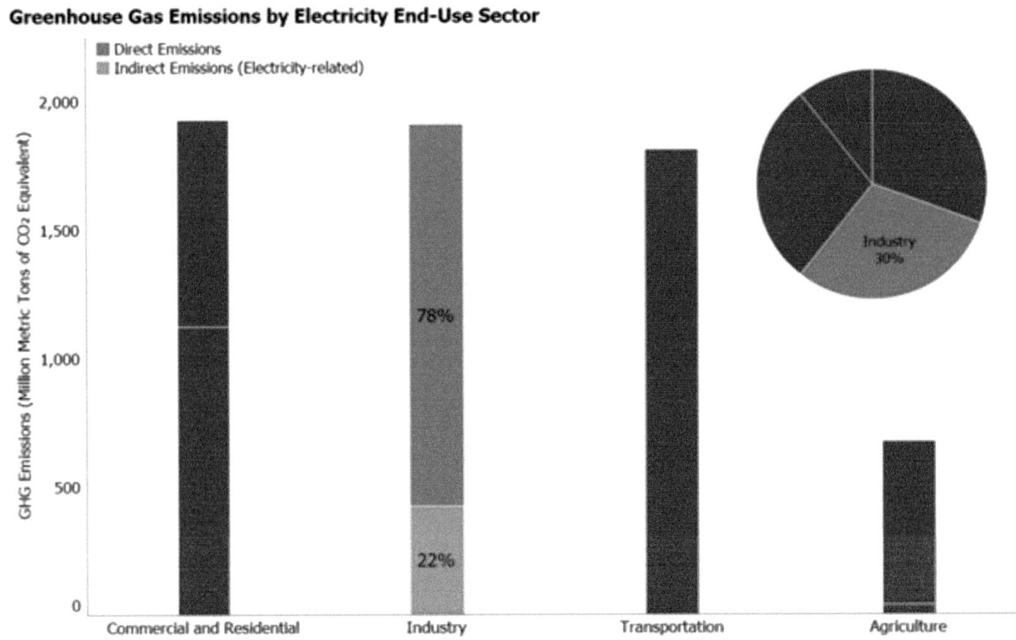

FIGURE 2.2 Total U.S. Greenhouse Gas Emissions by economic sector in 2021.

30% when including electricity-related emissions, making it the largest source of emissions (Isella & Manca, 2022; Schoeneberger, 2022). This is shown in Figure 2.3. The chemical industry, particularly processes like ammonia synthesis and steam cracking, is a major emitter, contributing significantly to global emissions (Isella & Manca, 2022). Efforts to reduce these emissions have included improvements in energy efficiency and fuel switching, which have led to a 14% decline in industrial emissions since 1990 (Isella & Manca, 2022). However, the global trend shows limited progress in reducing carbon footprints due to high consumer demands and insufficient shifts to low-carbon technologies, especially in rapidly industrializing regions like Asia (Bahman et al., 2022).

Electrification of industrial processes, such as boilers, presents a potential strategy for reducing emissions, though its effectiveness depends on the decarbonization of the electricity grid (Schoeneberger, 2022). Additionally, regional differences in industrial activities, such as those in China, highlight the importance of targeted policies that consider local industrial driving effects to reduce emissions effectively (Zheng et al., 2021). As demonstrated in Poland, high-resolution spatial modeling of emissions can improve the accuracy of emission inventories and support more effective mitigation strategies (Charkovska et al., 2019). These insights underscore the complexity and regional variability of industrial emissions, necessitating tailored approaches to achieve significant reductions.

There is a wide variety of industrial activities that cause GHG emissions, and there are many opportunities to reduce them. Table 2.2 below provides some examples of opportunities for industry to reduce emissions. For a more comprehensive list, see Chapter 11 of the *Contribution of Working Group III to the Sixth Assessment Report of the Intergovernmental Panel on Climate Change.*

2.3 TRANSPORTATION

The transportation sector, responsible for 28% of U.S. emissions in 2021, has seen a 19% increase since 1990 due to rising travel demand. However, advancements in vehicle efficiency and alternative fuels present significant opportunities for emission reductions. The adoption of zero-emission

TABLE 2.1
Example Reduction Opportunities for the Electric Power Sector

Type	How Emissions Are Reduced	Examples
Increased Efficiency of Fossil-fired Power Plants and Fuel Switching	Improving efficiency and switching to less carbon-intensive fuels or power plants.	In Great Britain, a rapid switch from coal to natural gas, facilitated by effective carbon pricing, led to a notable decrease in carbon emissions, highlighting the potential for similar strategies in other coal-dependent countries. One approach is gradually replacing carbon-rich fuels with carbon-free energy carriers, such as hydrogen, in modern power plants. As demonstrated in Cyprus, this transition can significantly reduce production costs and eliminate carbon dioxide emissions by 2050, with natural gas serving as an effective intermediate step toward achieving these goals by 2035. Additionally, using supercritical CO_2 power cycles offers a promising method to enhance thermal power plant efficiency. These cycles leverage the thermophysical properties of CO_2 to reduce compression and friction losses, thereby increasing efficiency and reducing greenhouse gas emissions compared to traditional cycles. Furthermore, the organic rankine cycle (ORC) technology is being utilized in energy-intensive industries to convert residual low-grade heat into electricity, thereby improving energy efficiency and reducing CO_2 emissions.
Renewable Energy	Generating electricity using renewable sources instead of fossil fuels.	Solar and wind power are prominent examples, with solar energy harnessed through photovoltaic cells and solar thermal collectors, and wind energy captured via turbines that convert kinetic energy into electricity. Hydropower, another significant renewable source, utilizes the energy of flowing water, while geothermal energy taps into the Earth's internal heat, providing a stable and reliable power supply. Biomass energy, derived from organic materials, is also gaining traction, with advancements in anaerobic digestion, pyrolysis, and gasification enhancing its efficiency and applications across power generation, transportation, and industrial processes. Innovative approaches, such as integrating renewable sources with waste tire recycling through pyrolysis, thermal degradation, and mechanical shredding, demonstrate the potential for reducing greenhouse gas emissions and fossil fuel dependency. Additionally, hybrid systems combining tidal and wind energy with diesel backup, as implemented on Stewart Island, illustrate the practical benefits of renewable energy in reducing fuel consumption and emissions while maintaining cost-effectiveness.

(*Continued*)

TABLE 2.1 (*Continued*)
Example Reduction Opportunities for the Electric Power Sector

Type	How Emissions Are Reduced	Examples
Increased End-Use Energy Efficiency	Reducing electricity use and peak demand through energy efficiency in homes, businesses, and industries.	Highly energy-efficient housing can significantly reduce peak load demands, as demonstrated in scenarios where efficient electrification cuts peak demand by about half compared to low-efficiency alternatives, thus minimizing the need for expanded electricity supply infrastructure and enhancing resilience to extreme weather condition. In residential settings, the implementation of home energy management systems (HEMS) can optimize energy use by managing essential and non-essential loads, integrating local solar photovoltaic systems, and facilitating load shifting during off-peak periods. Demand side management (DSM) techniques, such as end-use equipment control and peak clipping, are crucial for managing peak loads in both industrial and domestic contexts, ensuring optimal resource utilization and energy conservation (Kumar, 2021). Additionally, the use of algorithms like ant colony optimization in HEMS can further reduce energy consumption costs and peak demand by scheduling appliance use in response to time-of-use pricing, thereby enhancing user comfort and reducing the peak average ratio.
Nuclear Energy	Generating electricity with nuclear energy instead of fossil fuels.	Nuclear energy offers a reliable and substantial source of baseload power, which is crucial for reducing carbon emissions and combating climate change. Advanced nuclear technologies, including fourth-generation reactors and small modular reactors (SMRs), promise improvements in safety, economics, and sustainability, making nuclear energy a more attractive option for future energy systems. Additionally, the management of nuclear waste and the potential for nuclear proliferation remain significant concerns.
Carbon Capture and Sequestration (CCS)	Capturing and securely storing CO_2 before it enters the atmosphere.	CCS is a critical technology for mitigating climate change by capturing CO_2 emissions before they enter the atmosphere and securely storing them underground. Once captured, the CO_2 is purified, dehydrated, and compressed into a supercritical liquid for transportation, typically via pipelines, to geological storage sites. These sites, often deep rock formations, have historically contained other gases and are considered safe for long-term CO_2 storage). Recent advancements in CCS technologies, such as supercritical CO_2 injection and intelligent monitoring systems, have enhanced the efficiency and safety of these projects, ensuring secure containment and building public trust.

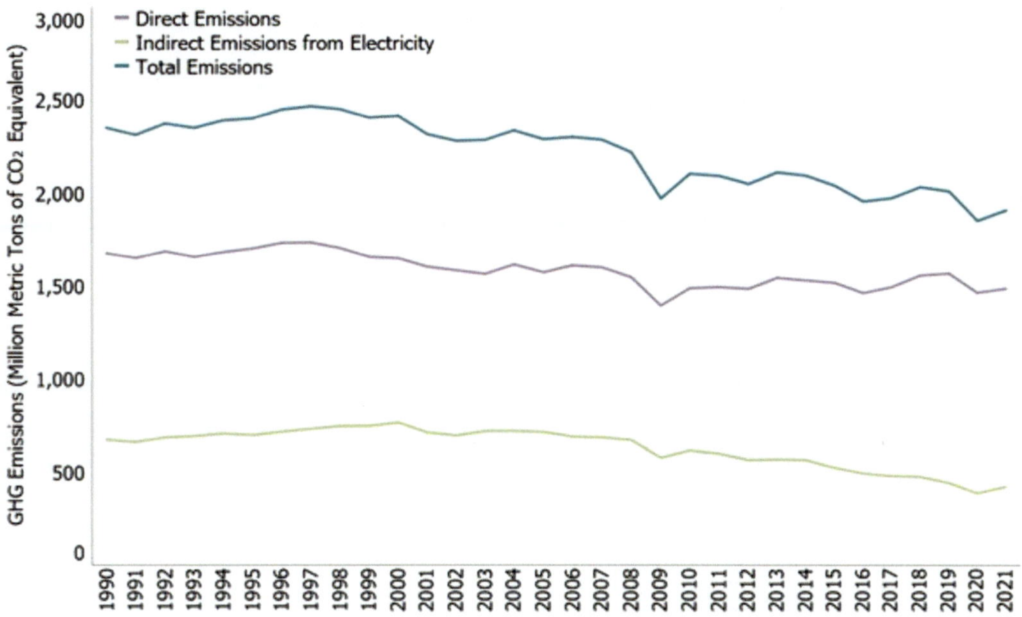

FIGURE 2.3 Greenhouse gas emissions from industry, 1990–2021. All emission estimates are from the *Inventory of U.S. Greenhouse Gas Emissions and Sinks: 1990–2021.*

vehicles (ZEVs) is crucial, with projections indicating that ZEV sales could reach 89% for passenger vehicles and 69% for freight trucks by 2030, achieving 100% by 2040, contingent on a clean electric grid (Hoehne et al., 2023). Electrification is a key strategy, with electric vehicles (EVs) significantly impacting the U.S. electrical grid, necessitating regional strategies to maximize decarbonization benefits (Moura et al., 2023). Hydrogen and biofuels are also pivotal, particularly for aviation and shipping, where electrification is less feasible (Speizer et al., 2023). The transition to a zero-emission transportation sector requires substantial technological shifts, with electrification and hydrogen playing central roles, though the latter is less cost-optimal for long-term electricity storage (Rinaldi et al., 2023). Research trends emphasize low-carbon travel and green supply chain management, highlighting the importance of managing travel demand to ease the transition (Fan et al., 2023). Overall, achieving deep decarbonization in transportation involves a multifaceted approach, integrating technological advancements, policy support, and strategic resource management to meet climate goals (Hoehne et al., 2023; Speizer et al., 2023; Moura et al., 2023).

Emissions involved in the use of electricity for transportation activities are included in Figure 2.4 above but are not shown separately (as was done for other sectors). These indirect emissions are negligible, accounting for less than 1% of the total emissions shown in the graph. All emission estimates are from the *Inventory of U.S. Greenhouse Gas Emissions and Sinks: 1990–2021.*

Relatively small amounts of methane (CH_4) and nitrous oxide (N_2O) are emitted during fuel combustion. In addition, hydrofluorocarbon (HFC) emissions also occur from the transportation sector. These emissions result from the use of mobile air conditioners and refrigerated transport.

In 2021, GHG emissions from transportation accounted for 28% of total U.S. GHG emissions, making it the largest contributor of U.S. GHG emissions. From 1990 to 2021, total transportation emissions from fossil fuel combustion increased by 19%. In 2021, emissions increased by 12%, which followed a decline of 13% in 2020 due to reduced travel demand during the height of the COVID-19 pandemic. In terms of the overall trend, from 1990 to 2021, total transportation emissions have increased due largely to increased travel demand. The number of vehicle miles traveled (VMT) by light-duty motor vehicles (passenger cars and light-duty trucks) increased by 45% from 1990 to 2021 as a result of a confluence of factors, including population growth, economic growth,

TABLE 2.2
Examples of Reduction Opportunities for the Industry Sector

Type	How Emissions Are Reduced	Examples
Energy Efficiency	Upgrading to more efficient industrial technology. EPA's Energy Star® program helps industries become more energy-efficient.	In the pulp and paper industry, the Energy Star® program has developed specific benchmarking tools and energy guides that highlight best practices and energy-efficient technologies, facilitating the adoption of more efficient processes. These technologies enhance efficiency by minimizing energy losses and improving the power factor, essential for effective energy utilization and reducing distribution network losses Furthermore, the potential for energy efficiency improvements in the industrial sector is significant, with estimates suggesting 10%–40% reductions in energy use across various sub-sectors.
Fuel Switching	Switching to fuels that result in less CO_2 emissions but the same amount of energy when combusted.	E-fuels, such as synthetic methanol, methane, and ammonia, are promising alternatives as they are produced using hydrogen from renewable sources and captured CO_2, making them approximately carbon neutral. These fuels can be used in combustion engines or fuel cells, with the latter offering higher efficiency but requiring more infrastructure changes. Ammonia, in particular, stands out due to its lower total emissions compared to other fuels like hydrogen and hydrocarbons, despite its lower energy density. It generates fewer CO_2 emissions and performs with high energy efficiency, making it a viable option for reducing carbon emissions (Wang et al., 2018). Furthermore, technologies like bioenergy with carbon capture and storage (BECCS) and the use of solid oxide fuel cells (SOFCs) in vehicles can achieve negative emissions by capturing and reusing CO_2, thus closing the carbon loop and potentially achieving significant reductions in CO_2 emissions.
Recycling	Producing industrial products from materials that are recycled or renewable, rather than producing new products from raw materials.	For steel, the use of scrap in electric arc furnaces (EAF) can reduce CO_2 emissions by 80% compared to traditional blast furnace-basic oxygen furnace (BF-BOF) methods, making it a crucial component of the steel industry's strategy to achieve net-zero emissions by 2050. In the aluminum industry, recycling has been integral due to the high cost of primary production and the ease of melting aluminum for reuse. The use of aluminum scrap reduces production costs and emissions by preserving raw materials, and innovative solutions are being developed to integrate recycled content into primary aluminum products without compromising quality. Companies like Rio Tinto are leading efforts to incorporate end-of-life recycling into their processes, further reducing greenhouse gas emissions and enhancing the circular economy in aluminum production.
Training and Awareness	Making companies and workers aware of the steps to reduce or prevent emissions leaks from equipment.	Training should emphasize the importance of continuous monitoring and predictive maintenance to reduce emissions, as this approach helps meet compliance and sustainability goals, enhances process efficiency, and reduces product losses. Understanding the economic implications of emissions leakage and the role of policy measures, such as output-based subsidies, can help workers appreciate the broader context of emissions management and the importance of adhering to reduction strategies.

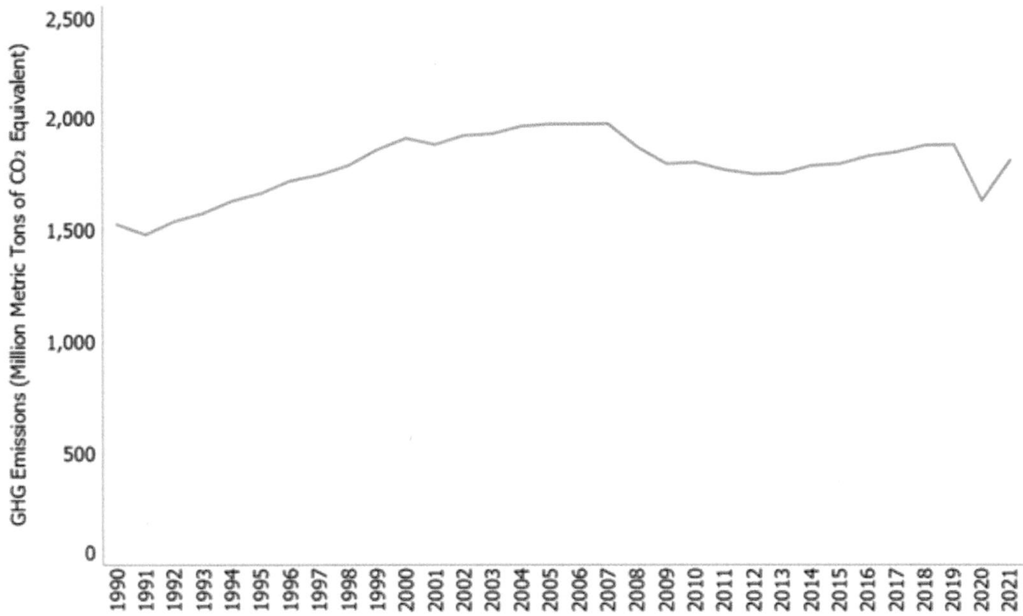

FIGURE 2.4 Greenhouse gas emissions from transportation, 1990–2021.

urban sprawl, and periods of low fuel prices. Between 1990 and 2004, average fuel economy among new vehicles sold annually declined, as sales of light-duty trucks increased. Starting in 2005, the average fuel economy of new vehicles began to increase, while light-duty VMT grew only modestly for much of the period. The average new vehicle fuel economy has improved almost every year since 2005, slowing the rate of increase in CO_2 emissions.

There are a variety of opportunities to reduce greenhouse gas emissions associated with transportation. Table 2.3 categorizes these opportunities and provides examples.

2.4 COMMERCIAL AND RESIDENTIAL SECTOR

In 2021, emissions from the commercial and residential sectors accounted for 13% of total U.S. emissions, primarily due to activities such as heating, cooking, and electricity use. Over the decades, trends in energy efficiency and renewable energy integration have contributed to a gradual decline in these emissions. In the commercial sector, HVAC systems are significant energy consumers, accounting for about 40% of total energy use, making them a focal point for energy efficiency improvements. Technological advancements in HVAC, lighting, and building insulation, along with the integration of renewable energy sources like solar and wind, have been pivotal in reducing energy consumption and emissions in commercial buildings (Anarene, 2024). In the residential sector, emissions are driven by factors such as large home sizes and on-site fossil fuel combustion. Decarbonization strategies, including the electrification of heating through heat pumps and improvements in thermal envelopes, are essential for reducing emissions. These strategies, combined with the decarbonization of electricity supply, could lead to a 91% reduction in residential emissions by 2050 (Berrill et al., 2022).

Strategic energy auditing and demand side management have also proven effective in both sectors, allowing for better energy management and significant reductions in electricity consumption (Kumar et al., 2019). The residential sector, accounting for 33% of global energy-related GHG emissions, requires rapid emissions reductions to support sustainability goals. Methodologies for modeling and spatial mapping of emissions are crucial for developing effective urban climate action plans

TABLE 2.3
Examples of Reduction Opportunities in the Transportation Sector

Type	How It Reduces Emissions	Examples
Fuel Switching	Switching to fuels that emit less CO_2 or using renewable energy.	Electric vehicles are increasingly seen as a viable alternative to traditional internal combustion engines, with their integration into the transportation sector expected to persist alongside renewable liquid fuels like bioethanol and biodiesel for the next 50 years, aiding in the reduction of carbon footprints during this transition period (Gutierrez & Ladisch, 2024). Buses powered by compressed natural gas (CNG) and hybrid CNG-pneumatic systems offer a cleaner alternative to diesel, with the hybrid CNG-pneumatic buses demonstrating significant operational reliability and reduced CO_2 emissions compared to diesel counterparts (Karaca & Dincer, 2021). However, the effectiveness of CNG in reducing emissions is limited, as evidenced by a mere 6.3% reduction in CO_2 emissions in a case study of a bus fleet in Oporto, suggesting that electric buses and other fuel alternatives may offer greater potential for decarbonization (Fancellu, 2023). The use of biofuels, such as biohydrogen and biomethane, can significantly reduce emissions, with reductions of up to 70% compared to fossil-based fuels, while also supporting renewable energy integration and enhancing engine performance (Bello et al., 2023). Despite the higher initial costs associated with electric buses, their deployment can be economically viable with government subsidies and infrastructure investments, although improvements in power generation efficiency and renewable energy adoption are necessary to maximize their environmental benefits (Yusof et al., 2021). Overall, a combination of these technologies, supported by policy and financial incentives, is essential for achieving sustainable transportation solutions.
Improving Fuel Efficiency	Using advanced technologies and designs to make vehicles more fuel-efficient.	The pursuit of energy efficiency in the automotive industry is being addressed through several innovative strategies, including lightweighting, aerodynamic improvements, and hybrid technology. Lightweighting involves using advanced materials and forming technologies to reduce vehicle mass, enhancing performance and reducing energy consumption. Techniques such as using hybrid materials like fiber-metal composites and integrating additive manufacturing with forming processes are pivotal in achieving these goals (Rosenthal et al., 2020). Additionally, lightweighting is recognized as a crucial measure for improving energy efficiency and reducing emissions, with the potential to significantly impact vehicle dynamics and energy usage throughout a vehicle's lifecycle (Candela et al., 2024). Aerodynamic improvements, such as the application of dimpled surfaces on vehicle roofs, have reduced drag coefficients, thereby enhancing fuel economy and vehicle stability (Saxena et al., 2022). Similarly, the concept of bionic groove surfaces, inspired by killer whale skin, has demonstrated substantial drag reduction in marine applications, suggesting potential applications in automotive design to further improve energy efficiency (Zheng et al., 2023). Hybrid technology, particularly in truck chassis design, is also being optimized through advanced modeling and optimization techniques, achieving significant mass reductions while maintaining structural integrity under various load conditions (De et al., 2020). Together, these strategies represent a comprehensive approach to reducing energy consumption and emissions in the automotive sector, highlighting the importance of integrating material science, aerodynamics, and hybrid technology in vehicle design.

(Continued)

TABLE 2.3 (*Continued*)
Examples of Reduction Opportunities in the Transportation Sector

Type	How It Reduces Emissions	Examples
Improving Operating Practices	Adopting better driving and vehicle maintenance practices to save fuel.	Avoiding rapid acceleration, reducing idling, and better planning for ships and aircraft are crucial strategies for enhancing safety and efficiency in maritime and aviation operations. The integration of variable speed route planning models, as proposed by Qian et al., can significantly mitigate risks associated with sudden maritime hazards by incorporating spatial and temporal dimensions into route planning. This approach minimizes risk and aligns with International Maritime Organization norms, demonstrating superior performance over traditional uniform speed schemes (Qian et al., 2023). In the context of dynamic ocean management, Hausner et al. highlight the effectiveness of adaptive strategies that adjust vessel speeds based on real-time environmental conditions, such as whale habitats, to reduce lethal ship collisions. These strategies, particularly dynamic ones, have shown increased efficacy during anomalous conditions like marine heatwaves, underscoring the importance of flexible management in response to climate change (Hausner et al., 2021). Furthermore, collision avoidance systems for both ships and aircraft, as described by Philippe, utilize trajectory prediction and lateral peripheral envelopes to preemptively address potential conflicts, enhancing navigational safety (Philippe, 2019). These advancements collectively emphasize the importance of integrating technology and adaptive strategies to optimize the planning and operation of ships and aircraft, thereby reducing unnecessary acceleration and idling while enhancing safety and environmental compliance.
Reducing Travel Demand	Urban planning to reduce the need for driving and encourage low-emission transport options.	Building public transport, bike paths, and mixed-use areas is a strategic approach to reducing driving distances and promoting sustainable urban development. Mixed-use urbanism, which integrates diverse land uses within pedestrian-accessible contexts, has decreased driving and air pollution while enhancing active transportation and public health (Simons, 2021). The relationship between transportation supply and mixed land-use (MLU) is crucial, as areas with high accessibility and density, particularly near city centers, tend to exhibit more effective MLU patterns (Almansoub et al., 2022). This integration supports urban sustainability by fostering walking-friendly, economically, and socially vibrant communities (Zagorskas, 2016, 2017). The use of advanced modeling techniques, such as GIS-based methods and neural networks, allows for precise analysis and forecasting of MLU patterns at both macro and micro scales, providing valuable insights for urban planners and policymakers. Moreover, mixed-use developments (MXDs) can significantly reduce traffic impacts by capturing a large share of trips internally, thus minimizing external traffic generation (Tian et al., 2015). These methods emphasize the importance of scale in planning, ensuring that urban environments are designed to support sustainable transportation and land-use practices effectively.

(Purcell et al., 2024). Overall, the integration of energy-efficient technologies, renewable energy, and strategic management practices is key to reducing emissions in the commercial and residential sectors (Martínez et al., 2019).

Total residential and commercial GHG emissions, including direct and indirect emissions, in 2021 have decreased by 2% since 1990. GHG emissions from on-site direct emissions in homes and businesses have increased by 2% since 1990. Additionally, indirect emissions from electricity use by homes and businesses increased from 1990 to 2007 but have decreased since then to approximately 4% below 1990 levels in 2021. Table 2.4 provides opportunities to reduce emissions from homes and businesses.

2.5 AGRICULTURE SECTOR

Agriculture's contribution to U.S. GHG emissions, primarily through soil management, enteric fermentation, and manure management, is a significant concern, accounting for 10% of the nation's emissions. This sector's emissions have increased by 7% since 1990, highlighting the need for effective mitigation strategies. Soil management is a major contributor to global emissions, with carbon dioxide, nitrous oxide, and methane being the primary gases released from soil activities (Kopittke et al., 2024). Nitrous oxide, in particular, has a global warming potential 265 times greater than carbon dioxide, and agriculture is responsible for approximately 60% of its global anthropogenic emissions (Júnior et al., 2021). Strategies to reduce these emissions include improving nitrogen fertilizer efficiency and adopting sustainable agricultural practices such as precision farming, crop rotation, and organic farming, which enhance soil fertility and reduce synthetic pesticide use (Saikanth et al., 2023). Additionally, reducing the carbon footprint of agricultural products through life cycle analysis can help optimize production methods and lower emissions by minimizing inputs like fertilizers and fuel (Litskas et al., 2020). Effective mitigation practices also involve enhancing soil carbon sequestration, diversifying cropping systems, and limiting deforestation (Jaiswal & Agrawal, 2020). These strategies not only aim to reduce emissions but also ensure food security and environmental conservation, making them essential for sustainable agricultural development.

Agricultural soil management activities, such as the application of synthetic and organic fertilizers, deposition of livestock manure, and growing nitrogen-fixing plants, were the largest contributors to U.S. N_2O emissions in 2021, accounting for 75% of total N_2O emissions. Emissions from other agricultural sources have generally remained flat or changed by a relatively small amount since 1990.

All emission estimates from the *Inventory of U.S. Greenhouse Gas Emissions and Sinks: 1990–2021* as shown in the above Figure 2.5.

Table 2.5 shown below provides examples of opportunities to reduce emissions from agriculture.

2.6 LAND USE, LAND-USE CHANGE, AND FORESTRY SECTOR EMISSIONS AND SEQUESTRATION

Land Use, Land-Use Change, and Forestry (LULUCF) activities in the U.S. have historically functioned as a significant carbon sink, offsetting a substantial portion of the nation's GHG emissions. However, this capacity has diminished by 14% since 1990, underscoring the critical need for effective forest preservation and improved land management practices. The decline in carbon sink capacity is influenced by various factors, including historical land-use changes and the legacy of past land-use practices, which affect the age distribution of forests and soil nutrient levels, impacting future carbon dynamics (Felzer, 2023). Globally, forests have been a significant carbon sink, although their strength has decreased over time, with net emissions from forest conversion partially offset by removals on forest land (Tubiello et al., 2021). In the U.S., forests, harvested wood products, and urban trees collectively represent the largest net carbon sink, offsetting over 11% of total

TABLE 2.4
Examples of Reduction Opportunities in the Residential and Commercial Sector

Type	How It Reduces Emissions	Examples
Homes and Commercial Buildings	Improving energy efficiency to use less energy for heating, cooling, lighting, and appliances.	Integrating better insulation, efficient LED lighting, energy-efficient appliances, and passive heating using sunlight is a multifaceted approach to enhancing building energy efficiency. Improved insulation techniques, such as high-performance glazing and advanced thermal mass materials, are crucial in minimizing energy loss and optimizing passive solar heating systems, which capture and utilize solar energy effectively to reduce reliance on mechanical heating and cooling systems (Patel, 2024). Energy-efficient appliances and LED lighting further reduce energy consumption by optimizing electricity use, as highlighted in the Green Ship Homes criteria by the Green Building Council Indonesia, which emphasizes the importance of energy-saving home appliances and artificial lighting (Rahmawati et al., 2018). Passive solar design features, such as Trombe walls and sunspaces, are effective in maintaining stable indoor temperatures and reducing heating energy consumption, as demonstrated in various climate regions in China (Gu & Xiermaimaiti, 2024). Additionally, shading devices play a significant role in controlling solar heat gain, thereby reducing cooling loads and enhancing daylight utilization throughout the year (Babota, 2014). The implementation of these strategies is supported by energy policies and infrastructure that promote the use of renewable energy sources and sustainable materials, as seen in notable green building projects like the Pearl River Tower and the Bullitt Center (Shafik, 2024). Collectively, these measures improve energy efficiency and contribute to environmental sustainability and occupant comfort in residential and commercial buildings.
Wastewater Treatment	Making water and wastewater systems use less energy.	Implementing energy-efficient systems in water and wastewater treatment plants can lead to significant energy savings, ranging from 15% to 30%, as demonstrated by various strategies and technologies discussed in the literature. One effective approach is the enhancement of energy efficiency through resource recovery techniques such as anaerobic digestion and advanced oxidation processes, which can transform wastewater treatment plants (WWTPs) into energy-efficient resource recovery hubs (Twi-Yeboah et al., 2024). Waste heat recovery (WHR) processes, particularly when combined with technologies like the Organic Rankine Cycle (ORC), have shown promise in maximizing energy efficiency. For instance, the As-Samra WWTP in Jordan demonstrated the economic feasibility of such systems, achieving a net present value of 2.09 million USD and a payback period of 4.8 years (Al-Dahidi et al., 2024). Optimization of aeration systems, a major energy consumer in WWTPs, can also contribute to energy savings, as evidenced by interventions that resulted in significant reductions in greenhouse gas emissions and favorable economic outcomes (Campo et al., 2023). Furthermore, advanced control frameworks, such as economic dynamic matrix control (EDMC), have been developed to optimize energy consumption, offering higher efficiency compared to traditional methods (Protoulis et al., 2023). Intelligent operation management systems using genetic algorithms and machine learning models, like the XGB-SEGA method, have been shown to reduce energy consumption by 13%–27% by optimizing key operational parameters without the need for complex data acquisition systems (Wang et al., 2024). These diverse strategies highlight the potential for substantial energy savings in WWTPs by adopting innovative technologies and optimization techniques.

(Continued)

TABLE 2.4 (*Continued*)
Examples of Reduction Opportunities in the Residential and Commercial Sector

Type	How It Reduces Emissions	Examples
Waste Management	Reducing waste sent to landfills and capturing methane from existing landfills.	Recycling, waste reduction programs, and capturing methane from landfills are critical components in managing landfill emissions and enhancing environmental sustainability. Effective landfill management can significantly mitigate methane emissions, a potent greenhouse gas. Early gas recovery and reducing biodegradable organic waste in landfills are identified as effective strategies for methane mitigation, though the latter can be challenging due to cost and infrastructure limitations in many regions (Scharff et al., 2023). In the European Union, adopting 'best-in-class' landfill practices, which include advanced gas capture systems and bioreactor operations, could reduce emissions by 36% and enhance energy recovery, aligning with EU environmental goals (Lair et al., 2024). In South America, the Villavicencio Landfill in Colombia exemplifies the potential of methane recovery for energy production, with projections indicating substantial electricity and thermal energy generation, supporting regional energy self-sufficiency and sustainability (Ríos et al., 2024). In developing countries, like Pakistan, conventional anaerobic landfills with methane capture and post-aeration have shown a 65% reduction in methane emissions, highlighting the importance of adopting sustainable waste management practices to address local and global environmental challenges (Sohoo et al., 2020). Collectively, these strategies underscore the importance of integrating recycling, waste reduction, and methane capture to enhance landfill management and reduce environmental impacts.
Air Conditioning and Refrigeration	Reducing leaks from cooling systems and using refrigerants with lower global warming potential.	Upgrading cooling technology and utilizing refrigerants with lower GWP are critical steps in reducing the environmental impact of refrigeration systems. The transition from high GWP refrigerants like R22 to alternatives such as R444B and R454C has shown promising results. R444B, for instance, not only improves the coefficient of performance (COP) and cooling capacity by 15% and 9%, respectively, but also reduces the total equivalent warming impact (TEWI) by 55% compared to R22, making it a viable option for retrofitting existing systems without significant modifications (Martins et al., 2024). The use of fourth-generation refrigerants like R1234ze(E) and R1225ye(Z) in optimized cooling cycles has demonstrated the potential to enhance energy and exergy efficiency, leading to significant reductions in CO_2 emissions and operational costs (Albà et al., 2024). In cryogenic applications, mixed refrigerants with a GWP of 150 or less have been shown to increase the COP by 52% and reduce carbon emissions by 60%, highlighting their effectiveness in meeting stringent environmental standards (Jung et al., 2024). Furthermore, integrating eco-friendly refrigerants such as R1234ze(Z) in vapor compression systems has been identified as a superior choice due to its low environmental impact and high performance under various temperature conditions (Mishra et al., 2024). These advancements underscore the importance of adopting low GWP refrigerants and optimizing refrigeration systems to mitigate environmental footprint while maintaining or enhancing performance (Hsieh & Yeh, 2024).

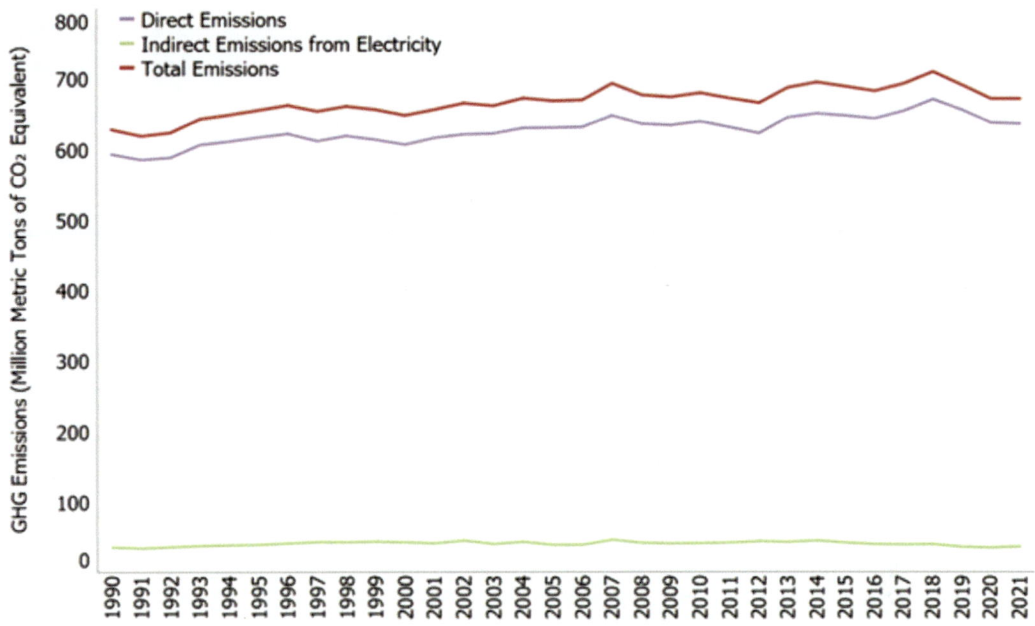

FIGURE 2.5 Greenhouse gas emissions from agriculture, 1990–2021.

GHG emissions annually (Domke et al., 2019). The role of forests in the carbon cycle is crucial, as they account for a significant portion of global carbon uptake. However, their effectiveness is heavily influenced by human land use, climate change, and disturbances (Hurteau, 2021). Land-use changes, such as converting forests to agricultural lands, significantly impact GHG emissions and the global carbon cycle, contributing to climate change (Gries et al., 2019). Therefore, maintaining and enhancing the carbon sink capacity of U.S. forests requires addressing these complex interactions through informed land management and conservation strategies. Additionally, while episodic in nature, increased CO_2, CH_4, and N_2O emissions from forest fires have also occurred over the time series (Figure 2.6).

Table 2.6 below provides some examples of opportunities for reducing emissions and enhancing sinks.

2.7 CONCLUSION

Greenhouse gas emissions come from nearly every part of modern life—power generation, industry, transport, buildings, agriculture, and land use. This chapter broke down how each sector contributes, with fossil fuel burning, industrial processes, and deforestation as major drivers. We also saw how emissions are deeply interconnected: electricity use in homes and factories, for example, links back to power plant emissions. Agriculture not only emits methane and nitrous oxide but is also tied to land-use change and deforestation. While land and forests can act as carbon sinks, their capacity is shrinking due to human activity.

Solutions exist, including cleaner energy, smarter industry, efficient transportation, sustainable farming, and improved land management, all of which help reduce emissions. However, progress requires a coordinated approach to policy, investment, and technology—and a focus on both reducing emissions and enhancing carbon sinks.

TABLE 2.5
Examples of Reduction Opportunities for the Agriculture Sector

Type	How It Reduces Emissions	Examples
Land and Crop Management	Changing how land and crops are managed to reduce emissions.	Several strategies have been identified across various studies to effectively manage nitrogen fertilizer and water in rice fields and minimize greenhouse gas emissions. Intermittent wetting and drying (IWD) is a promising water management technique that significantly reduces methane (CH_4) emissions compared to continuous flooding, as it alters the soil's redox potential, thereby affecting the microbial processes responsible for CH_4 production (Rajbonshi et al., 2024). Mid-summer drainage, particularly when prolonged and combined with slow-release fertilizers, has been shown to reduce both CH_4 and nitrous oxide (N_2O) emissions while slightly increasing grain yield (Jeong et al., 2023). The type and method of nitrogen fertilizer application also play crucial roles. For instance, deep placement of ammonium sulfate and tablet urea can lower CH_4 emissions and enhance rice yield compared to surface broadcasting (Wihardjaka et al., 2023). Moreover, replacing a portion of chemical fertilizers with organic alternatives or optimizing the timing and type of nitrogen application can reduce N_2O emissions significantly. However, this may sometimes lead to a slight increase in CH_4 emissions (Zheng et al., 2022). Balanced fertilizer application and intermittent irrigation have decreased N_2O emissions by 54% compared to conventional practices, highlighting the importance of integrated management approaches for sustainable rice production (Lina et al., 2022). These findings collectively suggest that a combination of judicious nitrogen management and strategic water control can effectively mitigate greenhouse gas emissions from rice paddies while maintaining or even enhancing crop yields.
Livestock Management	Adjusting feeding and care practices to cut methane emissions from animals.	Improving pasture quality and implementing better breeding practices are crucial strategies for enhancing livestock productivity and reducing methane emissions. Sustainable pasture management, such as the use of well-managed Brachiaria cultivars, has been shown to increase herbage mass and animal weight gain while reducing enteric methane emissions per unit of product, thus contributing to a more efficient forage-livestock system (Silva et al., 2024). Nutritional interventions, including the use of high-quality pastures with increased crude protein and digestibility, can significantly lower methane emissions from cattle, as demonstrated by a reduction of 14% in emissions when animals grazed on high-quality pastures (Dini et al., 2017). Integrating diverse forage species with complementary biochemical properties can enhance animal nutrition and welfare while mitigating environmental impacts (Distel et al., 2020). Breeding practices, such as genetic selection and biotechnology tools like RNA interference and marker-assisted selection, can further improve forage quality by increasing nutritional value and reducing harmful components (Sandal et al., 2023). These strategies, combined with effective management practices like monitoring forage mass and adjusting stocking rates, optimize forage quality and intake, reduce methane emissions, and improve sustainability in pasture-based systems (Galati et al., 2023). A holistic approach combining pasture quality improvement with advanced breeding techniques can significantly enhance livestock productivity and environmental sustainability.

(Continued)

TABLE 2.5 (*Continued*)
Examples of Reduction Opportunities for the Agriculture Sector

Type	How It Reduces Emissions	Examples
Manure Management	Controlling manure decomposition to cut methane and nitrous oxide emissions.	Storing manure as solids or using lagoons to capture methane for renewable energy presents a viable strategy for reducing greenhouse gas emissions and enhancing sustainability in agriculture. Mechanical solid-liquid separation technologies have shown significant potential in reducing methane emissions from manure storage. For instance, studies on dairy farms have demonstrated that using a two-stage sloped dual-screen separator can achieve methane emission reductions of up to 55.8% by effectively separating solids from liquids, thus preventing methane generation in lagoons (Edalati et al., 2023). Similarly, in swine production, retrofitting conventional lagoon systems with solid-liquid separation modules has resulted in a 65.5% reduction in methane emissions, highlighting the effectiveness of this approach in mitigating environmental impacts (Sohoulande et al., 2024). Beyond separation, integrating biogas plants to process manure into renewable energy offers additional benefits. This approach reduces waste and generates biogas, which can be used for energy production, thereby contributing to a closed nutrient cycle and reducing greenhouse gas emissions (Pata et al., 2024). Furthermore, the closed-loop dairy system concept, which includes converting manure into biochar and syngas, exemplifies how manure can be transformed into valuable resources, enhancing soil health and providing renewable energy (Muell et al., 2022). These strategies underscore the importance of innovative waste management practices in livestock farming, which can lead to economic benefits, improved environmental quality, and a sustainable agricultural paradigm (Singh & Singh, 2024; Pata et al., 2024).

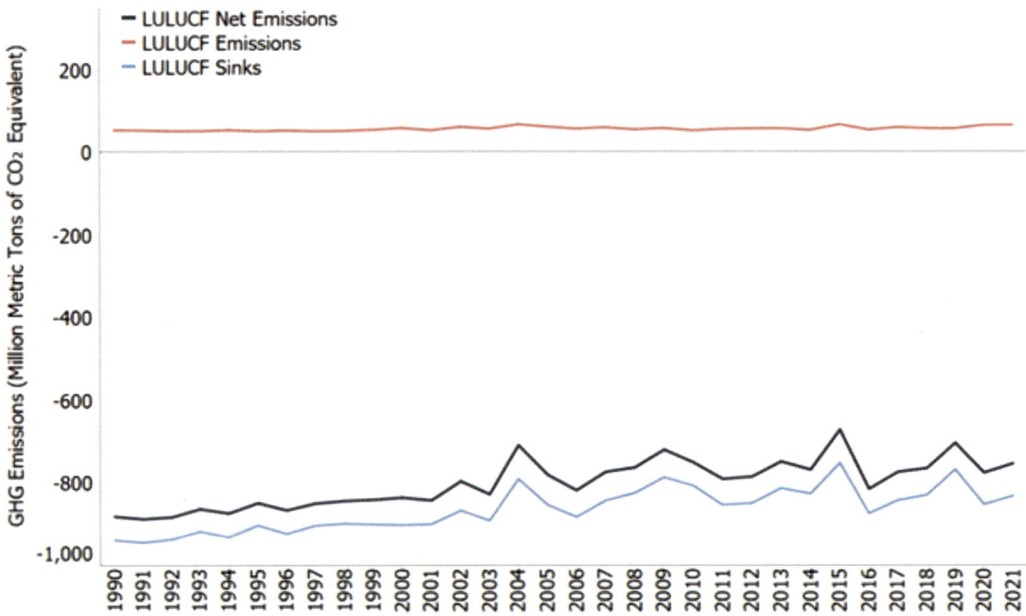

FIGURE 2.6 Greenhouse gas emissions and removals from U.S. Land Use, Land-Use Change, and Forestry, 1990–2021. Note: The LULUCF sector is a net "sink" of emissions in the United States (e.g., more greenhouse gas emissions are sequestered than emitted from land-use activities), so net greenhouse gas emissions from LULUCF are negative. All emission estimates are from the *U.S. Greenhouse Gas Emissions and Sinks Inventory: 1990–2021.*

TABLE 2.6
Examples of Reduction Opportunities in the LULUCF Sector

Type	How Emissions Are Reduced or Sinks Are Enhanced	Examples
Change in Uses of Land	Increasing carbon storage by using land differently or maintaining carbon storage by avoiding land degradation.	Forests are the largest terrestrial carbon pools, holding significant amounts of above-ground and soil organic carbon (SOC), crucial for reducing atmospheric CO_2 levels. In the Indian sub-Himalayan region, tree-based land uses, particularly mixed-species forests, have shown the highest carbon accumulation, significantly surpassing agricultural landscapes in carbon storage capacity. Afforestation, especially with species like black pine, has been demonstrated to significantly increase carbon stocks compared to bare lands, highlighting its potential in carbon sequestration efforts. Moreover, reforestation in the U.S. has been identified as a key strategy for enhancing carbon sequestration, with soils being the dominant carbon pool across ecosystems.
Changes in Land Management Practices	Improving management practices on existing land-use types.	Land and water management strategies, such as land allocation for climate-resilient crops, soil health management, and efficient water use technologies, are essential for maximizing agricultural productivity integrating agroecological, agrochemical, and agrotechnical approaches, such as soil assessment, rational fertilization, and erosion protection, further supports sustainable development in agriculture. Sustainable land management practices, including cover crops, mulching, and water harvesting, enhance yields, contribute to carbon sequestration, and reduce greenhouse gas emissions, particularly in areas with low and variable rainfall.

The next chapter moves from "where emissions come from" to "how we measure and compare them." Understanding tools such as carbon dioxide equivalent (CO_2e) and global warming potential (GWP) will be crucial for tracking progress, shaping policy, and holding sectors accountable as we work toward achieving net zero.

REVIEW QUESTIONS

1. What are the primary sectors contributing to GHG emissions?
2. How does carbon dioxide (CO_2) differ from other GHGs like methane (CH_4) in its impact on global warming?
3. What are the specific sources of emissions in the transportation sector, and how can they be reduced?
4. How do agricultural practices contribute to GHG emissions, and what solutions exist?
5. Why is it important to measure emissions in carbon dioxide equivalent (CO_2e) across sectors?
6. What role does industrial activity play in GHG emissions, and what steps can industries take to mitigate this?
7. How do land-use changes, like deforestation, contribute to climate change?
8. How can renewable energy sources replace fossil fuels in the power sector?
9. What are carbon sinks, and how do they help offset emissions?
10. How can governments and corporations work together to reduce emissions in high-impact sectors?

REFERENCES

Al-Dahidi, S., Alrbai, M., Al-Ghussain, L., & Alahmer, A. (2024). Maximizing energy efficiency in wastewater treatment plants: A data-driven approach for waste heat recovery and an economic analysis using Organic Rankine Cycle and thermal energy storage. *Applied Energy*. https://doi.org/10.1016/j.apenergy.2024.123008

Albà, C. G., Alkhatib, I. I. I., Llovell, F., & Vega, L. F. (2024). A novel approach for designing efficient and sustainable cooling cycles with low global warming refrigerants. *Applied Thermal Engineering*. https://doi.org/10.1016/j.applthermaleng.2024.122895

Almansoub, Y. A. M., Zhang, M., Raza, A. A., Safdar, M. N., Dahou, A., & Al-qaness, M. A. A. (2022). Exploring the effects of transportation supply on mixed land-use at the parcel level. *Land*. https://doi.org/10.3390/land11060797

Anarene, B. (2024). Revolutionizing energy efficiency in commercial and institutional buildings: A complete analysis. *International Journal of Scientific Research and Management*. https://doi.org/10.18535/ijsrm/v12i09.em12

Arif, A., Rizwan, M., Elkamel, A., Hakeem, L., & Zaman, M. (2020). Optimal selection of integrated electricity generation systems for the power sector with low greenhouse gas (GHG) emissions. *Energies*. https://doi.org/10.3390/EN13174571

Babatunde, O. M., Munda, J. L., & Hamam, Y. (2019). Decarbonisation of Electricity Generation: Efforts and Challenges. In S. Muthu (Ed.), *Carbon Footprints. Environmental Footprints and Eco-design of Products and Processes* (pp. 47–77). Springer. https://doi.org/10.1007/978-981-13-7912-3_3

Babota, F. (2014). *Increase Energy Efficiency and Comfort in Homes by Incorporating Passive Solar Design Features*. The Bulletin of the Polytechnic Institute of Jassy, Construction. Architecture Section.

Bahman, N., Alalaiwat, D., Abdulmohsen, Z., Khalifa, M. A., Baharna, S. A., Al-Mannai, M., & Younis, A. (2022). A critical review on global CO_2 emission: where do industries stand? *Reviews on Environmental Health*. https://doi.org/10.1515/reveh-2022-0105

Bello, K. A., Awogbemi, O., & Kanakana-Katumba, M. G. (2023). Assessment of alternative fuels for sustainable road transportation. In 2023 IEEE 11th International Conference on Smart Energy Grid Engineering (SEGE), Oshawa, ON, Canada, 2023, pp. 7–15. https://doi.org/10.1109/sege59172.2023.10274583

Berrill, P., Wilson, E. J. H., Reyna, J., Fontanini, A. D., & Hertwich, E. (2022). *Decarbonization Pathways for the Residential Sector in the United States*. https://doi.org/10.21203/rs.3.rs-1199406/v1

Bhardwaj, A. (2024). Impact of global warming. *International Journal of Recent Trends in Multidisciplinary Research*. https://doi.org/10.59256/ijrtmr.20240405002

Campo, G., Miggiano, A. G., Panepinto, D., & Zanetti, M. (2023). Enhancing the energy efficiency of wastewater treatment plants through the optimization of the aeration systems. *Energies*. https://doi.org/10.3390/en16062819

Candela, A., Sandrini, G., Gadola, M., Chindamo, D., & Magri, P. (2024). Lightweighting in the automotive industry as a measure for energy efficiency: Review of the main materials and methods. *Heliyon*. https://doi.org/10.1016/j.heliyon.2024.e29728

Charkovska, N., Halushchak, M., Halushchak, M., Bun, R., Nahorski, Z., Oda, T., Oda, T., Jonas, M., & Topylko, P. (2019). A high-definition spatially explicit modelling approach for national greenhouse gas emissions from industrial processes: reducing the errors and uncertainties in global emission modelling. *Mitigation and Adaptation Strategies for Global Change*. https://doi.org/10.1007/S11027-018-9836-6

Chowdhuri, A., & Gupta, C. K. (2023). Gauging the assessment of some anthropogenic factors driving climate-change. *Current World Environment*. https://doi.org/10.12944/cwe.18.2.16

De, S., Singh, K., Seo, J., Kapania, R. K., Ostergaard, E., Angelini, N., & Aguero, R. (2020). Lightweight chassis design of hybrid trucks considering multiple road conditions and constraints. *World Electric Vehicle Journal*. https://doi.org/10.3390/WEVJ12010003

Dehm, J. (2018). One tonne of carbon dioxide equivalent (1tCO_2e). In J. Hohmann, D. Joyce (Eds.), *International Law's Objects* (pp. 305–318). Oxford Academic. https://doi.org/10.1093/OSO/9780198798200.003.0026

Dini, Y., Gere, J., Cajarville, C., & Ciganda, V. S. (2017). Using highly nutritious pastures to mitigate enteric methane emissions from cattle grazing systems in South America. *Animal Production Science*. https://doi.org/10.1071/AN16803

Distel, R. A., Arroquy, J. I., Arroquy, J. I., Lagrange, S., Lagrange, S., & Villalba, J. J. (2020). Designing diverse agricultural pastures for improving ruminant production systems. *Frontiers in Sustainable Food Systems*. https://doi.org/10.3389/FSUFS.2020.596869

Domke, G. M., Domke, G. M., Walters, B. F., Walters, B. F., Nowak, D. J., Nowak, D. J., Smith, J. E., Smith, J. E., Ogle, S. M., Coulston, J. W., Coulston, J. W., Coulston, J. W., & Wirth, T. C. (2019). Greenhouse gas emissions and removals from forest land, woodlands, and urban trees in the United States, 1990–2018. In Resource Update FS–307. U.S. Department of Agriculture, Forest Service, Northern Research Station.

Edalati, A., Chen, Y. M., Barzee, T. J., El-Mashad, H. M., & Zhang, R. (2023). Effect of mechanical solids separators on potential reduction of methane emissions from dairy manure storage. *Journal of the ASABE*. https://doi.org/10.13031/ja.15371

Fan, J., Meng, X., Tian, J., Xing, C., Wang, C., & Wood, J. (2023). A review of transportation carbon emissions research using bibliometric analyses. *Journal of Traffic and Transportation Engineering*. https://doi.org/10.1016/j.jtte.2023.09.002

Fancellu, A. (2023). The impact of CNG on buses fleet decarbonization: A case study. In E.G. Nathanail, N. Gavanas, G.Adamos (Eds.), *Smart Energy for Smart Transport. CSUM 2022. Lecture Notes in Intelligent Transportation and Infrastructure* (pp. 606–620). Springer. https://doi.org/10.1007/978-3-031-23721-8_50

Felzer, B. S. (2023). Effect of land-use legacy on the future carbon sink for the conterminous US. *Biogeosciences*. https://doi.org/10.5194/bg-20-573-2023

Galati, R. L., Silva, L. C. R. P., Cabral, L. da S., & Cerom, M. F. S. Q. (2023). Challenges and strategies for reducing enteric methane emissions in sustainable animal production. *Concilium*. https://doi.org/10.53660/clm-1691-23j63

Ghosh, T., Ingwersen, W. W., Jamieson, M., Hawkins, T. R., Cashman, S., Hottle, T. A., Carpenter, A., & Richa, K. (2022). Derivation and assessment of regional electricity generation emission factors in the USA. *International Journal of Life Cycle Assessment*. https://doi.org/10.1007/s11367-022-02113-1

Gries, T., Redlin, M., & Ugarte, J. E. (2019). Human-induced climate change: The impact of land-use change. *Theoretical and Applied Climatology*. https://doi.org/10.1007/S00704-018-2422-8

Gu, W., & Xiermaimaiti, A. (2024). Evaluation of energy-saving potential and indoor thermal environment of passive solar houses in various heating climate regions of China. *Energy Sources Part A-Recovery Utilization and Environmental Effects*. https://doi.org/10.1080/15567036.2024.2339528

Gutierrez, D. M. R., & Ladisch, M. R. (2024). Electric vehicles, biofuels, and transitions in transportation energy. *Industrial Biotechnology*. https://doi.org/10.1089/ind.2023.0021

Hassan, N. E. (2024). Global warming: Causes, impacts and urgent strategies for a sustainable future: A review. *GSC Advanced Research and Reviews*. https://doi.org/10.30574/gscarr.2024.20.3.0338

Hausner, A., Samhouri, J. F., Hazen, E. L., Delgerjargal, D., Abrahms, B., & Abrahms, B. (2021). Dynamic strategies offer potential to reduce lethal ship collisions with large whales under changing climate conditions. *Marine Policy*. https://doi.org/10.1016/J.MARPOL.2021.104565

Hoehne, C. G., Muratori, M., Jadun, P., Bush, B., Yip, A. H., Ledna, C., Vimmerstedt, L., Podkaminer, K., & Ma, O. (2023). Exploring decarbonization pathways for USA passenger and freight mobility. *Nature Communications*. https://doi.org/10.1038/s41467-023-42483-0

Hsieh, C., & Yeh, R. (2024). Performance enhancement analysis of environmentally friendly refrigerants. *Processes*. https://doi.org/10.3390/pr12050904

Hurteau, M. D. (2021). The role of forests in the carbon cycle and in climate change. In T.M. Letcher (Ed.), *Climate Change: Observed Impacts on Planet Earth* (pp. 561–579). Elsevier. https://doi.org/10.1016/B978-0-12-821575-3.00027-X

Isella, A., & Manca, D. (2022). GHG emissions by (petro)chemical processes and decarbonization priorities—A review. *Energies*. https://doi.org/10.3390/en15207560

Jaiswal, B., & Agrawal, M. (2020). Carbon footprints of agriculture sector. In S. Muthu (Eds.), Carbon Footprints. Environmental Footprints and Eco-design of Products and Processes (pp. 81–99). Springer. https://doi.org/10.1007/978-981-13-7916-1_4

Jeong, H. C., Gwon, H. S., Lee, H.-S., Park, H., Lee, J. M., Oh, T. K., & Lee, S.-I. (2023). Effect of water management on greenhouse gas emissions from rice paddies using a slow-release fertilizer. *Korean Journal of Environmental Agriculture*. https://doi.org/10.5338/kjea.2023.42.2.14

Jung, H., Son, C.-H., & Lee, J.-H. (2024). Proposals for next-generation eco-friendly non-flammable refrigerants for a −100 °C semiconductor etching chiller based on 4E (energy, exergy, environmental, and exergoeconomic) analysis. *Energies*. https://doi.org/10.3390/en17194969

Júnior, V. M., Comin, J. J., Ferreira, G. W., Tavares, J. M. R., Couto, R. da R., & Filho, P. B. (2021). Nitrous oxide emissions in soils fertilized with pig manure: soil processes and strategies of control and mitigation. *Research, Society and Development*. https://doi.org/10.33448/RSD-V10I2.12427

Kabeyi, M. J. B., & Olanrewaju, O. A. (2020). Managing sustainability in electricity generation. *Industrial Engineering and Engineering Management*. https://doi.org/10.1109/IEEM45057.2020.9309994

Karaca, A. E., & Dincer, I. (2021). A hybrid compressed natural gas-pneumatic system as a powering option for buses: A comparative assessment. *Energy*. https://doi.org/10.1016/J.ENERGY.2021.120865

Kopittke, P. M., Dalal, R. C., McKenna, B. A., Smith, P., Wang, P., Weng, Z., Bom, F. J. T. van der, & Menzies, N. W. (2024). Soil is a major contributor to global greenhouse gas emissions and climate change. *EGUsphere*. https://doi.org/10.5194/egusphere-2024-1782

Kumar, D. S. (2021). Demand Side Management Techniques for Peak Reduction. *International Journal for Research in Applied Science and Engineering Technology*, 9, 2911–2913. https://doi.org/10.22214/IJRASET.2021.36979

Kumar, P., Brar, G. S., Singh, S., Nikolovski, S., Baghaee, H. R., & Balkić, Z. (2019). Perspectives and intensification of energy efficiency in commercial and residential buildings using strategic auditing and demand-side management. *Energies*. https://doi.org/10.3390/EN12234539

Kyung, J. Y., Phil, P. K., Beom, L. J., Jin, L. H., & Ra, J. A. (2017). Ships using the aircraft safe operation support systems.

Lair, A., Mansuy, M., Romand, C., Oberti, O., Pradels, C., Barina, G., Denoun, T., Venturini, M., & Trommsdorff, C. (2024). Enhancing landfill efficiency to drive greenhouse gas reduction: A comprehensive study on best practices and policy recommendations. *Waste Management & Research*. https://doi.org/10.1177/0734242x241270951

Lina, S., Hesti, Y., & Ali, P. J. (2022). Managing fertilizers and water treatments on rice production and N_2O emission. *IOP Conference Series*. https://doi.org/10.1088/1755-1315/1039/1/012012

Litskas, V. D., Litskas, V. D., Platis, D. P., Anagnostopoulos, C. D., Tsaboula, A. C., Menexes, G., Kalburtji, K. L., Stavrinides, M. C., & Mamolos, A. P. (2020). Climate change and agriculture: Carbon footprint estimation for agricultural products and labeling for emissions mitigation. In N. Betoret, E. Betoret (Eds.), *Sustainability of the Food System: Sovereignty, Waste, and Nutrients Bioavailability* (pp. 33–49). Elsevier. https://doi.org/10.1016/B978-0-12-818293-2.00003-3

Martínez, D. M., Ebenhack, B. W., & Wagner, T. P. (2019). Residential and commercial sector energy efficiency. In D.M. Martínez, B.W. Ebenhack, T.P. Wagner (Eds.), *Energy Efficiency: Concepts and Calculations* (pp. 227–269). Elsevier. https://doi.org/10.1016/B978-0-12-812111-5.00008-1

Martins, L. V. S., Braga, C. H. M., Pabón, J. J. G., Machado, L., & Duarte, W. M. (2024). Assessment of total equivalent warming impact (TEWI) of alternative refrigerants for retrofit of R22 in single split air conditioning system. *Journal of Building Engineering*. https://doi.org/10.1016/j.jobe.2024.109085

Mishra, R. S., Agarwal, A., Dixit, J., & Kadam, S. (2024). Methods of improving thermal performance of vapour compression based refrigeration system through eco friendly refrigerants to reduce their environmental impact. *International Journal of Emerging Science and Engineering*. https://doi.org/10.35940/ijese.f9569.12050424

Moura, P., Mohan, A., Lau, S., & Mohammadi, J. (2023). Maximizing decarbonization benefits of transportation electrification in the U.S. *Electricity*. https://doi.org/10.3390/electricity4010004

Muell, J., Mohtar, R. H., Kan, E., Assi, A. T., & Pappa, V. A. (2022). Farm-scale water-energy-food-waste nexus analysis for a closed-loop dairy system. *Frontiers in Environmental Science*. https://doi.org/10.3389/fenvs.2022.880839

Nunes, L. J. R. (2023). The rising threat of atmospheric CO2: A review on the causes, impacts, and mitigation strategies. *Environments*. https://doi.org/10.3390/environments10040066

Pata, B. A., Patel, S., & Lokendra, L. (2024). Optimizing livestock farm waste management: Effective strategies for efficiency and sustainability. In *Futuristic Trends in Agriculture Engineering & Food Sciences Volume 3 Book 21* (pp. 138–154). IIP. https://doi.org/10.58532/v3bcag21p2ch8

Patel, A. K. (2024). Energy efficient building design with solar passive heating and ventilation systems. *Global Journal of Engineering and Technology Advances*. https://doi.org/10.30574/gjeta.2024.20.2.0144

Philippe, R. (2019). Systems *and Method for Collision Avoidance between Aircraft or Ships*. US Patent: 1192897932 (Extrated from Google Patent).

Protoulis, T., Kalogeropoulos, I., Kordatos, I., Kapnopoulos, A., Zervas, P. L., Vangelatos, G., Sarimveis, H., & Alexandridis, A. (2023). An identification and control framework for optimizing the energy consumption of a wastewater treatment plant. In *2023 IEEE 6th International Conference and Workshop Óbuda on Electrical and Power Engineering (CANDO-EPE)*, Budapest, Hungary, pp. 000043–000048. https://doi.org/10.1109/cando-epe60507.2023.10418031

Purcell, L., O'Regan, A., McGookin, C., & Nyhan, M. M. (2024). Modelling & spatial mapping of residential-sector emissions for sub-national & urban areas. *MethodsX*. https://doi.org/10.1016/j.mex.2024.102617

Qian, L., Hong, M., Qi, Y., & Guo, Z. (2023). Avoiding sudden maritime risk: A new variable speed route planning model by integrating Spatio-temporal dimension. *Ocean Engineering*. https://doi.org/10.1016/j.oceaneng.2023.115950

Rahmawati, A., Sucipto, T. L. A., & Aripin, M. S. (2018). Implementing energy efficiency and conservation in green homes as learning media for vocational building engineering in Indonesia. *AIP Conference Proceedings*. https://doi.org/10.1063/1.5043024

Rajbonshi, M. P., Mitra, S., & Bhattacharyya, P. (2024). Agro-technologies for greenhouse gases mitigation in flooded rice fields for promoting climate smart agriculture. *Environmental Pollution*. https://doi.org/10.1016/j.envpol.2024.123973

Rinaldi, A., Patel, M. K., & Parra, D. (2023). Optimal pathways for the decarbonisation of the transport sector: Trade-offs between battery and hydrogen technologies using a whole energy system perspective. *Cleaner Production Letters*. https://doi.org/10.1016/j.clpl.2023.100044

Ríos, L. F. R., Becerra, D., & Ortega-Contreras, J. Y. (2024). Potential use of methane gas from the Villavicencio sanitary landfill, Colombia. *Ingeniería y Competitividad*. https://doi.org/10.25100/iyc.v26i2.14019

Rosenthal, S., Maaß, F., Kamaliev, M., Hahn, M., Gies, S., & Tekkaya, A. E. (2020). Lightweight in automotive components by forming technology. *Automotive Innovation*. https://doi.org/10.1007/S42154-020-00103-3

Sahoo, S. K. (2019). Energy and electricity. In S.N. Kundu, M. Nawaz (Eds.), *Sustainable Energy and Environment: An Earth System Approach*. Apple Academic Press. https://doi.org/10.1201/9780429430107-7

Saikanth, D. R. K., Supriya, S., Singh, B., Rai, A. K., Bana, S. R., & Singh, B. (2023). Advancing sustainable agriculture: A comprehensive review for optimizing food production and environmental conservation. *International Journal of Plant and Soil Science*. https://doi.org/10.9734/ijpss/2023/v35i163169

Sandal, S. S., Walia, P., & Vidya, A. S. (2023). Quality enhancement in forage crops of cereals and legumes. *International Journal of Plant and Soil Science*. https://doi.org/10.9734/ijpss/2023/v35i163142

Saxena, S., Mahesha, G. T., & Srinivas, G. (2022). Aerodynamic drag reduction of commercial ground vehicles using numerical techniques. *IOP Conference Series: Earth and Environmental Science*. https://doi.org/10.1088/1755-1315/1100/1/012013

Scharff, H., Soon, H.-Y., Taremwa, S. R., Zegers, D., Dick, B., Zanon, T. V. B., & Shamrock, J. (2023). The impact of landfill management approaches on methane emissions. *Haikibutsu Gakkaishi*. https://doi.org/10.1177/0734242x231200742

Schoeneberger, C. A. (2022). Electrification potential of U.S. industrial boilers and assessment of the GHG emissions impact. *Advances in Applied Energy*. https://doi.org/10.1016/j.adapen.2022.100089

Seftiani, M. A., Sudarti, S., & Yushardi, Y. (2024). Mekanisme Dampak Pemanasan Global Oleh Gas Rumah Kaca. *Jurnal Pendidikan, Sains Dan Teknologi*. https://doi.org/10.47233/jpst.v3i2.1708

Shafik, W. (2024). Energy policies and infrastructure requirements for efficient green buildings. In C.S. Meena, A. Kumar, V.P. Singh, A. Ghosh (Eds.), *Sustainable Technologies for Energy Efficient Buildings*. CRC Press. https://doi.org/10.1201/9781003496656-10

Silva, D. S. M., Monteiro, A., Pedreira, B. C. E., Mombach, M. A., Pereira, D. H., Rodrigues, R. de A. R., & Matos, E. da S. (2024). Enhancing forage–livestock system productivity and mitigating greenhouse gas emissions via sustainable pasture management of two Brachiaria cultivars. *Crop & Pasture Science*. https://doi.org/10.1071/cp24054

Simons, G. D. (2021). The application of mixed-use measures at the pedestrian-scale. *arXiv: Physics and Society*.

Singh, N., & Singh, D. (2024). Agricultural waste management. In *Futuristic Trends in Agriculture Engineering & Food Sciences Volume 3 Book 22* (pp. 311–319). IIP. https://doi.org/10.58532/v3bcag22ch25

Sohoo, I., Ritzkowski, M., Kuchta, K., & Cinar, S. Ö. (2020). Environmental sustainability enhancement of waste disposal sites in developing countries through controlling greenhouse gas emissions. *Sustainability*. https://doi.org/10.3390/SU13010151

Sohoulande, C. D. D., Vanotti, M. B., & Szogi, A. A. (2024). Estimating methane emissions from swine waste treatment lagoons and the reduction through solid-liquid separation: A multiscale case evaluation. *Cleaner Waste Systems*. https://doi.org/10.1016/j.clwas.2024.100133

Speizer, S., Fuhrman, J., Aldrete, L. E., George, M., Kyle, P., Monteith, S., & McJeon, H. (2023). *A Zero-Emissions Global Transportation Sector: Advanced Technologies and Their Energy and Environmental Implications*. Research Square Platform LLC. https://doi.org/10.21203/rs.3.rs-2921936/v1

Tian, G., Ewing, R., White, A., Hamidi, S., Walters, J., Goates, J. P., & Joyce, A. (2015). Traffic generated by mixed-use developments: Thirteen-region study using consistent measures of built environment. *Transportation Research Record*. https://doi.org/10.3141/2500-14

Tubiello, F. N., Conchedda, G., Wanner, N., Federici, S., Rossi, S., & Grassi, G. (2021). Carbon emissions and removals from forests: new estimates, 1990–2020. *Earth System Science Data*. https://doi.org/10.5194/ESSD-13-1681-2021

Twi-Yeboah, N., Osei, D., Dontoh, W. H., Asamoah, G. A., Baffoe, J., & Danquah, M. K. (2024). Enhancing energy efficiency and resource recovery in wastewater treatment plants. *Energies*. https://doi.org/10.3390/en17133060

U.S. Energy Information Administration. (2022). *Electricity Explained - Basics*. U.S. Energy Information Administration.

Wang, Y., Zheng, S., & Wang, Z. (2018). Energy Efficiency and Emissions Analysis of Ammonia, *Hydrogen and Hydrocarbon Fuels*, 7(1), 47. https://doi.org/10.11648/J.JENR.20180701.17

Wang, Z., Zhou, X., Wang, H., Huang, Z., Ji, J., Peng, Z., & Jiang, K. (2024). XGB-SEGA coupled energy saving method for wastewater treatment plants. *Applied Water Science*. https://doi.org/10.1007/s13201-023-02081-3

Wihardjaka, A., Harsanti, E. S., Viandari, N. A., & Zu'amah, H. (2023). Effect of soil tillage and nitrogen fertilizer management on methane emissions from irrigated rice fields in Central Java, Indonesia. *Chilean Journal of Agricultural Research*. https://doi.org/10.4067/s0718-58392023000300347

Yusof, N. K., Abas, P. E., Mahlia, T. M. I., & Hannan, M. A. (2021). Techno-economic analysis and environmental impact of electric buses. *World Electric Vehicle Journal*. https://doi.org/10.3390/WEVJ12010031

Zagorskas, J. (2016). GIS-based modelling and estimation of land use mix in urban environment. *International Journal of Education, 1*.

Zagorskas, J. (2017). Gis-based estimation of function mix in urban environment at neighbourhood scale. In *10th International Conference "Environmental Engineering" (ICEE-2017)*. Vilnius, Lithuania, 27–28 April, 2017. https://doi.org/10.3846/ENVIRO.2017.129

Zheng, H., Gao, X., Zhang, J., Liu, D., Xi, X., & Sun, X. (2021). *Characteristics of Industrial Driving Effects Network and Impacts on Carbon Emissions*. Taylor & Francis. https://doi.org/10.6084/m9.figshare.14987981

Zheng, M.-Q., Liu, J., Jiang, P., Wu, J., Li, Y., & Li, S. (2022). [Effects of nitrogen fertilizer management on CH_4 and N_2O emissions in paddy field]. *Huan jing ke xue= Huanjing kexue*. https://doi.org/10.13227/j.hjkx.202107260

Zheng, S., Liang, X., Li, J., Liu, Y., & Tang, J. (2023). Drag reduction using bionic groove surface for underwater vehicles. *Frontiers in Bioengineering and Biotechnology*. https://doi.org/10.3389/fbioe.2023.1223691

3 The Human and Environmental Toll of Global Warming

This chapter explores the profound consequences of global warming, driven primarily by human activities, on human societies and the environment. It introduces the concept of the green economy as a potential solution, emphasizing the critical need to harmonize economic development with environmental sustainability. Key topics include the escalating disruption of ecosystems, mounting threats to food security, and the growing frequency and severity of natural disasters. The discussion also highlights the human toll of global warming, focusing on public health crises, exacerbating economic inequalities, and amplifying social injustices. Drawing on insights from the latest Intergovernmental Panel on Climate Change (IPCC) report, this chapter underscores the urgency of immediate and ambitious measures to mitigate and adapt to the unfolding climate crisis.

3.1 THE EVIDENCE OF TEMPERATURE CHANGE

Climate change, driven by human activities, is causing significant and widespread disruption to natural ecosystems and human societies. The latest IPCC report, *Climate Change 2022: Impacts, Adaptation and Vulnerability*, highlights the severe consequences of inaction and underscores the urgency of immediate and ambitious measures to mitigate and adapt to the accelerating climate crisis.

The planet is experiencing an increase in extreme weather events such as heatwaves, droughts, and floods, which exceed the tolerance thresholds of many plant and animal species. These events are leading to mass mortality, especially in vulnerable species like trees and corals. Furthermore, these weather extremes often occur simultaneously, causing cascading impacts that are difficult to manage. As a result, millions of people—particularly in Africa, Asia, Central and South America, small islands, and the Arctic—face acute food and water insecurity.

Global warming of 1.5°C above preindustrial levels, projected over the next two decades, will bring multiple unavoidable climate hazards. Even temporarily surpassing this warming threshold will lead to severe, irreversible impacts. Society will face increased risks, including threats to infrastructure and low-lying coastal settlements. Adaptation efforts to cope with climate change are progressing unevenly, with significant gaps between the current actions and what is needed. These gaps are most pronounced among lower-income populations, exacerbating social inequalities. The IPCC report calls for urgent, ambitious, and accelerated actions to adapt to climate change while simultaneously making deep cuts in greenhouse gas emissions. Failure to act now will result in a substantial loss of life, biodiversity, and infrastructure.

Cities, where more than half of the world's population resides, are hotspots of climate impacts and risks. Urban areas face complex challenges from climate change, especially those with poorly planned growth, high poverty levels, and inadequate services. However, cities also offer opportunities for climate action through green buildings, renewable energy, and sustainable transportation systems.

To secure a livable future, the report advocates for the protection and restoration of healthy ecosystems, which are more resilient to climate change and provide essential services like food and clean water. By conserving and restoring 30%–50% of Earth's land, freshwater, and ocean habitats, society can leverage nature's capacity to absorb and store carbon, thereby accelerating progress toward sustainable development. Climate change poses a grave threat to human well-being and the health of the planet. The window for effective action is rapidly closing, and

DOI: 10.1201/9781003479666-3

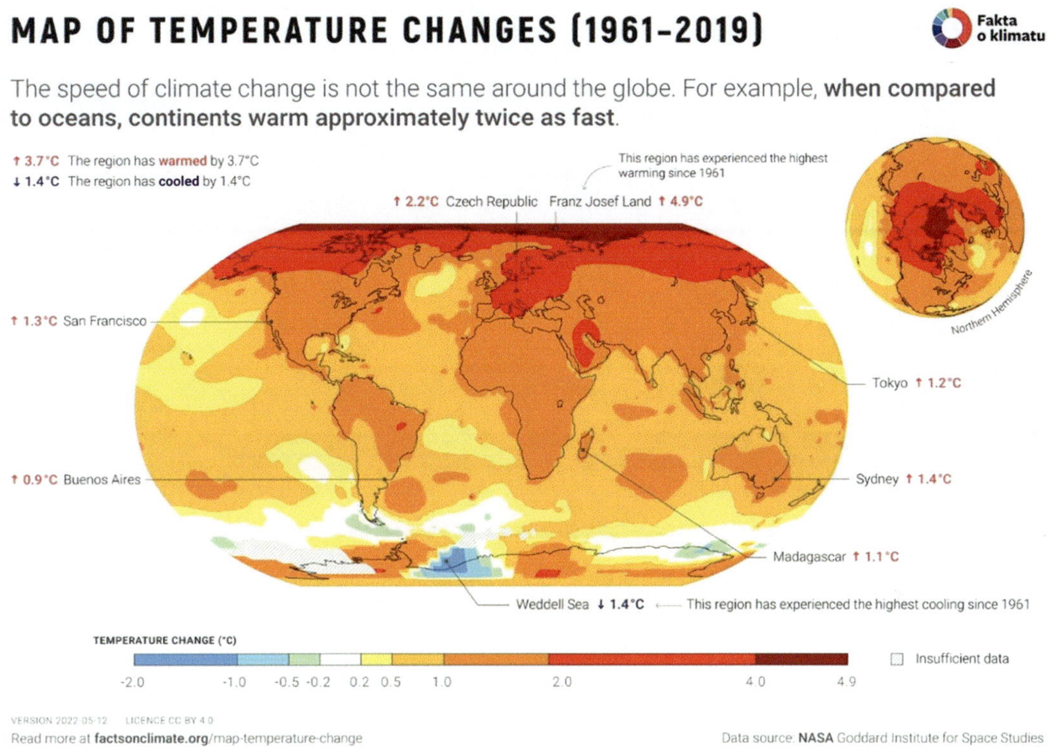

FIGURE 3.1 Map of temperature changes (1961–2019).

any delay will jeopardize our ability to secure a sustainable and equitable future. The IPCC report makes it clear that tackling the challenges of climate change requires a concerted global effort, emphasizing equity, justice, and the integration of diverse knowledge systems to build resilience and mitigate risk.

As seen in Figure 3.1, the global temperature increase since 1961 is approximately 1.1°C. However, the warming rate is not the same for all locations: while the temperature in most of the planet's oceans has not increased by more than 0.8°C, most of the world's land is warming considerably faster. The Arctic Ocean has the highest warming rate on Earth—over 4°C in the last 60 years.

Human activities, especially the burning of fossil fuels, result in significant CO_2 emissions—about 40 billion tons annually. This includes 35 billion tons from energy, transport, and industry, and 5 billion tons from deforestation. On average, each person contributes around 5 tons of CO_2 yearly. Besides CO_2, greenhouse gases like methane (CH_4) and nitrous oxide (N_2O), which come from agriculture, also contribute to global warming (Bouley et al., 2014).

Figure 3.2 shows the climate change scheme with the cascading effects of climate change through a systematic flow diagram, starting with primary human activities and their subsequent impacts on global systems. Four main human activities—fossil fuel burning, deforestation, cattle breeding, and industrial processes—contribute to increased greenhouse gas concentrations. These emissions trigger two parallel impact chains: atmospheric and oceanic. The atmospheric pathway shows how the enhanced greenhouse effect leads to air temperature increases and glacier melting, while the oceanic pathway demonstrates rising water temperatures and acidification. The diagram culminates in ten distinct ecosystem and societal impacts, including extreme weather events, shifting vegetation zones, sea-level rise, and coral reef degradation. The scheme effectively demonstrates how single human activities can propagate through the Earth system to produce multiple, interconnected environmental consequences.

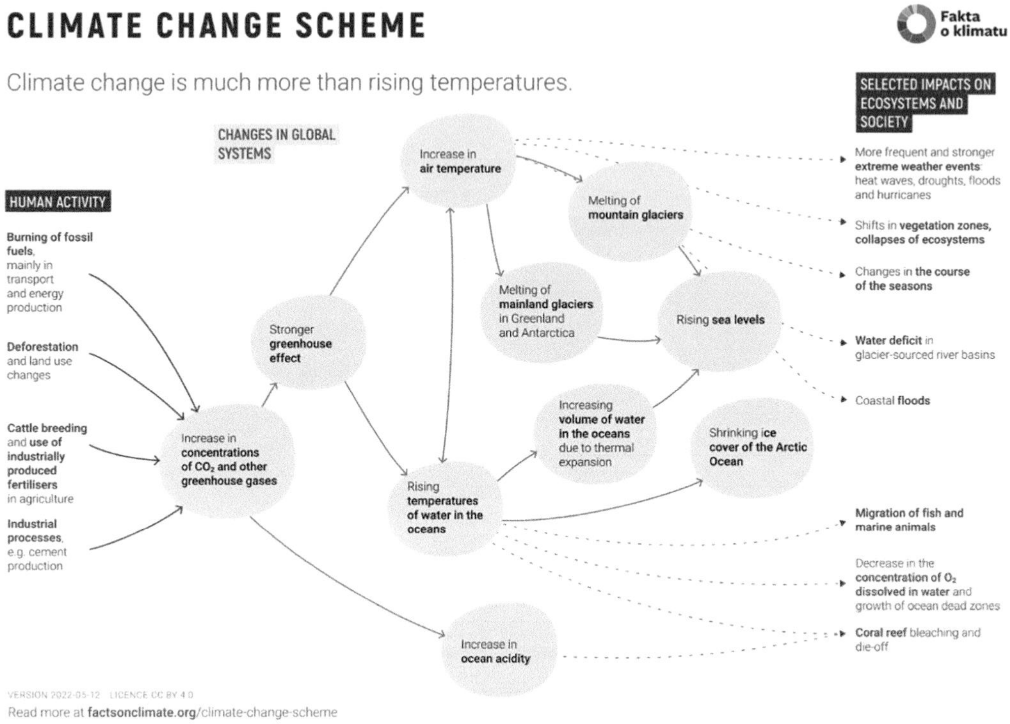

FIGURE 3.2 Climate change scheme (Source: Fakta o klimatu, 2022.).

The increased greenhouse gas concentrations enhance the greenhouse effect, trapping more heat on Earth and causing global warming. Since the Industrial Revolution, average air temperatures have risen by 1.2°C, with the ocean absorbing most of the heat. The Arctic is warming four times faster than the oceans, leading to significant changes, such as the reduction of sea ice. Projections indicate ice-free Arctic summers around 2050.

This warming trend results in more frequent and intense extreme weather events, including heatwaves, hurricanes, droughts, and floods. Additionally, CO_2 dissolving in oceans causes acidification, harming marine life. Sea levels are rising by 3.3 cm per decade due to melting glaciers and the thermal expansion of seawater. Melting mountain glaciers in the Alps, Himalayas, and Andes will impact agriculture and water supplies by altering river flows.

3.2 THE THREAT TO FOOD SECURITY: FOOD SAFETY AND NUTRITION

Climate change poses significant threats to food security by affecting food safety and nutritional quality. Rising temperatures and altered precipitation patterns facilitate the proliferation of harmful bacteria such as *Salmonella*, *Campylobacter*, and *Vibrio*, increasing the risk of foodborne illnesses (Dietrich et al., 2023). Additionally, ocean warming contributes to elevated mercury levels in seafood, further compromising food safety (Dietrich et al., 2023). Extreme weather events disrupt food supply chains, exacerbating food shortages and price volatility, particularly in low- and middle-income countries where agricultural productivity is already challenged (Toromade et al., 2024; Fanzo et al., 2024). Elevated atmospheric carbon dioxide levels, while promoting plant growth, paradoxically reduce the protein and mineral content of staple crops like wheat, rice, and potatoes, thereby threatening nutritional quality and exacerbating malnutrition (Toromade et al., 2024; Rani & Reddy, 2023). These impacts are particularly severe in regions already vulnerable to food insecurity, where marginalized communities and smallholder farmers lack the resources to adapt (Toromade et al., 2024; *Climate Change and Food Insecurity*, 2023). To address these challenges, climate-smart agricultural practices, such as agroforestry, crop diversification, and soil conservation, are essential for enhancing the resilience of food systems. Furthermore, investing in sustainable infrastructure, promoting gender equity in agriculture, and fostering international collaboration are critical for building adaptive and inclusive food systems capable of withstanding the impacts of climate change (Toromade et al., 2024; Fanzo et al., 2024).

The complex interplay between rising CO_2 levels and climate change presents significant food quality and safety challenges throughout the farm-to-table supply chain, as illustrated in Figure 3.3. The system demonstrates multiple critical impact pathways: temperature and precipitation extremes increase pathogen loads in agricultural production, while climate variations alter pest populations, potentially leading to increased pesticide usage in farming practices.

Rising atmospheric CO_2 concentrations directly influence the nutritional content of food products, adding another layer of complexity to food security concerns. Furthermore, warmer temperatures accelerate food spoilage during storage and transportation phases, while extreme climate events can disrupt the entire food distribution network, affecting the journey from

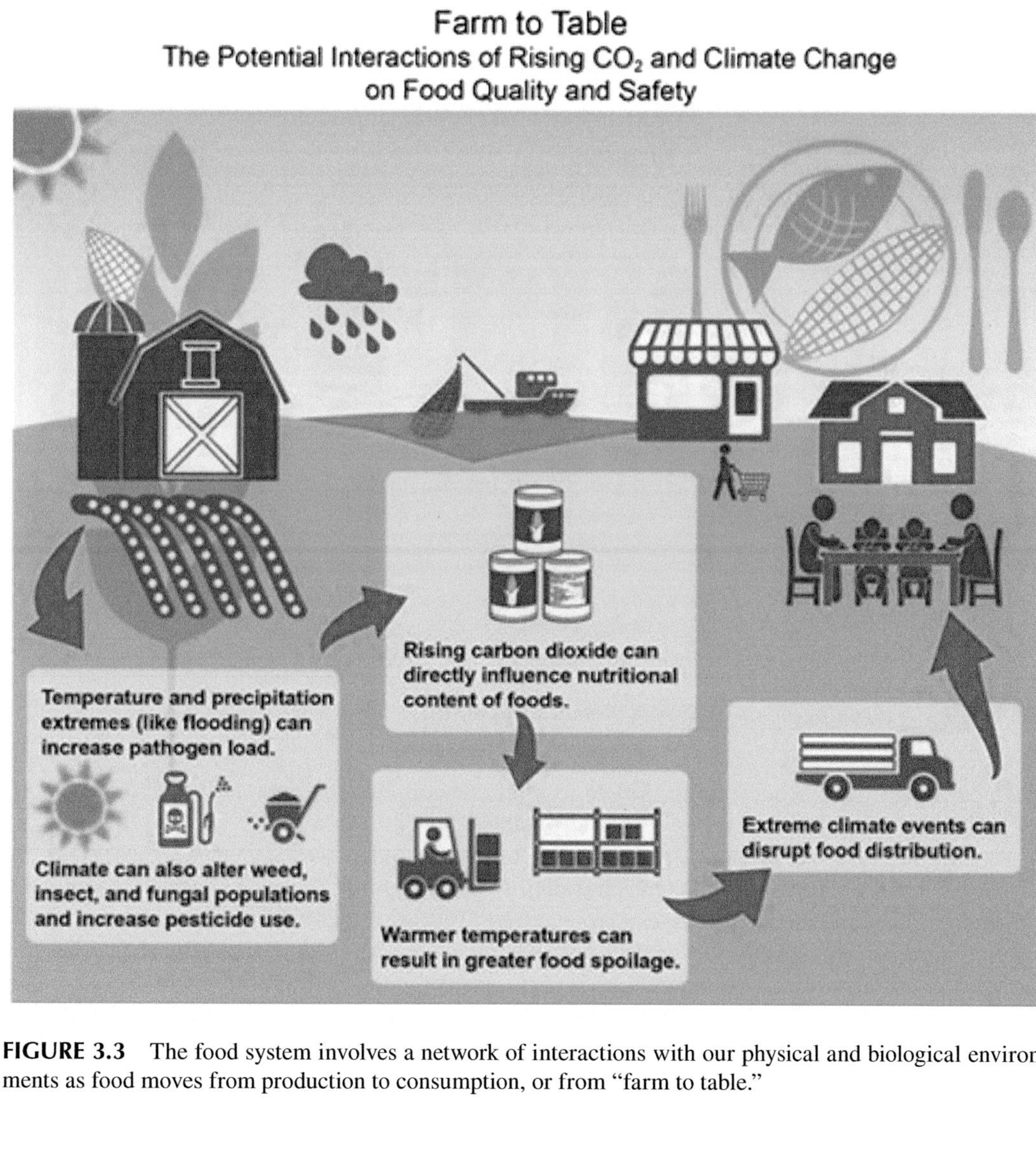

FIGURE 3.3 The food system involves a network of interactions with our physical and biological environments as food moves from production to consumption, or from "farm to table."

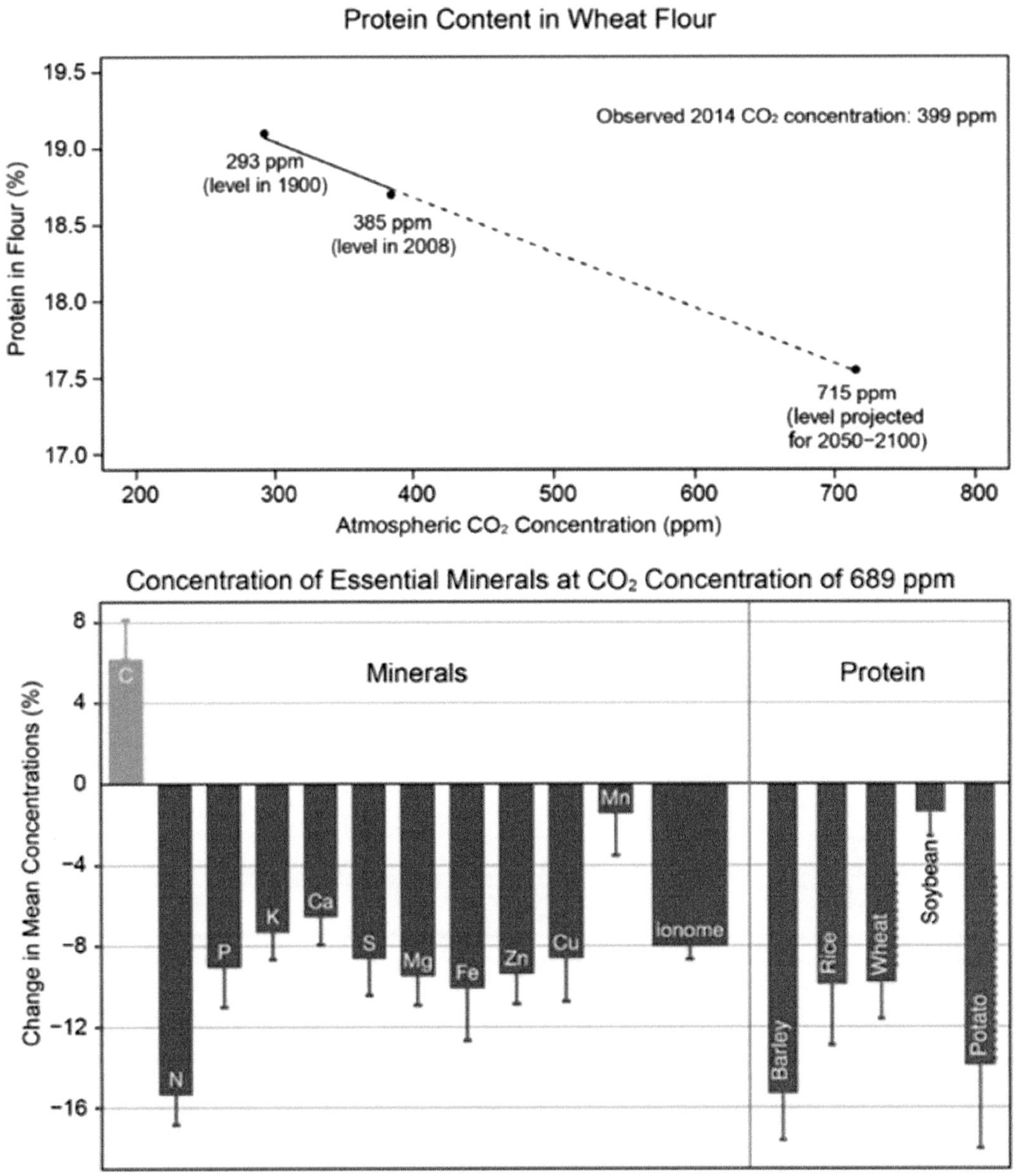

FIGURE 3.4 Effects of carbon dioxide on protein and minerals.

production facilities through retail outlets to final consumption, ultimately impacting the quality and safety of food that reaches consumers' tables.

Rising atmospheric CO_2 levels directly impact the protein and mineral content in crops. Figure 3.4 illustrates that the increase in CO_2 concentration from 293 ppm at the start of the 20th century to 385 ppm in 2008, and further to 715 ppm projected by 2100 under the RCP8.5 and RCP6.0 pathways, leads to a noticeable reduction in protein concentrations in wheat flour, averaging data from four varieties of spring wheat. The second graph highlights the broader implications of elevated CO_2, showing that doubling CO_2 levels from preindustrial times reduces essential mineral concentrations in 125 plant species and cultivars. This includes declining ionome levels and protein concentrations in staple crops such as barley, rice, wheat, and potato.

Elevated levels of CO_2 in the atmosphere paradoxically affect plant growth and nutrition. While CO_2 stimulates plant growth and boosts carbohydrate production, it negatively impacts the nutritional content of most food crops, particularly in terms of protein and minerals. As CO_2 levels rise, plants require less protein for photosynthesis, leading to a decrease in protein concentration within plant tissues. This trend is evident in wheat flour produced from various wheat varieties grown under conditions simulating the increase in atmospheric CO_2 since the year 1900. Major food crops like barley, wheat, rice, and potato are projected to exhibit 6%–15% lower protein concentrations when grown at the CO_2 levels anticipated for the year 2100. However, protein content in corn and sorghum may not be significantly affected.

Rising CO_2 levels also impact micronutrient levels in plants. Essential minerals like calcium, copper, iron, magnesium, and zinc are projected to decrease by 5%–10% in most plants due to elevated CO_2. This decline is attributed to the increased carbohydrate accumulation in plant tissues, which can dilute mineral content, and the reduced water uptake by plants, resulting in fewer nutrients being absorbed.

The combined effect of increased carbohydrates and decreased protein in plants exposed to higher CO_2 levels leads to a significant increase in the carbohydrate-to-protein ratio. This dietary shift can have adverse effects on human metabolism and body composition. While the exact impact of reduced protein and mineral content in crops on human health is not fully understood, the decline in nutritional value poses a potential threat, especially for populations with existing nutritional deficiencies. It is important to note that this overview is a synthesis of available information and should not be taken as a definitive assessment of the impact of rising CO_2 on human health. Further research and investigation are necessary to fully comprehend the long-term consequences of this phenomenon.

3.3 EXTREME WEATHER EVENTS

Global warming has significantly intensified the frequency and severity of extreme weather events, with profound impacts on both human and natural systems. The Arctic, warming at a rate more than twice the global average, exemplifies this trend through rapid sea ice loss and increased precipitation, primarily driven by local evaporation due to sea ice retreat (Bintanja & Selten, 2016; Cohen et al., 2014). This warming contributes to a more pronounced Arctic hydrological cycle, which affects global sea levels and potentially alters global climate patterns through changes in oceanic circulation (Bintanja & Selten, 2016). Concurrently, the Northern Hemisphere has experienced unprecedented heat extremes, such as those in 2018, which were virtually impossible without human-induced climate change (Vogel et al., 2019). These heatwaves, alongside other extreme events like droughts and floods, are becoming more frequent and intense, particularly in Europe, where they often co-occur and persist over consecutive years (Pradhan et al., 2022).

Figure 3.5 illustrates anticipated shifts in key climate factors—by mid-century, the coldest night is predicted to warm considerably across most of the U.S. At the same time, the hottest day will also see a significant temperature rise. Extreme precipitation events are expected to intensify, and the longest dry spells will lengthen slightly. The increase in average temperatures translates to hotter days and more frequent, extended heatwaves. This may result in thousands to tens of thousands more heat-related deaths annually by the year 2100 during summer months. While cold-related deaths might decrease in winter, this won't compensate for the rise in heat-related mortality. However, adaptation strategies like increased air conditioning usage could potentially mitigate these impacts.

Figure 3.6 illustrates a significant correlation between high temperatures and increased mortality rates during the 1995 Chicago heatwave. The peak in deaths observed in mid-July far exceeded the typical death rate for that time of year, highlighting the detrimental impact of

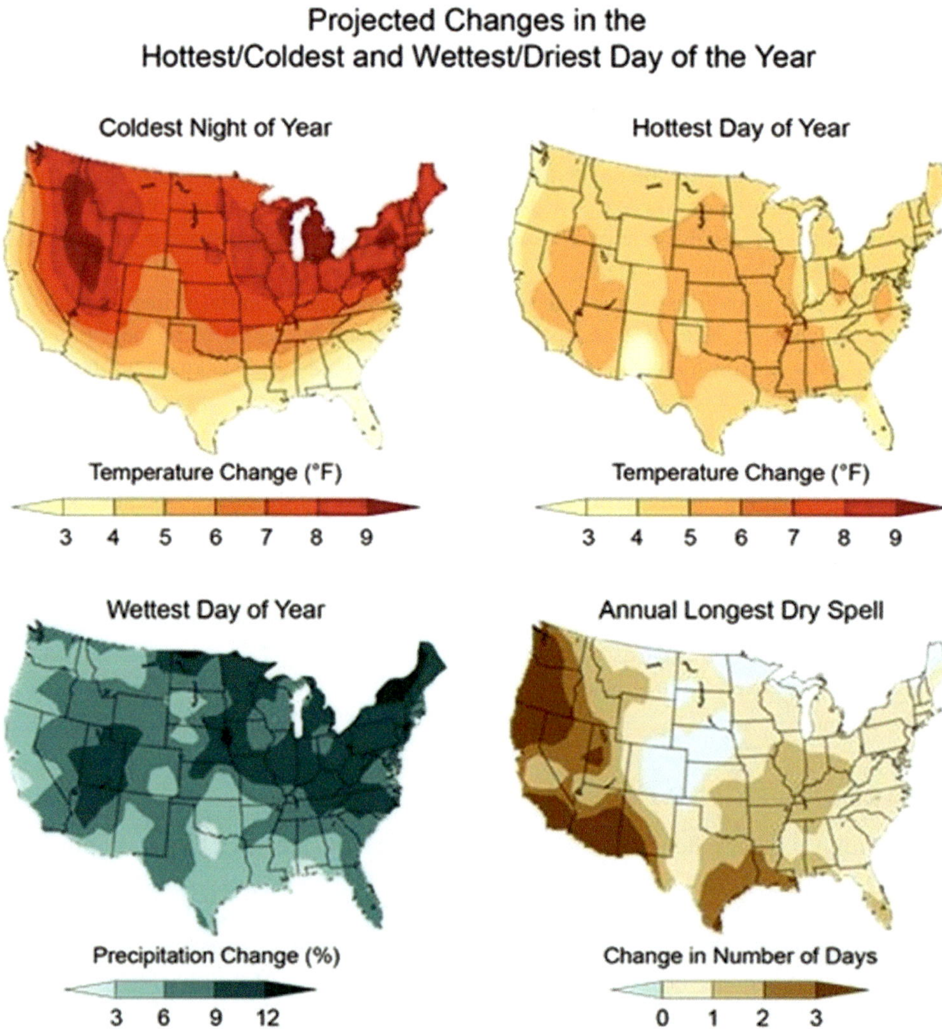

FIGURE 3.5 Projected changes in several climate variables for 2046–2065 with respect to the 1981–2000 average for the RCP6.0 scenario.

extreme heat. This phenomenon is not unique to Chicago, as major urban centers like St. Louis, Philadelphia, and Cincinnati have also experienced similar spikes in mortality during heat waves. This heightened vulnerability of urban populations is expected to worsen with climate change, further amplifying the health risks associated with extreme heat events. Additionally, heatwaves often coincide with periods of stagnant air, exacerbating air pollution and its adverse effects on human health.

A study led by Monash University found that 9.43% of global deaths from 2000 to 2019 were attributable to hot and cold temperatures, as shown in Figure 3.7. Although cold-related deaths decreased slightly, heat-related deaths increased, indicating that climate change may worsen this issue. This study used a broader dataset than previous research, providing a more comprehensive understanding of the global impact of temperature-related mortality. The data showed geographical variations in the global death trend: more than half occurred in Asia, particularly in East and South Asia; Europe had the highest excess death rates per 100,000 due to heat exposure; and Sub-Saharan Africa had the highest death rates per 100,000 due to exposure to cold. The study revealed that

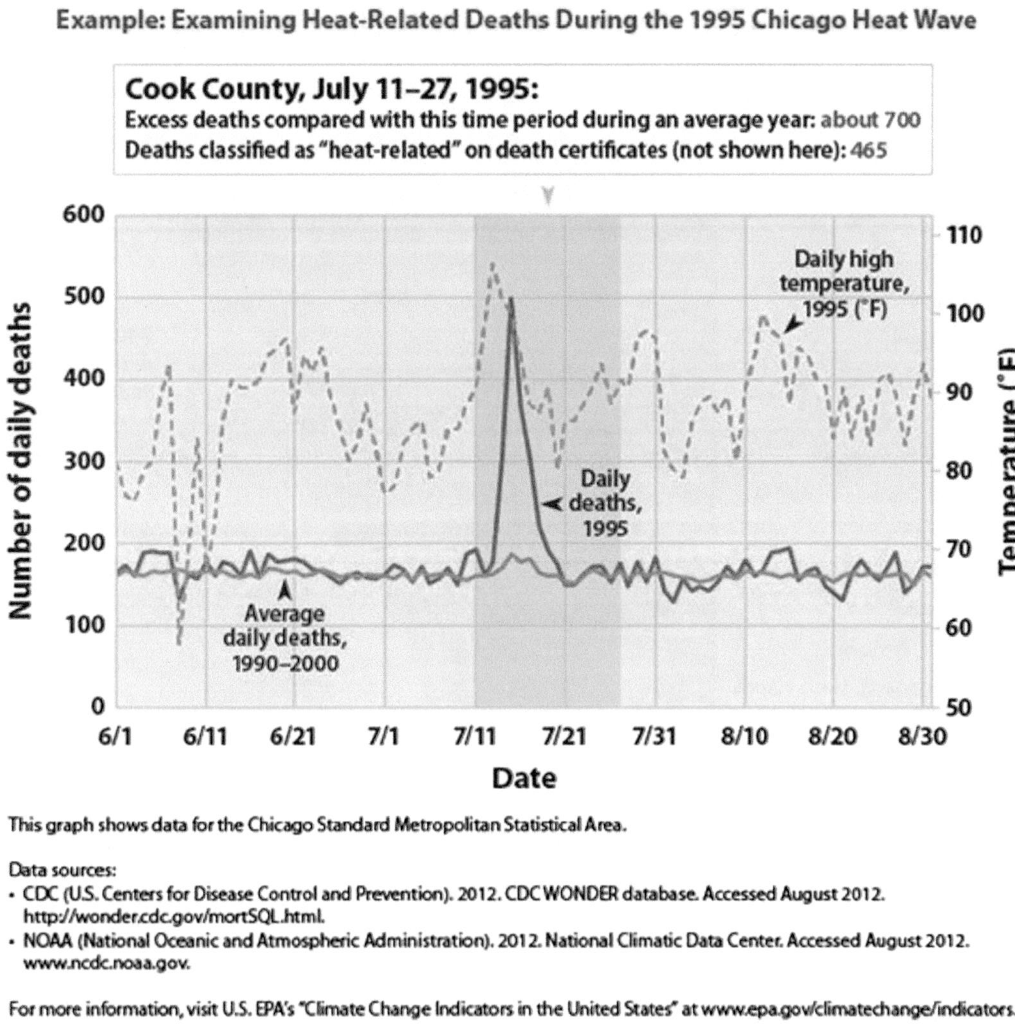

FIGURE 3.6 Heat-related death in Chicago in 1995.

over five million annual deaths are attributed to abnormal temperatures, with Asia experiencing the highest number (2.6 million), predominantly due to cold exposure (2.4 million), while Europe has the highest heat-related death rates (178,700).

The essay details the significant impact of global heating on extreme weather events across five continents. In Europe, the 2019 heatwave in France saw temperatures rise by 4°C due to climate change, and extreme rains in Germany and Belgium in 2021 became nine times more likely. In North America, Texas' 2011 heat and California's 2012–2014 drought were ten times and substantially more likely due to global warming, while Canada's 2017 wildfires burned areas up to 11 times larger. Asia faced unparalleled warmth in 2016, Super Typhoon Haiyan in 2013, and various extreme events in China, all exacerbated by climate change. Australia's "Black Summer" of 2019–2020 bushfires and the 2016 Great Barrier Reef bleaching event were both significantly intensified by global heating. Africa's 2022 South African floods and the 2015 Ethiopian drought were likewise made more likely, highlighting compounding effects with other crises. Latin America saw Brazil's 2022 floods and Argentina's 2013 heatwave, both influenced by climate change, with fewer studies but notable impacts like the 2010 Amazon drought.

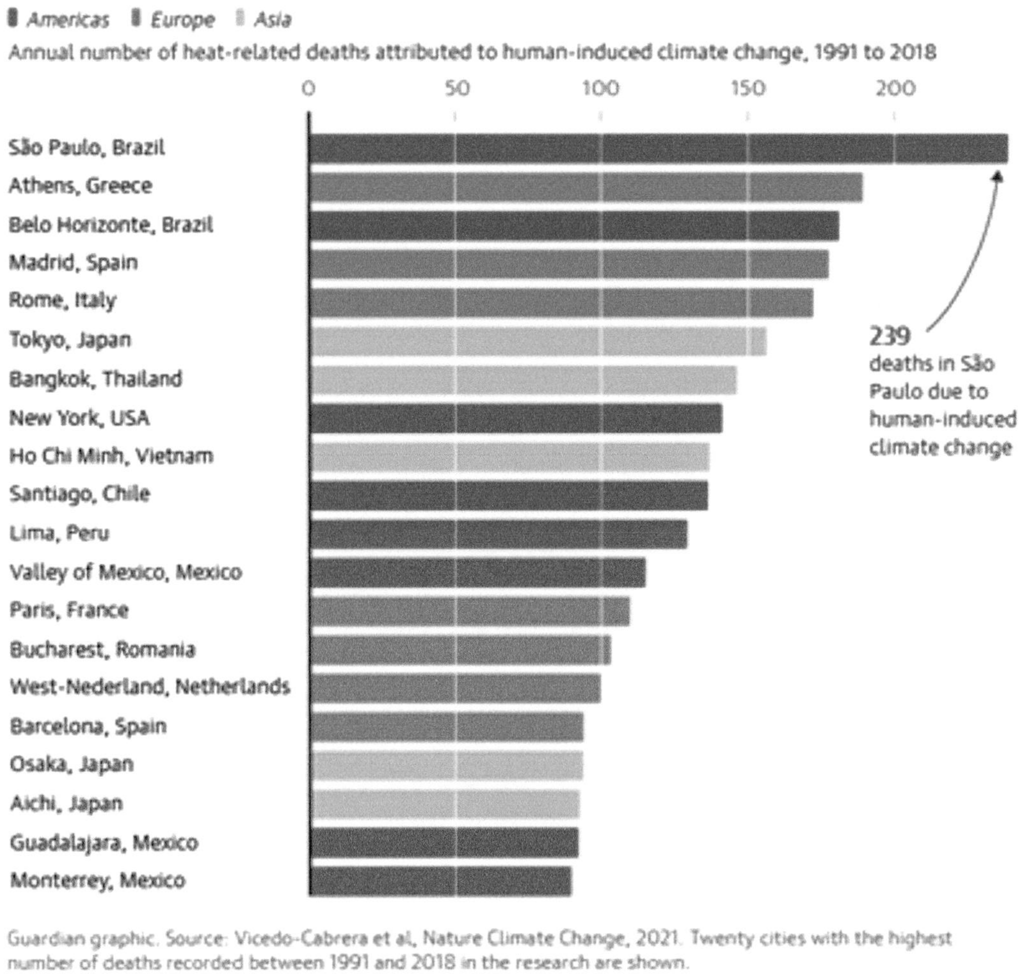

FIGURE 3.7 Heat-related deaths related to anthropogenic climate change.

3.4 VECTOR-BORNE DISEASES

The shifting climate significantly impacts the distribution and activity of disease vectors such as mosquitoes and ticks, spreading vector-borne illnesses like Lyme disease and West Nile virus (WNV). Climate change influences the habitat suitability for vectors and hosts, altering disease risk patterns. For instance, in the UK, the potential emergence of WNV is linked to climate-driven changes in the distribution of Culex mosquitoes and avian hosts, with future scenarios predicting increased risk in the south-east and decreased risk in the north by 2100 (Withers et al., 2024). Similarly, in the United States, the relationship between climate variables and WNV cases varies significantly when analyzed across ecological versus geopolitical regions, suggesting that ecological frameworks may better capture the environmental dynamics of disease transmission (Moser et al., 2024). In the United States, over three million people were estimated to have been infected between 1999 and 2010.

The IPCC has highlighted the increasing prevalence of vector-borne diseases due to climate change, emphasizing the need for enhanced monitoring and public health strategies (Parums, 2023). In West Africa, climate models predict shifts in malaria transmission patterns, with high transmission periods potentially expanding due to climate change, although overall prevalence may decrease in some scenarios (Fall et al., 2023). Additionally, the interaction between parasites and their hosts, such as the mosquito parasite *Lambornella clarki*, is influenced by local adaptation rather than temperature, indicating complex ecological interactions that may be affected by climate change (Lyberger et al., 2024). These findings underscore the necessity for adaptive public health policies and disease surveillance systems to mitigate the impacts of climate change on vector-borne diseases globally. While developed nations like the United States have public health systems in place to monitor and manage these diseases, poorer countries with limited resources face greater challenges in prevention and treatment.

3.5 WATER-RELATED ILLNESSES

Rising temperatures and extreme rainfall, exacerbated by climate change, significantly elevate the risks of waterborne diseases by creating environments conducive to harmful bacterial and algal growth. This intersection of climate and health is particularly pronounced in vulnerable regions with limited healthcare and water management system access. In developing countries like Pakistan, chaotic urbanization and inadequate water management have led to increased incidences of diseases such as hepatitis, cholera, and typhoid, with morbidity rates exacerbated by insufficient healthcare infrastructure (Ahmed et al., 2016). Similarly, in the Hoshangabad district of India, variations in temperature and precipitation have been linked to increased public health risks from waterborne diseases, particularly during the monsoon season (Jain & Mishra, 2016). The broader implications of climate change on water quality and disease transmission are evident, as warmer temperatures and altered rainfall patterns affect the incidence and transmission of waterborne diseases, impacting vulnerable populations disproportionately. Furthermore, the persistence of pathogens like *Cryptosporidium* and *Giardia* in water systems poses significant public health challenges, as these organisms are resistant to conventional treatment technologies (Hipsey & Brookes, 2013).

Exposure to contaminated water can have severe health consequences, ranging from gastrointestinal illnesses like diarrhea to damage to the nervous, respiratory, and other organ systems. However, public health agencies in the United States have implemented safeguards to mitigate these risks. These include water quality monitoring, treatment standards, beach closures, and advisories for boiling water and harvesting shellfish.

3.6 MENTAL HEALTH AND VULNERABLE POPULATIONS

The physical and psychological impacts of climate change are indeed disproportionately felt by vulnerable populations, including children, the elderly, and low-income communities, as evidenced by various studies. Climate change exacerbates mental health challenges, particularly through the stress induced by extreme weather events and pre-existing socioeconomic disparities. For instance, research highlights that younger populations are experiencing significant climate distress, with their mental health burden expected to increase as climate change progresses, yet their perspectives are often underrepresented in interventions (Dunk & Rhodes, 2023). In the aftermath of natural disasters like Cyclone Yash, children and adolescents in affected regions have shown high incidences of post-traumatic stress disorder (PTSD), underscoring the profound mental health impacts of such events (Dastidar, 2023). Socioeconomic factors also play a critical role in shaping perceptions of health risks and adaptive responses to climate change, as seen in semiarid regions of Brazil, where these factors influence how individuals perceive and respond to climate risks (de Junior et al., 2023).

Various groups are more susceptible to the health risks associated with climate change. Three key factors determine vulnerability: sensitivity (how affected individuals or groups are by stressors like high temperatures), exposure (the extent of contact with these stressors), and adaptive capacity (the ability to adjust to or avoid hazards). For instance, while older adults are sensitive to extreme heat, those with access to air conditioning and the means to use it can mitigate their exposure and thereby reduce their risk. In the United States, the Climate Vulnerability Index reveals that existing disparities are often exacerbated by climate change, with vulnerable groups facing greater exposure and lower capacity to adapt and recover from climate impacts (Cury, 2023). Similarly, in Korea, the relative risk of mortality due to heatwaves has increased among vulnerable groups, such as single-person households and the less educated, highlighting the need for targeted policies to address these disparities (Park & Chae, 2023). The mental health impacts of climate change are further illustrated by the increased levels of depression, anxiety, and stress among Australian youth following the Black Summer bushfires, indicating significant vulnerabilities in this demographic (Lykins et al., 2023). Overall, these findings emphasize the necessity for mental health support and targeted interventions to protect vulnerable populations from the compounded effects of climate change and socioeconomic disparities (Dastidar, 2023; Cury, 2023; Park & Chae, 2023).

3.7 CLIMATE CHANGE AND ENERGY DEMAND

Between 2010 and 2050, the largest rises in baseline final energy demand are expected in densely populated tropical and mid-latitude regions. These areas could experience up to 75 (50) more hot days and up to 40 (30) fewer cold days under severe (moderate) warming scenarios, as shown in Figure 3.8. In the figure, Representative Concentration Pathways (RCPs) and Shared Socioeconomic Pathways (SSPs) are complementary frameworks used to model future climate scenarios. RCPs focus specifically on greenhouse gas concentration trajectories and their resulting radiative forcing by 2100, measured in watts per square meter (e.g., RCP4.5 represents 4.5 W/m² of forcing). In contrast, SSPs expand upon RCPs by incorporating broader socioeconomic factors such as population growth, economic development, urbanization, and technological advancement to provide comprehensive storylines of how society might evolve. While RCPs tell us where global temperatures could end up, SSPs explain the societal pathways that could lead to those outcomes. The key distinction is that SSPs provide economic and social context for emission pathways, whereas RCPs focus solely on the physical climate response to greenhouse gas concentrations.

This increase is most pronounced in regions with minimal changes in hot or cold days (<10), highlighting that most areas will only see slight temperature changes by mid-century, particularly under the moderate RCP 4.5 scenario. According to SSP-specific multi-ESM means, socioeconomic trends consistently influence this pattern, with the highest absolute energy demand observed under the rapid economic growth SSP5 scenario and the lowest under the population growth SSP3 scenario. Exposure patterns also vary significantly among different climate model projections.

The total energy demand is exposed to changes in cold and hot days. The upper half displays the results for RCP 8.5, and the bottom half for RCP 4.5. In the detailed panels (a, b, e, f), the blue lines depict present-day energy demand, and the red lines depict SSP5 baseline energy demand for 2050; brown lines indicate energy use under SSP5 after the impacts of climate change (mean and all 21 individual ESMs) exposed to changes in hot and cold days. The aggregate panels (c, d, g, h) show the multi-ESM mean for all five SSPs. Impacts from climate change are only shown for changes in hot (b, d, f, h) or cold (a, c, e, g) days.

3.8 WATER AVAILABILITY AND ENERGY

Energy and water systems are connected. Energy is needed to pump, transport, and treat drinking water and wastewater. Cooling water is needed to run many of today's power plants. Hydroelectricity (electricity produced by running water) is itself an important source of power in parts of the United States. Changes in precipitation, increased risk of drought, reduced snowpack, and changes in the timing of snowmelt in spring will influence our patterns of energy and water use. For example, power plants can require large amounts of water for cooling. On average, a kilowatt-hour of electricity (enough power to run 400 typical compact fluorescent light bulbs for an hour) requires 25 gallons of water to be withdrawn from rivers or lakes. Parts of the Southeast and Southwest face increased competition for water to meet the demands of population and economic growth while also protecting natural ecosystems.

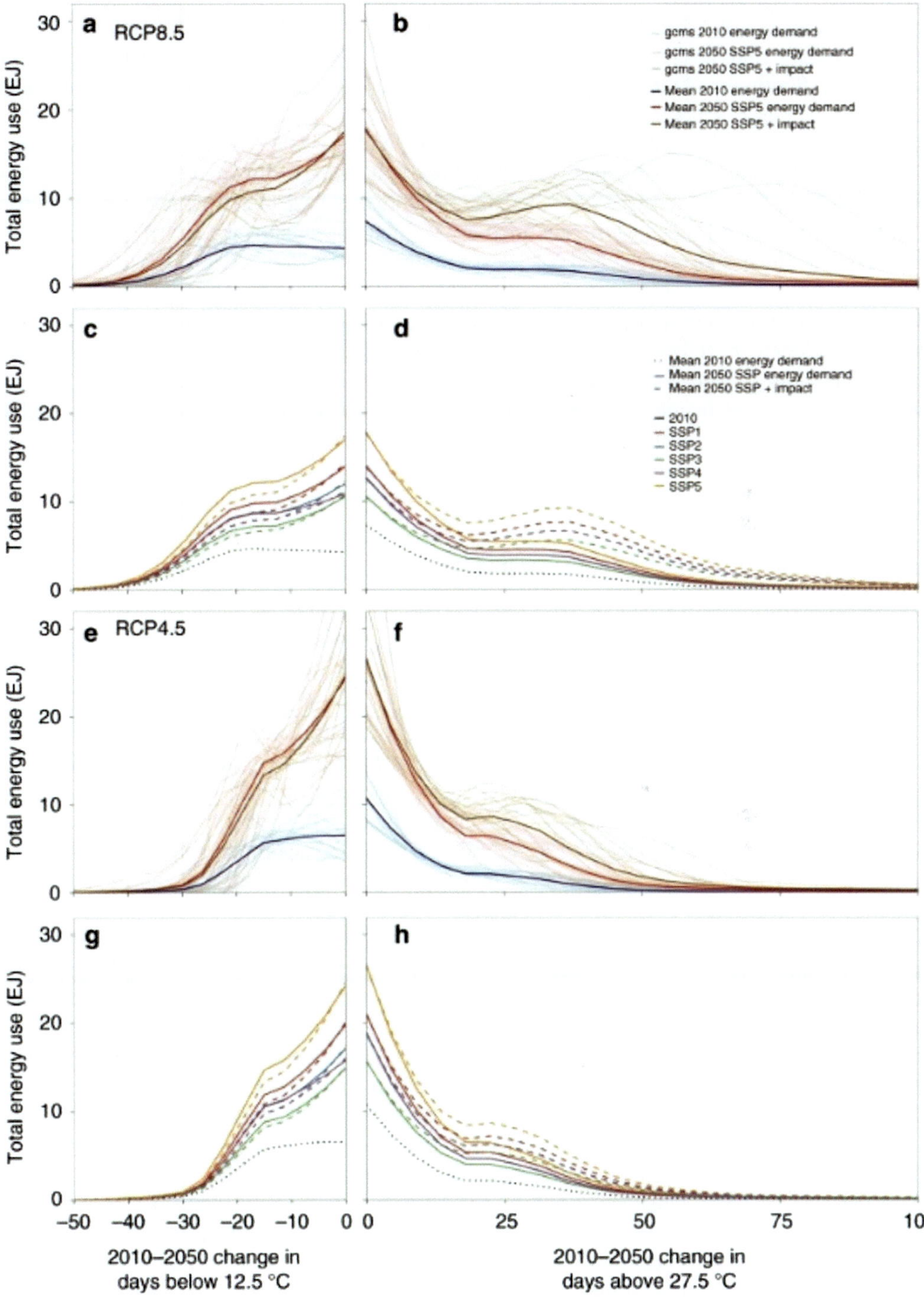

FIGURE 3.8 The total energy demand exposed to changes in cold and hot days under RCP 4.5 and 8.5. (Courtesy: van Ruijven et al., 2019, Available under CC BY 4.0 https://creativecommons.org/licenses/by/4.0/, no changes made.)

Many local governments in these regions have slowed or stopped plans for new power plants that require large withdrawals of water due to concerns about the adequate availability of cooling water.

More frequent and severe heat waves will likely increase the demand for electricity in the Southeast and Southwest. At the same time, these areas are likely to experience reduced water supplies due to increased temperatures and evaporation, as well as possible decreased rainfall. Since water is necessary for electricity production, these combined effects could stress water resources.

Hydroelectric power plants are sensitive to the volume and timing of stream flows. In some regions, especially during times of increased rainfall, dam operators may have to allow some water to bypass the electric turbines to prevent downstream flooding. Maintaining stream flow for hydroelectric dams could present conflicts with other activities, such as salmon habitat restoration in the Pacific Northwest.

Growing crops for biomass and biofuel energy could stress water resources in certain regions, depending on the type of crop, where it is grown, agricultural production in the region, and current water and nutrient management practices. Given the many factors involved, more research is needed to understand how climate change may affect these resources.

Rising temperatures, increased evaporation, and drought may increase the need for energy-intensive methods of providing drinking and irrigation water. For example, desalinization plants can convert salt water into freshwater but consume a lot of energy. Climate change may also require irrigation water to be pumped over longer distances, particularly in dry regions across the Western United States.

3.9 SEA-LEVEL RISE NEGATIVELY INFLUENCES COASTAL AREAS

Sea-level rise poses a significant threat to coastal regions worldwide, with projections indicating substantial impacts by 2050. This phenomenon is primarily driven by global warming, which causes the thermal expansion of ocean waters and the melting of glaciers and ice sheets, notably in Greenland and Antarctica (Mottram et al., 2022; Siegert & Pearson, 2021). The potential for the Antarctic Ice Sheet to contribute significantly to sea-level rise, with estimates suggesting up to 2.5 m by the end of the century, underscores the urgency of addressing this issue (Phipps et al., 2020). Coastal areas, which are ecologically rich and socioeconomically vital, face diverse challenges due to sea-level rise, including permanent inundation, increased storm surges, and groundwater salinization (Maliva, 2021; Pednekar & Raju, 2020). These changes threaten biodiversity, disrupt ecosystems, and jeopardize infrastructure, potentially displacing millions and causing significant economic losses (Siegert & Pearson, 2021; Pednekar & Raju, 2020).

In California, for instance, sea-level rise could displace over 100,000 people and put billions in property value at risk by 2050 (Matella, 2022). Effective management of these risks requires a transdisciplinary approach, integrating climate science with local stakeholder engagement to develop adaptive strategies tailored to specific regional needs (Mottram et al., 2022). Legal frameworks, such as the United Nations Convention on the Law of the Sea, need to evolve to address the challenges posed by sea-level rise, promoting international cooperation and domestic legislation improvements (Xiao, 2022). Additionally, innovative planning and infrastructure adaptation, as seen in California's Local Coastal Programs, can enhance resilience through phased adaptation, nature-based solutions, and asset relocation (Matella, 2022). Despite the uncertainties in sea-level rise projections, particularly concerning ice sheet dynamics, it is crucial to implement both mitigation and adaptation measures to minimize future impacts (Siegert & Pearson, 2021; Oddo et al., 2020).

Urban areas like Miami and New Orleans are particularly vulnerable to the compounded risks of rising tides, storm surges, and saltwater intrusion due to their geographical locations and low-lying topographies. Miami, for instance, faces immediate and long-term threats from sea-level rise and erosion, which are exacerbated by its hurricane-prone environment. Historical perspectives on flood resilience can offer insights into developing spatial strategies to enhance Miami's future resilience (Firley & Deupi, 2017). Similarly, New Orleans, situated in a region with high relative sea-level rise and erosion, is at significant risk due to its low elevation and subsidence issues (Gornitz et al., 1991).

The threat of sea-level rise is not isolated to these cities; it is a global challenge that demands both mitigation and adaptation strategies. Coastal cities worldwide, including those in the

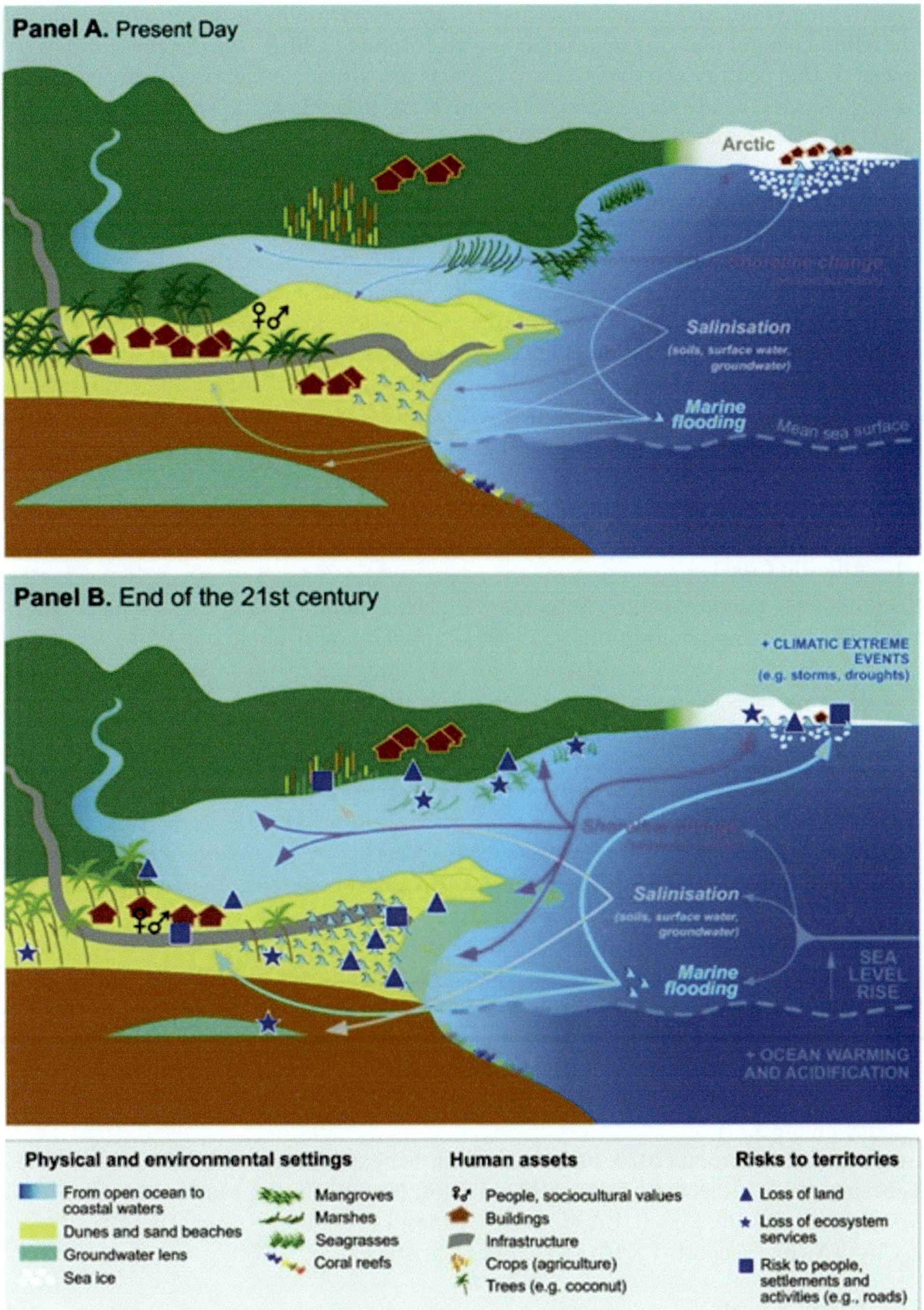

FIGURE 3.9 Schematic visualization of SLR impacts by the end of the 21st century. (Courtesy: Magnan et al., 2022, https://www.nature.com/articles/s41598-022-14303-w, Available under CC BY 4.0 https://creativecommons.org/licenses/by/4.0/, no changes made.)

United States, are increasingly focusing on integrated coastal management to address these threats. This involves a combination of protection, accommodation, and retreat strategies to mitigate impacts (Nicholls, 2011). Figure 3.9 represents a hypothetical coastal area composed of tropical, temperate, and polar coasts in the face of marine hazards (flooding, salinization, and shoreline change) and impacts, including the effects of SLR, for the present day (A) and by the end of this century (B). (A) Coastal impacts are already occurring, especially shoreline retreat (e.g., shaded sandy shores and yellow arrows), flooding (blue wave symbols), and ecosystem degradation (shaded coastal vegetation and coral reefs). Future impacts are represented with colored arrows and dark blue triangles, stars, and squares (loss of land, loss of ecosystem services, and risk to human assets, respectively). The thickening of the arrows from (A) to (B) illustrates the increasing influence of SLR. The comparison between (A) and (B) shows SLR risks in terms of the reconfiguration of the coast (mainly coastal recession here) in all environments; the reduction in size and salinization of groundwater lenses; the salinization of soils used for coconut trees and crops (shaded trees and yellow and brown icons); the degradation of coastal vegetation and coral reefs (shaded green and coral icons), partly also due to ocean warming and acidification; and the loss of human assets (e.g., houses and roads). Finally, it shows the expected decrease in sea ice that will amplify SLR effects through reduced physical protection of the land from wave action.

3.9.1 Projected Sea-Level Rise Impact on U.S. Coastal Cities by 2050

3.9.1.1 Atlantic Coast

By 2050, sea-level rise is expected to increase high-tide flooding in 11 U.S. Atlantic coast cities (Figure 3.10), exposing an additional 773–951 km². This will impact 59,000–263,000 people and 32,000–163,000 properties. Miami, with an average elevation below 2 m, will bear the brunt of this exposure, accounting for 38%–44% of the affected area, 38%–46% of the population, and 41%–49% of the properties. The estimated property value at risk is between $14 and 64 billion.

Areas exposed to current (2020) high tide and further exposed areas by 2050, considering VLM and geocentric SLR projections under SSP2–4.5 for: Boston, MA (**a**); New York City, NY and Jersey City, NJ (**b**); Atlantic City, NJ (**c**); Virginia Beach, VA (**d**); Wilmington, NC (**e**); Myrtle Beach, SC (**f**); Charleston, SC (**g**); Savannah, GA (**h**); Jacksonville, FL (**i**); and Miami, FL (**j**).

3.9.1.2 Gulf Coast

Similarly, 11 cities along the U.S. Gulf Coast will see an additional 528–826 km² exposed to flooding, affecting 110,000–225,000 people and 58,000–109,000 properties, valued between $14 and $21 billion. Notably, New Orleans already has substantial exposure due to areas below sea level and existing flood defenses. However, the effectiveness of these defenses is uncertain, as highlighted by the aftermath of Hurricane Katrina. Key infrastructure like airports, roads, and refineries are also at risk. (See Figure 3.11)

Areas exposed to current (2020) high tide and further exposed areas by 2050, considering VLM and geocentric SLR projections under SSP2–4.5 for: Naples, FL (a); Mobile, AL (b); Biloxi, MS (c); New Orleans, LA (d); Slidell, LA (e); Lake Charles, LA (f); Port Arthur, TX (g); Galveston and Texas City, TX (h); Freeport, TX (i); and Corpus Christi, TX (j)

3.9.1.3 Pacific Coast

Projections for ten cities on the U.S. Pacific coast indicate a significantly lower impact, with an additional 20–40 km² exposed, affecting 6,000–30,000 people and 3,000–15,000 properties valued between $4.5 and $22 billion, as illustrated in Figure 3.12. This reduced risk is attributed to higher topographic elevations, lower land subsidence rates, and slower sea-level rise compared to the Atlantic and Gulf coasts. Nonetheless, factors like cliff retreat and increased high-tide flooding will still pose challenges to coastal residents and properties.

Areas exposed to current (2020) high tide and further exposed areas by 2050, considering VLM and geocentric SLR projection under SSP2–4.5 for: Richmond, CA (a); Oakland, CA (b);

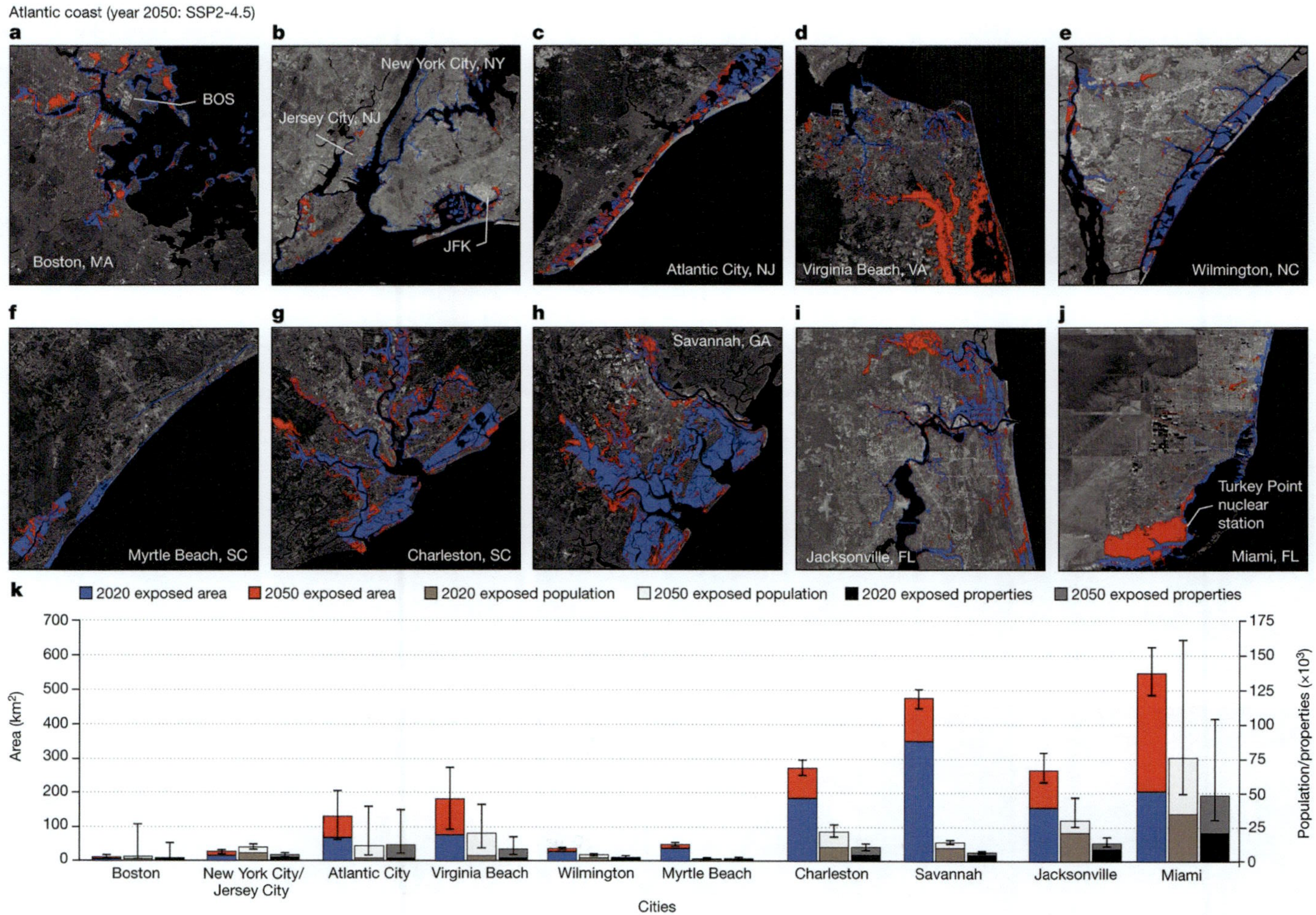

FIGURE 3.10 Inundation maps for cities on the US Atlantic coast. (Courtesy: Ohenhen, 2024, https://www.nature.com/articles/s41586-024-07038-3, Available under CC BY 4.0: https://creativecommons.org/licenses/by/4.0/, no changes made.)

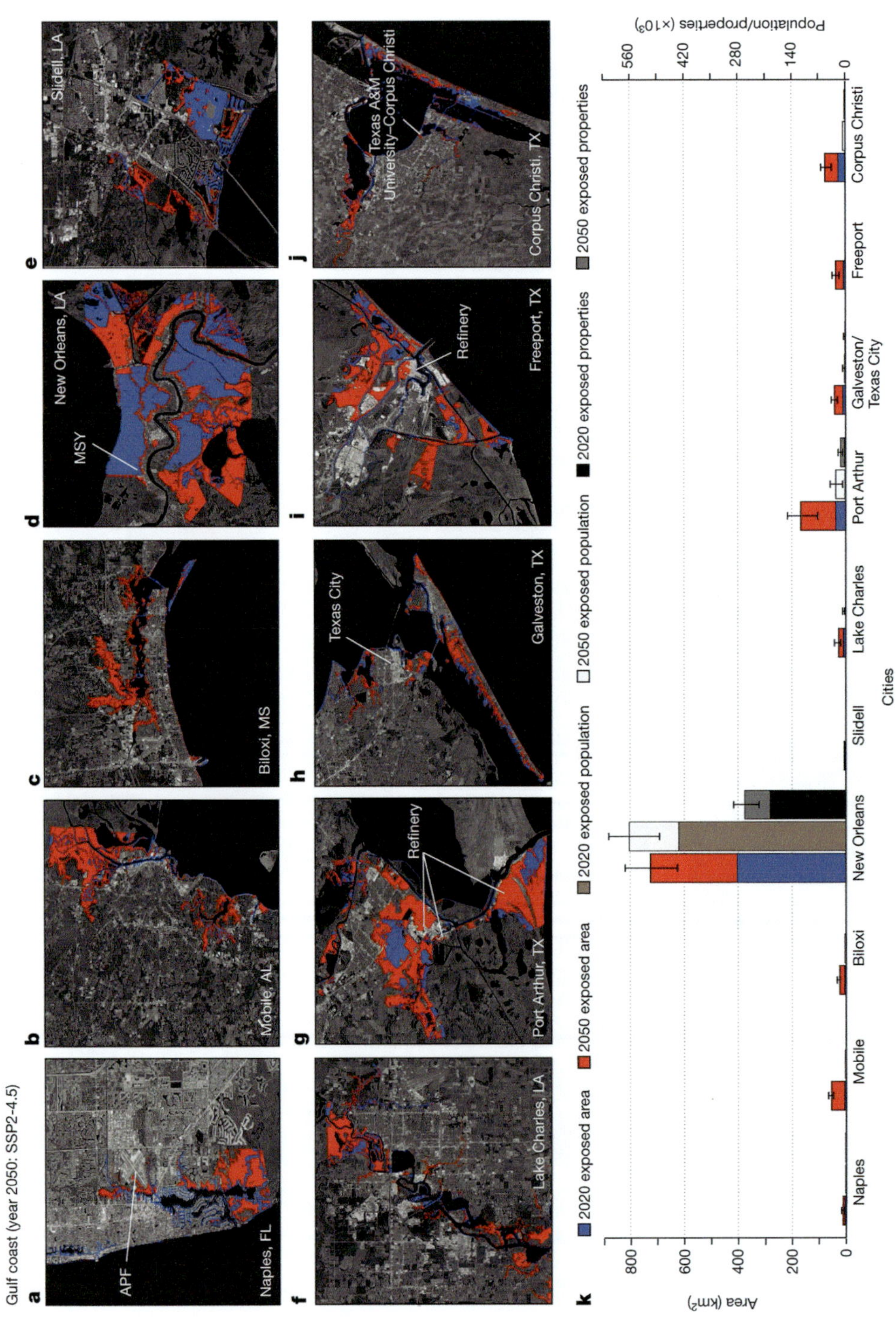

FIGURE 3.11 Inundation maps for cities on the US Gulf coast. (Courtesy: Ohenhen, 2024, https://www.nature.com/articles/s41586-024-07038-3, Available under CC-BY-4.0 https://creativecommons.org/licenses/by/4.0/.)

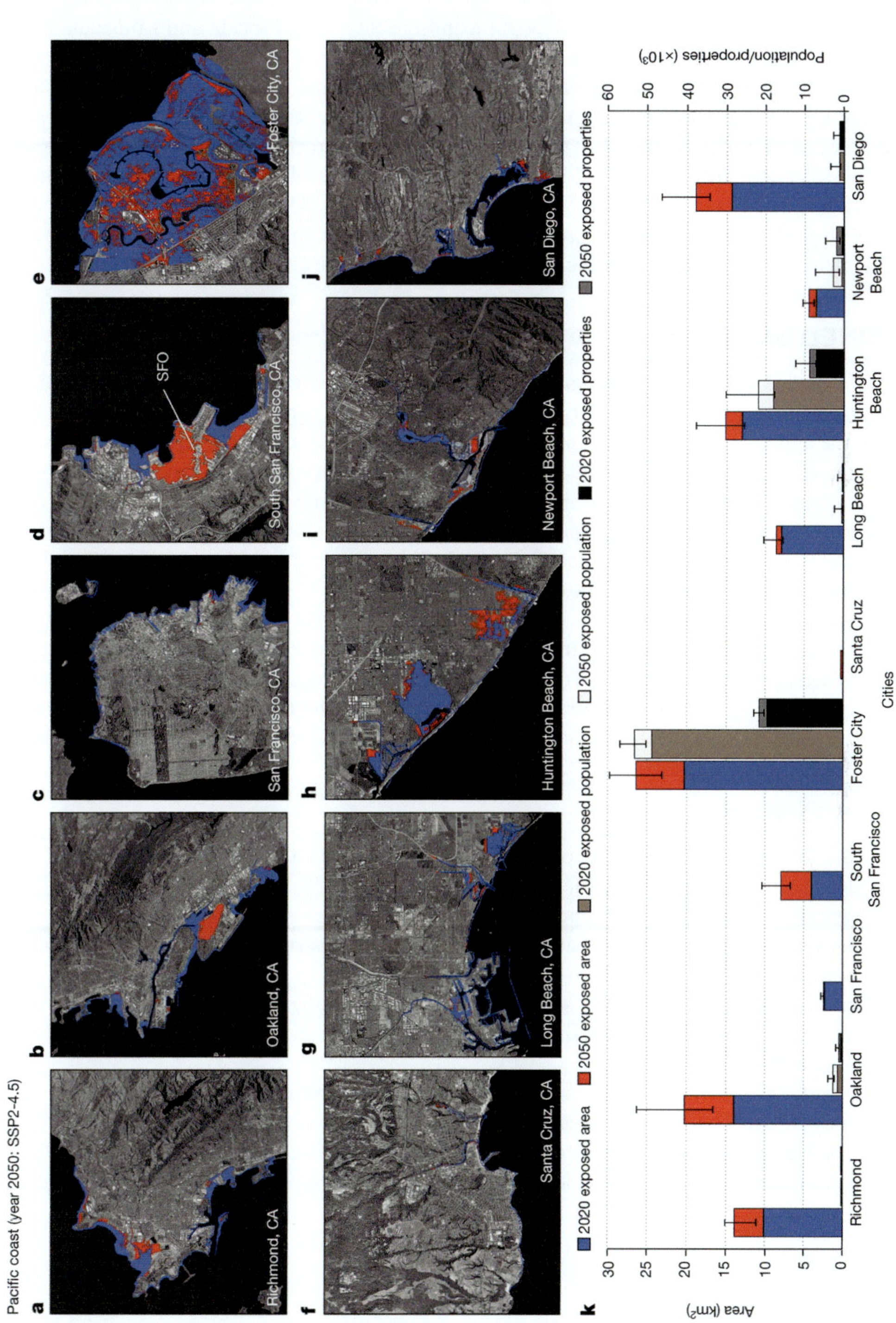

FIGURE 3.12 Inundation maps for cities on the US Pacific coast. (Courtesy: Ohenhen, 2024 https://www.nature.com/articles/s41586-024-07038-3, Available under CC-BY-4.0 https://creativecommons.org/licenses/by/4.0/.)

San Francisco, CA (c); South San Francisco, CA (d); Foster City, CA (e); Santa Cruz, CA (f); Long Beach, CA (g); Huntington Beach, CA (h); Newport Beach, CA (i); and San Diego, CA (j).

Furthermore, the vulnerability of U.S. coastal areas is highlighted by the fact that over 22.9 million Americans live on land within 6 m of local mean high tide, with Florida and Louisiana being among the most at-risk states (Strauss et al., 2012). Urban resilience to flooding requires a multi-sectoral approach that considers the interdependence of urban functions and services, as demonstrated in studies of cities like Lisbon (Almeida et al., 2020). Adaptable planning strategies, such as those employed in Boston, emphasize the use of greenways as resilient infrastructure to protect against rising seas and storm surges, showcasing the importance of proactive urban planning in mitigating the impacts of climate change (Tang & Lukenda, 2019). These strategies underscore the necessity for coastal cities to adopt comprehensive and adaptable measures to safeguard their populations and infrastructures against the increasing threats of climate change.

3.10 CONCLUSION

This chapter clarifies how greenhouse gas emissions are measured and compared using standardized metrics, such as carbon dioxide equivalent (CO_2e) and global warming potential (GWP). These tools enable the aggregation of diverse gases with varying lifetimes and heat-trapping abilities into a single, comparable figure, which is essential for tracking emissions, setting targets, and shaping policy. We explored the complexities behind these metrics, including their limitations and the need for transparency in their application, especially when addressing short-lived versus long-lived climate pollutants. Understanding these measurement frameworks is crucial for accurately assessing emissions across various sectors, regions, and timeframes.

As we advance, the next chapter will build on this foundation by examining the economic and policy instruments designed to reduce emissions. It will explore carbon pricing, emissions trading systems, subsidies, and regulations that incentivize decarbonization and support the transition to a low-carbon economy. By linking measurement to management, the upcoming chapter bridges the science of emissions with the practical tools needed to drive effective climate action worldwide.

REVIEW QUESTIONS

1. What are the main health risks associated with increased global temperatures?
2. How does climate change impact food security and nutrition?
3. What are vector-borne diseases, and how does climate change influence their spread?
4. How does climate change exacerbate mental health issues, particularly among vulnerable populations?
5. What are the projected impacts of sea-level rise on coastal cities by 2050?
6. How are extreme weather events linked to climate change, and what are their effects on communities?
7. Why is water availability a concern in the context of global warming?
8. How does climate change affect energy demand, particularly in hotter regions?
9. What are the socioeconomic impacts of climate change on disadvantaged communities?
10. How can countries prepare for and adapt to the health challenges posed by climate change?

REFERENCES

Map of Temperature Changes (1961–2019) infographic by Fakta o klimatu, licenced under CC BY 4.0. https://factsonclimate.org/infographics/map-temperature-change#data-source

Ahmed, T., Scholz, M., Al-Faraj, F. A. M., & Niaz, W. (2016). Water-related impacts of climate change on agriculture and subsequently on public health: A review for generalists with particular reference to Pakistan. *International Journal of Environmental Research and Public Health*. https://doi.org/10.3390/IJERPH13111051

Almeida, M. do C., Telhado, M. J., Morais, M., Barreiro, J., & Lopes, R. (2020). Urban resilience to flooding: Triangulation of methods for hazard identification in urban areas. *Sustainability*. https://doi.org/10.3390/SU12062227

Bintanja, R., & Selten, F. (2016). Future increases in Arctic precipitation linked to local evaporation and sea ice retreat. *Nature*, *509*, 479–482.

Bouley, T., Gilbert, M., Whung, P.-Y., Gall, F. L., & Plante, C. A. (2014). *Reducing climate-sensitive disease risks (English).* Agriculture and environmental services discussion paper; no. 7, Washington, DC; World Bank Group. http://documents.worldbank.org/curated/en/486511468167944431.

Bryant, C.R., Vitali, A., Madani, A. (2023). *Climate Change and Food Insecurity*. MDPI. https://doi.org/10.3390/books978-3-0365-7440-0

Cohen, J., Screen, J. A., Furtado, J. C., Barlow, M., Whittleston, D., Coumou, D., Francis, J. A., Dethloff, K., Entekhabi, D., Overland, J. E., & Jones, J. (2014). Recent Arctic amplification and extreme mid-latitude weather. *Nature Geoscience*. https://doi.org/10.1038/NGEO2234

Cury, R. G. (2023). Characterizing vulnerabilities to climate change across the United States. *Environment International*. https://doi.org/10.1016/j.envint.2023.107772

Dastidar, B. G. (2023). The incidence of post traumatic stress disorder amongst cyclone survivors in a rural hamlet of West Bengal. *British Journal of Psychiatry Open*. https://doi.org/10.1192/bjo.2023.186

Dietrich, J. S., Hammerl, J. A., Johne, A., Kappenstein, O., Loeffler, C. R., Nöckler, K., Rosner, B., Spielmeyer, A., Szabó, I., & Richter, M. (2023). Impact of climate change on foodborne infections and intoxications. *Journal of Health Monitoring*. https://doi.org/10.25646/11403

Dunk, J., & Rhodes, P. (2023). Climate distress: A review of current psychological research and practice. *Sustainability*. https://doi.org/10.3390/su15108115

Fakta o klimatu. (2022, May 12). *Climate change scheme* [Infographic]. https://factsonclimate.org/infographics/climate-change-scheme

Fall, P., Diouf, I., Deme, A., Diouf, S. M., Sene, D., Sultan, B., & Janicot, S. (2023). Enhancing understanding of the impact of climate change on malaria in West Africa using the vector-borne disease community model of the International Center for Theoretical Physics (VECTRI) and the bias-corrected phase 6 coupled model intercomparison project data (CMIP6). *Microbiology Research*. https://doi.org/10.3390/microbiolres14040145

Fanzo, J., Schneider, K., & Wood, S. (2024). Food insecurity and malnutrition. In B.S. Levy, & J.A. Patz (Eds.), *Climate Change and Public Health*. Oxford Academic, New York. https://doi.org/10.1093/oso/9780197683293.003.0008

Firley, E., & Deupi, V. (2017). Miami rising: Historical perspectives on sea level rise as a view into the future. *The Plan Journal*. https://doi.org/10.15274/TPJ.2017.02.02.10

Gornitz, V. M., White, T. W., & Cushman, R. M. (1991). Vulnerability of the U.S. to future sea level rise. In *Proceedings of the Coastal Zone 91 Proceedings of the 7th Symposium on Coastal & Ocean Management, ASCE*, Oak Ridge, pp. 1345–1359.

Hipsey, M. R., & Brookes, J. D. (2013). *Pathogen Management in Surface Waters: Practical Considerations for Reducing Public Health Risk*. InTech. https://doi.org/10.5772/55367

Jain, R., & Mishra, S. (2016). Potential implications of climate change on water quality and water-borne diseases in Hoshangabad District (Madhya Pradesh, India). *Indian Journal of Geo-Marine Sciences*, 45, 639–652.

de Junior, V. M. B., de Magalhães, H. F., & Albuquerque, U. P. (2023). Perception of health risks in contexts of extreme climate change in semiarid Northeastern Brazil: an analysis of the role of socioeconomic variables. *Journal of Ethnobiology and Ethnomedicine*. https://doi.org/10.1186/s13002-023-00597-1

Lyberger, K., Farner, J. E., Couper, L. I., & Mordecai, E. A. (2024). A mosquito parasite is locally adapted to its host but not temperature. *The American Naturalist*, 204(2). https://doi.org/10.1086/730522

Lykins, A. D., Parsons, M., Craig, B. M., Cosh, S., Hine, D. W., & Murray, C. V. (2023). Australian youth mental health and climate change concern after the black summer bushfires. *Ecohealth*. https://doi.org/10.1007/s10393-023-01630-1

Magnan, A. K., Oppenheimer, M., Garschagen, M., Buchanan, M. K., Duvat, V. K. E., Forbes, D. L., Ford, J. D., Lambert, E., Petzold, J., Renaud, F. G., Sebesvari, Z., van de Wal, R. S. W., Hinkel, J., Pörtner, H.-O. (2022). Sea level rise risks and societal adaptation benefits in low-lying coastal areas. *Science Reports*, 12, 10677. https://doi.org/10.1038/s41598-022-14303-w

Maliva, R. (2021). *Sea Level Rise and Groundwater*. Springer. https://doi.org/10.1007/978-3-030-66813-6_6

Matella, M. K. (2022). Sea level rise planning for resilient coastal infrastructure in California's coastal zone. *Share & Beach*. https://doi.org/10.34237/1009026

Moser, S. K., Spencer, J. A., Barnard, M., Hyman, J. M., Manore, C. A., & Gorris, M. E. (2024). Exploring climate-disease connections in geopolitical versus ecological regions: The case of west Nile virus in the United States. *Geohealth*. https://doi.org/10.1029/2024gh001024

Mottram, R., Winkelmann, R., Edwards, T. L., Burgard, C., van Calcar, C., Rahiman, W., Wang, T., & Jiménez, P. J. (2022). Sea-level rise: From global perspectives to local services. In *GU General Assembly 2022*, Vienna, Austria, 23–27 May 2022, EGU22-13026. https://doi.org/10.5194/egusphere-egu22-13026

Nicholls, R. J. (2011). Planning for the impacts of sea level rise. *Oceanography*. https://doi.org/10.5670/OCEANOG.2011.34

Oddo, P., Lee, B. S., Garner, G. G., Srikrishnan, V., Reed, P. M., Forest, C. E., Keller, K., & Keller, K. (2020). Deep uncertainties in sea-level rise and storm surge projections: Implications for coastal flood risk management. *Risk Analysis*. https://doi.org/10.1111/RISA.12888

Ohenhen, L. O., Shirzaei, M., Ojha, C., Sherpa, S. F., & Nicholls, R. J. (2024). Disappearing cities on US coasts. *Nature*, 627, 108–115. https://doi.org/10.1038/s41586-024-07038-3

Park, J., & Chae, Y. (2023). Analysis of time-dependent effects of ambient temperatures on health by vulnerable groups in Korea in 1999–2018. *Dental Science Reports*. https://doi.org/10.1038/s41598-023-28018-z

Parums, D. V. (2023). Editorial: Climate change and the spread of vector-borne diseases, including dengue, malaria, lyme disease, and West Nile Virus infection. *Medical Science Monitor*. https://doi.org/10.12659/msm.943546

Pednekar, G., & Raju, S. S. (2020). Sea level rise and its socio-economic impacts: A case study in Mumbai, India. In R. Akhtar (Eds.), *Extreme Weather Events and Human Health* (pp. 155–169). Springer https://doi.org/10.1007/978-3-030-23773-8_11

Phipps, S. J., King, M. A., Roberts, J. L., & Zhang, X. (2020). The commitment to global sea level rise over the next 500 years: Exploring the threat of the antarctic ice sheet to coastal infrastructure. In *Australasian Coasts and Ports 2019*, Hobart, Tasmania, Australia, 10–13 September 2019. https://doi.org/10.5281/zenodo.3890244

Pradhan, P., Seydewitz, T., Zhou, B., Lüdeke, M. K. B., & Kropp, J. P. (2022). Climate extremes are becoming more frequent, co-occurring, and persistent in Europe. *Anthropocene Science*. https://doi.org/10.1007/s44177-022-00022-4

Rani, P., & Reddy, R. G. (2023). Climate change and its impact on food security. *International Journal of Enviornment and Climate Change*. https://doi.org/10.9734/ijecc/2023/v13i31687

van Ruijven, B. J., De Cian, E., Sue Wing, I. (2019). Amplification of future energy demand growth due to climate change. *Nature Communications*, 10, 2762. https://doi.org/10.1038/s41467-019-10399-3

Siegert, M. J., & Pearson, P. (2021). Reducing uncertainty in 21st century sea-level predictions and beyond. *Frontiers in Environmental Science*. https://doi.org/10.3389/FENVS.2021.751978

Strauss, B. H., Zielinski, R., Weiss, J., & Overpeck, J. T. (2012). Tidally adjusted estimates of topographic vulnerability to sea level rise and flooding for the contiguous United States. *Environmental Research Letters*. https://doi.org/10.1088/1748-9326/7/1/014033

Tang, H., & Lukenda, J. (2019). Rising seas: Adaptable planning strategies for coastal urban greenways - Case studies in the US and China. *Fábos Conference on Landscape and Greenway Planning*, 6. https://doi.org/10.7275/A811-1Q32

Toromade, A. S., Soyombo, D. A., Kupa, E., & Ijomah, T. I. (2024). Reviewing the impact of climate change on global food security: Challenges and solutions. *International Journal of Applied Research in Social Sciences*. https://doi.org/10.51594/ijarss.v6i7.1300

Vogel, M. M., Zscheischler, J., Zscheischler, J., Zscheischler, J., Wartenburger, R., Dee, D., & Seneviratne, S. I. (2019). Concurrent 2018 hot extremes across northern hemisphere due to human-induced climate change. *Earth's Future*. https://doi.org/10.1029/2019EF001189

Withers, A. J., Croft, S., Budgey, R., Warren, D., & Johnson, N. (2024). Using correlative and mechanistic species distribution models to predict vector-borne disease risk for the current and future environmental and climatic change: A case study of West Nile Virus in the UK. *bioRxiv*. https://doi.org/10.1101/2024.09.12.612656

Xiao, H. (2022). Suggestions on international law system to mitigate the adverse effects of sea-level rise caused by climate change. *Law and Economy*. https://doi.org/10.56397/le.2022.10.03

4 Scientific and Technical Analysis of Climate Change

This chapter provides a comprehensive overview of the scientific and technical aspects of climate change. It delves into the current status of climate change research, highlighting key milestones in the understanding of human influence on the climate system. This chapter explores the Earth's climate system, including natural and anthropogenic drivers, and examines observed changes in sea level, temperature, and other climate variables. It also discusses the progress made in climate science, focusing on observation-based products, climate model performance, and the understanding of human influence on climate change. This chapter further explores the causes of climate change, emphasizing the role of greenhouse gas emissions and other human activities in driving global warming. It also presents future projections of climate change, highlighting the potential risks and impacts associated with different emissions scenarios. Additionally, this chapter discusses the concept of climate-resilient development and the importance of mitigation and adaptation strategies in achieving a sustainable future.

4.1 CURRENT UNDERSTANDING OF EARTH'S CLIMATE SYSTEM

Earth's climate system has undergone significant changes over millions of years, with natural archives like ice cores, tree rings, and sediment layers providing valuable insights into past climates. These records reveal a strong correlation between atmospheric CO_2 concentrations and global surface temperatures, as shown in Figure 4.1. For instance, during the mid-Pliocene Warm

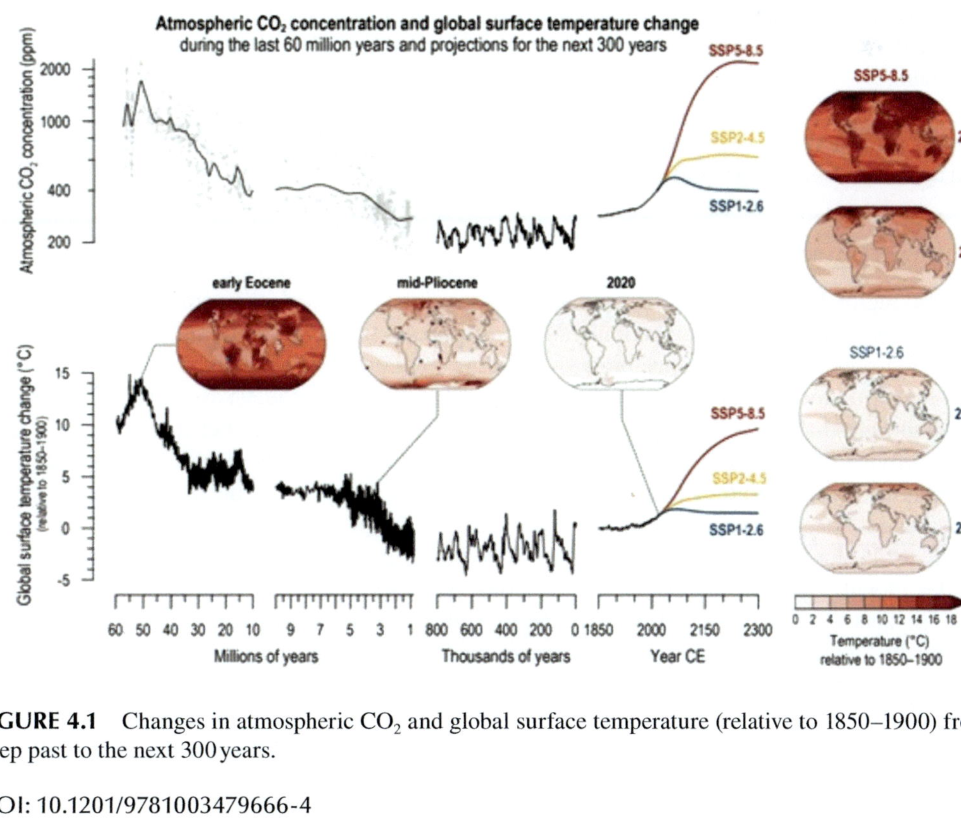

FIGURE 4.1 Changes in atmospheric CO_2 and global surface temperature (relative to 1850–1900) from the deep past to the next 300 years.

DOI: 10.1201/9781003479666-4

Period, about 3.2 million years ago, temperatures were 2.5°C–4°C higher than pre-industrial levels, aligning with CO_2 concentrations similar to those projected for an intermediate emissions scenario by 2300 (Chandan & Peltier, 2018). In the Early Eocene Climate Optimum, around 50 million years ago, temperatures were much higher, ranging from 10°C to 18°C above pre-industrial levels, corresponding to CO_2 levels projected under a high emissions scenario (Shared Socioeconomic Pathways – SSP5–8.5) by 2300 (Bess, 2022).

Future global warming levels depend on the emissions pathway followed. By 2300, under an intermediate emissions scenario, global surface temperatures are projected to rise by 2.3°C–4.6°C compared to 1850–1900 levels. This scenario mirrors conditions during the mid-Pliocene Warm Period (Chandan & Peltier, 2018). Conversely, a high CO_2 emissions scenario (Shared Socioeconomic Pathways – SSP5–8.5) could lead to temperatures increasing by 6.6°C–14.1°C, akin to the Early Eocene Climate Optimum (Bess, 2022). These projections underscore the significant impact of human activities on future climate conditions. The fundamental elements of the climate system are well understood, with major natural and anthropogenic factors identified since the 19th century. Natural drivers include:

- **Incoming Solar Radiation**: Variations in solar energy reaching Earth.
- **Volcanic Activity**: Emissions of particles and gases affecting atmospheric conditions.
- **Orbital Cycles**: Changes in Earth's orbit and tilt influencing climate patterns.
- **Biogeochemical Cycles**: Natural processes like the carbon cycle impacting greenhouse gas levels.

Anthropogenic drivers, which were better understood and attempts made to quantify by the 1970s, include:

- **CO_2 Emissions**: From fossil fuel combustion (petroleum, coal, natural gas).
- **Atmospheric Aerosols**: Fine solid particles or liquid droplets from human activities.
- **Land-Use Change**: Deforestation and urbanization altering land surfaces.
- **Non-CO_2 Greenhouse Gases**: Emissions of methane (CH_4), nitrous oxide (N_2O), and others.

Figure 4.2 shows that concentrations of the key greenhouse gases have all increased since the Industrial Revolution due to human activities.

4.1.1 Temperature Trends

The Arctic region has experienced a pronounced warming trend, with temperatures rising at more than twice the global average, a phenomenon known as Arctic Amplification. This trend is driven by increased greenhouse gas concentrations, leading to significant changes in the Arctic climate system, including reductions in snow and ice (Taylor et al., 2023; Gorodetskaya et al., 2022). Recent studies highlight that the Arctic is warming approximately three times faster than the global average, with projections indicating that this trend will continue, potentially reaching critical temperature thresholds by mid-century (Ma et al., 2022). The rapid warming in the Arctic is evidenced by unprecedented extreme events, such as record summer temperatures, heatwaves, and significant ice melt, which have no historical analogues (Overland, 2024). These changes are indicative of a broader global warming pattern, with the Arctic serving as an early warning sign of climate change impacts (Taylor et al., 2023). The melting of glaciers and sea ice, along with rising sea levels, further underscores the severity of warming in polar regions (Gorodetskaya et al., 2022). The Arctic's climate changes have been documented since the 1980s, with recent decades showing more pronounced effects, including extremes in sea ice levels and precipitation (Linden, 2022). These findings collectively confirm the unprecedented nature of current warming trends, aligning

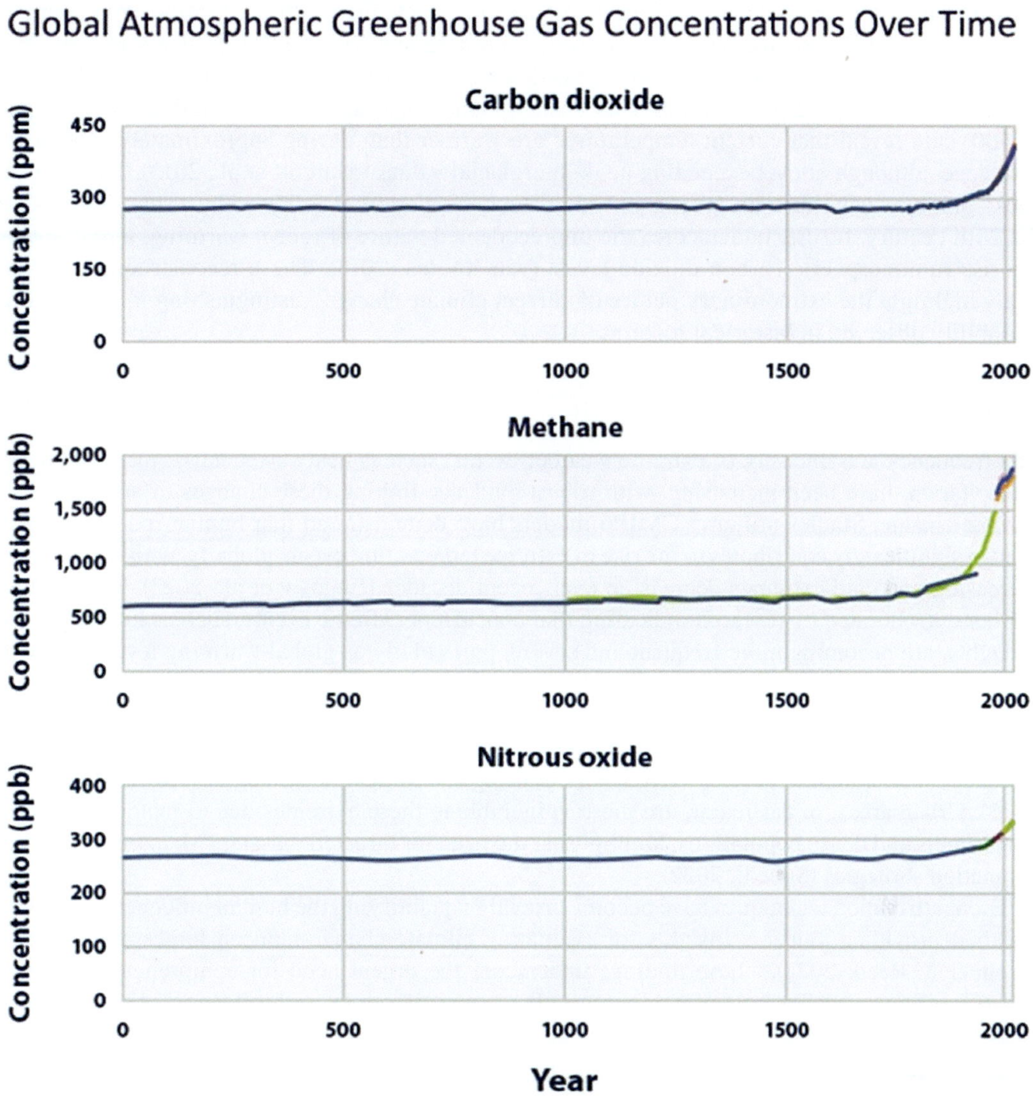

FIGURE 4.2 Graph showing the increase in atmospheric concentrations of three key greenhouse gases over the last 2000 years. Increases in concentrations of these gases since 1750 are due to human activities in the industrial era. (Source: *U.S. EPA, Climate Change Indicators in the United States: Atmospheric Concentrations of Greenhouse Gases, 2021*).

with historical temperature reconstructions from paleoclimate records and emphasizing the urgent need for global climate action (Linden, 2022).

Historical temperature reconstructions from paleoclimate records provide substantial evidence supporting the notion of unprecedented current warming. These reconstructions, derived from various natural proxies such as tree rings, ice cores, and sediment layers, reveal that the recent warming trend is exceptional when compared to the past millennia. For instance, the Intergovernmental Panel on Climate Change (IPCC) has concluded that the latest decade is warmer than any multi-century period over the past 125,000 years, a claim supported by the absence of elevated greenhouse gases during pre-instrumental periods of the Holocene, as indicated by ice core data (Esper et al., 2024). In Siberia, tree-ring records show that the current warming has disrupted a multi-millennial cooling trend, marking the industrial era warming as unprecedented in both rate and magnitude over the past 7,638 years (Hantemirov, 2022). Similarly, in southern China, temperature reconstructions using tree-ring data and historical documents indicate that the warming since the 1980s has been unparalleled in the past 160 years, despite a brief hiatus in the 2000s (Liu et al., 2017). On a broader scale, global temperature reconstructions over the past 11,300 years reveal that current temperatures are warmer than during approximately 75% of the Holocene, although not yet exceeding peak interglacial values (Marcott et al., 2013). The "hockey stick" graph, which illustrates a long period of stable temperatures followed by a sharp increase in the 20th century, further underscores the unprecedented nature of recent warming, aligning with the rise in atmospheric carbon dioxide levels (van Kooten, 2013). These reconstructions collectively highlight the extraordinary nature of current climate change, distinguishing it from natural variability observed in historical records.

4.1.2 Extreme Weather Events

The frequency and intensity of extreme weather events, such as heatwaves, hurricanes, and heavy precipitation, have been increasing, with robust evidence linking these changes to anthropogenic climate change. Studies utilizing CMIP6 models have demonstrated that human-induced climate change significantly contributes to the rise in extreme temperature events globally, with hot extremes increasing and cold extremes decreasing over recent decades (Engdaw et al., 2023). This trend is further corroborated by research indicating that concurrent extreme events, such as heatwaves and droughts, are becoming more frequent and severe, particularly as global warming levels rise, with some regions experiencing up to a 9.6-fold increase in such events at +3°C warming compared to pre-industrial levels (Batıbeniz et al., 2023). The increasing severity of heatwaves poses significant threats to public health and ecological diversity, necessitating multidisciplinary research and international collaboration to develop effective monitoring and mitigation strategies (Klingelhöfer et al., 2023). Urban areas, in particular, are more vulnerable to these extremes due to factors like urban heat islands and dense populations, highlighting the need for targeted disaster risk management and adaptation strategies (Subedi, 2022).

Event attribution techniques have become crucial in quantifying the human influence on extreme weather, providing valuable insights for evaluating climate change-induced losses and damages (Wehner & Reed, 2022). These findings underscore the urgent need for comprehensive climate action to mitigate the impacts of extreme weather events driven by anthropogenic climate change. The accuracy of attribution studies linking specific extreme weather events to anthropogenic climate change is enhanced by several key factors, as highlighted in recent research. One significant advancement is the use of machine learning techniques, such as convolutional neural networks, which allow for the creation of dynamically consistent counterfactual scenarios. This approach has been shown to provide rapid and cost-effective attribution of extreme events, aligning well with traditional methods and offering insights into the influence of global mean temperature on event intensity and frequency (Trok et al., 2024).

Additionally, the integration of probabilistic assessments with conditional methods helps in understanding how climate change alters the likelihood and intensity of extreme events, which is crucial for informing policy and adaptation strategies (Otto, 2023). The collective attribution of multiple events, such as historical heatwaves, also contributes to a more comprehensive understanding by systematically evaluating the influence of anthropogenic factors across numerous instances, thereby supporting climate litigation and policy development (Quilcaille et al., 2024). Furthermore, the use of storyline approaches, which focus on the dynamic and thermodynamic aspects of climate change, provides a quantifiable framework for understanding the impacts of climate change on specific events, reducing uncertainties and enhancing the robustness of attribution studies (Mindlin & Garderen, 2023). Finally, the emerging field of Extreme Event Impact Attribution (EEIA) combines climate data with socioeconomic impact assessments, offering a nuanced view of the consequences of climate change on extreme weather events and aiding in the quantification of climate change costs and adaptation planning (Noy et al., 2024). Together, these methodological advancements and integrative approaches significantly improve the precision and applicability of attribution studies in linking extreme weather events to human-induced climate change.

4.1.3 Sea-Level Rise

As noted in Figure 4.3, the global mean sea level (GMSL) has increased by approximately 0.20 m since 1901, with an accelerated rate of increase observed since the 1960s, primarily driven by the thermal expansion of seawater and the melting of glaciers and ice sheets (Durand et al., 2022; Scambos & Abdalati, 2022). Recent studies project that under high emissions scenarios, sea levels could rise by up to 1.01 m by 2100, posing significant risks to coastal communities and ecosystems (Scambos & Abdalati, 2022). The Antarctic and Greenland ice sheets are major contributors to this rise, with their future mass loss being a significant source of uncertainty in long-term projections (Durand et al., 2022). Projections using the Coupled Model Intercomparison Project Phase 6 (CMIP6) models under various climate targets suggest that global sea-level rise could range from 60 to 320 mm, depending on the extent of global warming (Sung et al., 2021). The Antarctic ice sheet alone could contribute between 0.86 and 1.44 m by 2500 under different emissions scenarios,

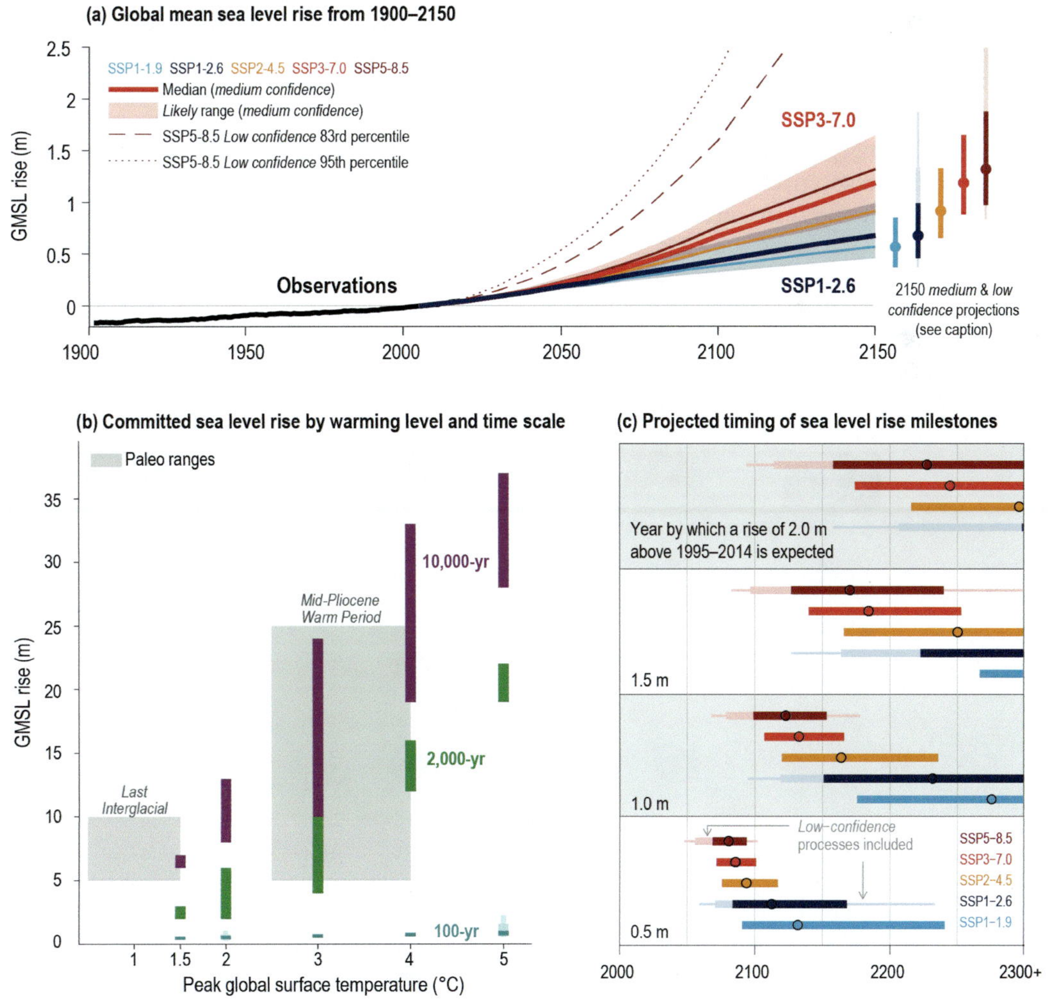

FIGURE 4.3 Global mean sea-level (GMSL) changes on different time scales and under different scenarios.

highlighting the critical role of ice sheet dynamics in future sea-level changes (Turner et al., 2023b). The variability in regional sea-level changes is influenced by factors such as ocean circulation, surface winds, and the exchange of heat and freshwater between the air and ocean (Scambos & Abdalati, 2022). Addressing these challenges requires a transdisciplinary approach, involving climate scientists, coastal impact specialists, and local stakeholders to develop effective adaptation strategies (Durand et al., 2022). Overall, the integration of multi-centennial projections and local adaptation measures is crucial to mitigate the impacts of sea-level rise on vulnerable coastal regions (Turner et al., 2023a,b).

4.1.4 The Climate System and Observed Changes

The climate system responses to global warming, as noted in Figure 4.4, demonstrate significant escalation across multiple indicators when compared to the 1850–1900 baseline period, as reported by the IPCC in 2023:

- Current observations at +1.1°C warming show the hottest days are already 1.2°C warmer (range: +0.7 to 1.5°C), with drought frequency increased by 1.7 times and extreme precipitation events occurring 1.3 times more frequently.
- As warming progresses to +1.5°C, these impacts intensify substantially, with temperature anomalies reaching +1.9°C (range: +1.3 to 2.2°C), drought frequency doubling, and extreme precipitation events increasing by 1.5 times.
- At +2°C warming, the system shows further deterioration, with temperature anomalies of +2.6°C (range: +1.8 to 3.1°C), while drought and precipitation extremes increase by 2.4 and 1.7 times, respectively.
- The most severe scenario at +4°C projects dramatic changes, including temperature anomalies of +5.1°C (range: +4.3 to 5.8°C), drought frequency increasing by 4.1 times, and extreme precipitation events occurring 2.7 times more often.

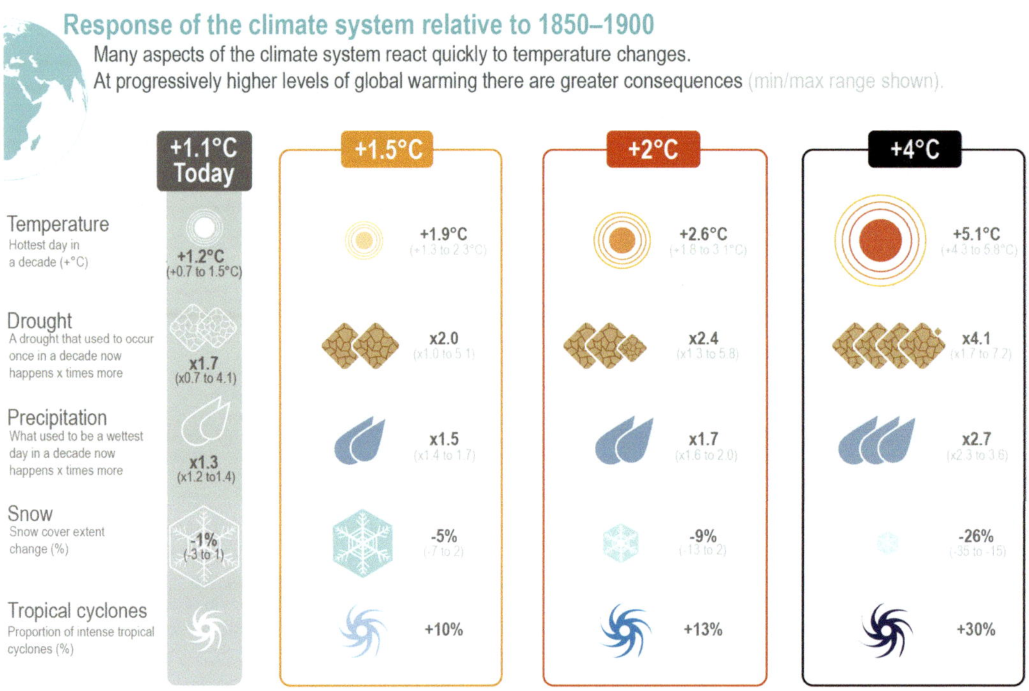

FIGURE 4.4 Response of selected climate variables to four levels of global warming (°C). Changes in the 'Today' column are based on a global warming level of 1°C.

- Snow cover exhibits a particularly concerning trend, declining from −5% at +1.5°C to −26% at +4°C, while the proportion of intense tropical cyclones shows a steady increase from +10% to +30% across the warming scenarios.

These findings underscore the non-linear nature of climate system responses and the cascading consequences of continued global warming.

4.2 PROGRESS IN RESEARCH AND STUDY OF CLIMATE SCIENCE

The evolution of climate science, particularly the understanding of the greenhouse effect, has been a gradual process spanning centuries. Early foundational work by scientists such as Fourier, Tyndall, and Arrhenius established the critical role of greenhouse gases like water vapor and carbon dioxide in regulating Earth's temperature. By the mid-19th century, it was recognized that industrial activities, notably coal combustion, could significantly alter the climate by increasing atmospheric CO_2 levels (Haszpra, 2022). This understanding was further solidified in the mid-20th century when empirical evidence, such as the Keeling Curve, demonstrated the accumulation of CO_2 due to human activities (Haszpra, 2022). The greenhouse effect, a process where gases trap heat in the atmosphere, is crucial for maintaining Earth's temperature but has been exacerbated by anthropogenic emissions, leading to global warming (Samoshkina, 2023; Schwartz, 2018). The industrial era warming, which began in the mid-19th century, was marked by significant temperature increases across oceans and continents, highlighting the early onset of human-induced climate change (Abram et al., 2016). Different greenhouse gases vary in their impact on the climate, with some being more effective than others in enhancing the greenhouse effect, thus contributing to the complexity of climate change challenges (Tavassoli & Kamran-Pirzaman, 2023). The historical progression of climate science underscores the shift from understanding the causes of climate change to addressing its rate and impacts, as current research focuses on the complex responses of the climate system to increased atmospheric energy forcing (Haszpra, 2022).

By the 21st century, the understanding of climate change had evolved significantly, with systematic assessments like those by the IPCC transforming it from a theoretical concept into an established fact. This transformation was largely due to the integration of complex feedback mechanisms and the impact of aerosols into climate models. Aerosols, for instance, play a crucial role in atmospheric processes, influencing air pollution and climate through their interactions with boundary-layer dynamics and turbulence, which are essential for understanding atmospheric feedback and interactions (Kulmala et al., 2023). The latest Earth system models, such as those in the Coupled Model Intercomparison Project Phase 6 (CMIP6), incorporate aerosol and chemistry components that interact with the biosphere, revealing a complex web of feedbacks that are crucial for quantifying climate responses to anthropogenic emissions (Thornhill et al., 2021). These models show that aerosol feedback generally results in negative forcing, which can be offset by positive feedbacks from methane emissions (Thornhill et al., 2021).

Additionally, aerosol–cloud interactions, as studied in models like the Community Atmosphere Model version 6 (CAM6), demonstrate that aerosol forcing is predominantly negative due to shortwave radiation effects, while feedbacks can vary regionally, highlighting the intricate balance between forcing and feedbacks (Gettelman et al., 2024). CAM6 is a state-of-the-art global climate model and the atmospheric component of the Community Earth System Model version 2 (CESM2). The interaction between forests, aerosols, and climate systems illustrates how increased emissions of monoterpenes from forests can enhance aerosol and cloud formation, thereby affecting climate feedback mechanisms (Hede et al., 2015). The terrestrial climate feedback loop, involving carbon dioxide, gross primary production, and organic aerosol formation, further underscores the complexity of climate feedbacks, emphasizing the need for comprehensive field measurements and model simulations to understand these interactions (Kulmala et al., 2014). These advancements in understanding feedback mechanisms and aerosol impacts have been pivotal in solidifying climate change as an established scientific fact.

The observed changes in climate, such as shrinking ice sheets, rising sea levels, and extreme weather events, are strongly linked to human activities, providing compelling evidence of anthropogenic influence on the climate system. Research has shown that human-induced climate warming significantly increases the likelihood of extreme glacier mass-loss events, as demonstrated in New Zealand's Southern Alps, where such events were found to be six to ten times more likely due to anthropogenic factors (Vargo et al., 2020). Additionally, the frequency and intensity of weather extremes, such as heat anomalies and natural disasters, have risen markedly since the industrial era, correlating with increased carbon dioxide emissions from fossil fuel combustion (Carr, 2013).

The impacts of climate change are not confined to temperature changes but also include alterations in precipitation patterns, affecting various regions and sectors globally (Rosenzweig & Neofotis, 2013; Neofotis, 2013). Statistical analyses further confirm that human factors are the primary drivers of the observed long-term trends in global temperatures, with significant warming beginning around 1955 (Perron et al., 2012). These findings underscore the urgent need for continued research and action to mitigate climate change impacts, emphasizing the importance of reducing greenhouse gas emissions and developing effective adaptation strategies (Rosenzweig & Neofotis, 2013; Neofotis, 2013). The evidence collectively highlights the critical role of human activities in shaping current climate dynamics and the necessity for global efforts to address this pressing challenge.

Some key milestones in the history of climate research, including major influences external to the science itself, are presented in Table 4.1.

Despite concerted efforts at regional and national levels, recent years have witnessed significant fluctuations and record-breaking levels in GHG emissions. In 2021, global energy-industry-related GHG emissions reached an unprecedented 40.8 Gt of CO_2-eq per year, marking the highest recorded level to date (Friedlingstein et al., 2022). This peak followed a temporary decrease in global CO_2 emissions during 2020, attributed to the COVID-19 pandemic. In 2021, boreal forest fires accounted for 23% (0.48 billion metric tons of carbon) of global fire CO_2 emissions, the highest proportion since 2000. Additionally, agricultural practices continue to be significant GHG contributors, with potato cultivation emitting up to 3,186.2 kg CO_2-eq/ha in some regions. These findings underscore the complex and multifaceted nature of GHG emissions trends, highlighting the need for continued monitoring and mitigation efforts across diverse sectors.

4.2.1 Climate Model Performance

The Coupled Model Intercomparison Project Phase 6 (CMIP6) has advanced climate modeling by incorporating higher-resolution simulations and improved representations of physical processes, which have led to better alignment with observed temperature trends. However, discrepancies remain, particularly in simulating regional precipitation and tropical ocean processes. CMIP6 models have shown improved performance over their predecessors, CMIP5, in simulating mean and extreme precipitation, especially in regions like East Africa, where they better capture the annual mean cycle and extreme indices, although challenges persist in simulating extremely wet days (Ayugi et al., 2021). In terms of global surface temperatures, CMIP6 models exhibit a range of equilibrium climate sensitivity (ECS) values, with low ECS models aligning more closely with observed warming trends, suggesting that actual ECS may be lower than previously estimated (Ayugi et al., 2021). Despite these advancements, biases remain in simulating extreme precipitation, with overestimations in arid regions and underestimations in northern areas, although the multimodal ensemble generally performs better than individual models (Zhang et al., 2024). In the tropical oceans, CMIP6 models face significant challenges, particularly in simulating precipitation and its variability, due to issues with deep convection and radiative interactions, leading to overestimations of precipitation and water vapor variability (Zhang et al., 2024). These findings underscore the need for continued refinement of climate models to address persistent biases and improve the accuracy of regional climate projections.

TABLE 4.1
Key Milestones of Climate Research and Major Conferences

Year/Period	Milestone	Significance
1824	Joseph Fourier proposes the greenhouse effect concept	Laid foundation for understanding Earth's heat balance
1859	John Tyndall discovers heat-trapping gases	Identified CO_2 and other gases as key to atmospheric warming
1896	Svante Arrhenius calculates CO_2 warming effect	First quantitative link between CO_2 levels and global temperature
1930s–1950s	Guy Stewart Callendar compiles global temperature data	Provided early evidence of global warming trend
1958	Charles David Keeling begins Mauna Loa CO_2 measurements	Resulted in the "Keeling Curve," showing rising CO_2 levels
1960s	Development of computer climate models begins	Enhanced ability to simulate and predict climate changes
1970	U.S. Clean Air Act	First major environmental law in the U.S. to include air quality standards
1980	World Climate Research Programme (WCRP) established	Promoted international cooperation in climate research
1987	Montreal Protocol signed	International treaty to phase out ozone-depleting substances
1988	Intergovernmental Panel on Climate Change (IPCC) formed	Marked significant step in global climate policy and research coordination
1990s–2000s	Rapid advancement in climate modeling and satellite observations	Improved understanding of climate systems and changes
1992	United Nations Convention on Biological Diversity	Global agreement on sustainable development and biodiversity conservation
2003	Numerous observations raise concern that collapse of ice sheets (West Antarctica, Greenland) can raise sea level faster than most had believed	Sea rise, ice, floods, deadly summer heat wave in Europe accelerates divergence between European and U.S. public opinion
2005	Kyoto treaty goes into effect, signed by major industrial nations except U.S.	Work to mitigate emissions accelerates in Japan, Western Europe, U.S. regional governments and corporations
2006	California's Global Warming Solutions Act (AB32)	Comprehensive program to reduce greenhouse gas emissions
2009	U.S. EPA declares CO_2 a pollutant	Allowed for regulation of greenhouse gases under the Clean Air Act
2015	Researchers find collapse of West Antarctic ice sheet may be irreversible	Predictions indicating meters of sea-level rise over future centuries
2015	Paris Agreement adopted	Nearly all nations pledge to set their own targets for greenhouse gas cuts and to report their progress
2017	New Zealand introduces regulations for earthquake-prone buildings	Set timeframes for buildings to achieve minimum structural standards
2021	The sixth IPCC report warns that catastrophic outcomes cannot be ruled out	The Glasgow conference spurs pledge to restrict emissions—not enough to avoid dangerous climate change, but risk of catastrophic change is reduced

Climate models have increasingly integrated biogeochemical feedback, enhancing their utility for projecting future climate scenarios. This integration is crucial for understanding the complex interactions between climate and biogeochemical cycles, such as the carbon cycle, which are essential for accurate climate projections. For instance, Earth system models now incorporate carbon cycle feedback, which is critical for assessing the impacts of carbon dioxide removal (CDR) strategies and temperature overshoot scenarios. Carbon cycle feedbacks, quantified by metrics like β (carbon-concentration feedback) and γ (carbon-climate feedback), play a crucial role in Earth system models and climate projections. These feedbacks describe how changes in atmospheric CO_2 concentration and climate affect the exchange of CO_2 between the atmosphere and carbon reservoirs in land and oceans. These models reveal significant hysteresis, indicating that feedback strengths change during phases of negative emissions, thus affecting the reliability of climate projections under such scenarios (Asaadi et al., 2024; Séférian et al., 2023). Additionally, the inclusion of feedback mechanisms involving biogenic volatile organic compounds (BVOCs) and their impact on methane lifetime further illustrates the complexity of biogeochemical interactions. This feedback can increase methane's radiative forcing, thereby influencing climate sensitivity and future climate projections (Boy et al., 2022).

Coupled ocean-sea biogeochemistry models, such as Finite-volumE Sea ice-Ocean Model and Regulated Ecosystem Model version 3 (FESOM2.1-REcoM3), have also been updated to better simulate the ocean's role in carbon cycling, improving the accuracy of climate projections on decadal to centennial timescales (Hauck et al., 2023). These advancements underscore the importance of integrating biogeochemical feedback into climate models to provide more comprehensive and reliable projections, which are essential for informing international climate policy and mitigation strategies (Jones et al., 2024).

4.2.2 Climate Variability and Emerging Changes

Observed changes in the global climate are unequivocal and increasingly evident on regional and local scales. The rate of long-term change and year-to-year variability differ between regions and climate variables, affecting the visibility of these changes amid natural variations. Temperature changes are more apparent in tropical regions due to smaller year-to-year variations, despite larger warming elsewhere. Long-term changes in variables like weather extremes and Arctic sea ice have emerged in many regions. Improved observational datasets since AR5 offer stronger evidence of climate change and better estimates of natural variability. The slower global surface temperature increase from 1998 to 2012 was temporary, driven by internal variability and external factors, with ocean heat continuing to rise. Large ensembles of climate model simulations have enhanced understanding of internal variability and forced change. While anthropogenic impacts on multi-decadal mean precipitation are evident in some regions, internal variability can obscure these signals. Most land regions show temperature changes exceeding natural variability, with many temperature-related changes attributable to human activities. Ocean acidification, deoxygenation, and Arctic sea ice reduction have also emerged globally.

4.2.3 Understanding of Human Influence

The evidence for human influence on recent climate change has progressively strengthened from the IPCC Second Assessment Report to AR5, and is even stronger in the latest assessment, particularly at regional scales and for extreme events. As shown in Figure 4.5, human activities, such as anthropogenic greenhouse gas emissions, aerosol emissions, and land-use changes like urbanization, significantly alter the climate system's radiative properties, leading to warming. Advances in observational datasets, paleoclimate information, climate models, and attribution techniques have increased confidence in attributing climate changes to human activities. Since AR5, the rate of energy accumulation in the Earth system has become a robust measure of global climate change, enhancing the accuracy of trend estimation. Regional climate changes are influenced by local factors like urbanization and aerosols, with new techniques allowing for better attribution of extreme weather events to human influence. Improved understanding and data accessibility have further solidified the link between human activities and climate change, particularly regarding regional precipitation changes and weather extremes.

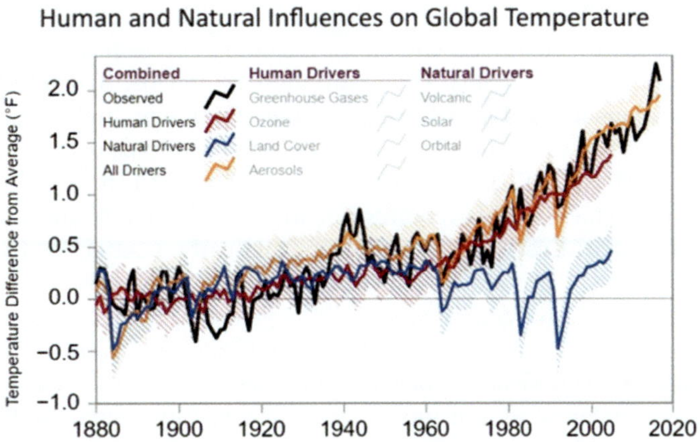

FIGURE 4.5 Human and natural factors both influence the Earth's climate, but the long-term trend observed over the past century can only be explained by the effect of human activities on climate.

4.3 ASSESSING FUTURE CLIMATE CHANGE

4.3.1 Climate Change Scenarios

The IPCC AR6 report utilizes a set of five illustrative scenarios, known as the Shared Socio-economic Pathways (SSPs), to explore a wide range of future greenhouse gas and air pollutant trajectories. These scenarios include SSP1-1.9, SSP1–2.6, SSP2–4.5, SSP3–7.0, and SSP5–8.5, each representing distinct socio-economic and climate mitigation pathways. SSP1-1.9 and SSP1–2.6 depict low emissions scenarios with stringent climate policies, aiming to limit warming to below 1.5°C and achieving net-zero CO_2 emissions around mid-century. SSP2–4.5 represents a scenario with moderate emissions reduction efforts, while SSP3–7.0 and SSP5–8.5 portray high emissions pathways, with SSP5–8.5 projecting the highest levels of greenhouse gas emissions and associated climate impacts. These scenarios are crucial for assessing climate responses under different socio-economic conditions and mitigation efforts, providing a comprehensive basis for understanding future climate change impacts and informing policy decisions.

4.3.2 Global Warming Levels and Cumulative CO_2 Emissions

The concept of global warming levels (GWLs), such as 1.5°C or 2.0°C above pre-industrial levels, serves as a crucial framework for evaluating climate change impacts across various scenarios and timeframes. These levels provide a standardized metric independent of emissions pathways, facilitating the integration of climate information across scientific disciplines and socio-economic sectors. For rapidly responding variables like temperature and precipitation, GWLs reliably predict changes in mean and extreme conditions. However, slow-responding factors such as ice sheet mass loss and deep ocean warming exhibit variability influenced by the trajectory and duration of warming. Despite these complexities, GWLs offer a robust tool for assessing climate impacts and guiding mitigation and adaptation strategies, aligning with global climate goals outlined in international agreements like the Paris Agreement.

4.4 FUTURE PROJECTIONS AND RISKS

Under intermediate emissions scenarios, global temperatures are projected to rise by 2.3°C–4.6°C by 2300, resembling the mid-Pliocene Warm Period, a time characterized by significant warming due to elevated greenhouse gas concentrations, particularly CO_2 and water vapor, which reduced planetary emissivity and albedo (Chandan & Peltier, 2018). This warming scenario underscores the profound risks posed by continued emissions, including severe heatwaves, biodiversity loss, and irreversible changes to ecosystems. High-**emissions scenarios, projecting increases up to 14.1°C, mirror conditions of the Early Eocene Climate** Optimum, highlighting the potential for catastrophic climate impacts such as self-reinforcing warming spirals, ocean acidification, and the collapse of agriculture (Bess, 2022). The Paris Agreement aims to limit global temperature rise to well below 2°C, but achieving this requires ambitious mitigation strategies, including carbon neutrality, reduction of short-lived climate pollutants, and carbon extraction (Xu & Ramanathan, 2017). Despite these efforts, there remains approximately a 50% probability that global warming will exceed 2°C, with cumulative emissions of 1,000 gigatons of carbon likely causing permanent shifts in climate regimes, particularly in the tropics (Diffenbaugh & Charland, 2016). The historical reliance on fossil fuels has significantly contributed to these emissions, necessitating global cooperation and innovative strategies to transition to renewable energy sources and achieve zero GHG emissions (Filonchyk et al., 2024). The urgency of these actions is underscored by the potential for dangerous to catastrophic warming within the next few decades if emissions remain unchecked (Xu & Ramanathan, 2017).

4.4.1 Projected Sea-Level Rise

Sea-level rise (SLR) driven by potential ice sheet collapse is a significant long-term concern, with projections indicating the critical importance of emissions reductions to mitigate global warming and its impacts. The uncertainty in SLR projections is largely due to the complex dynamics of ice sheets, particularly in Antarctica and Greenland, which are influenced by factors such as surface albedo effects and ice shelf buttressing (Broomfield, 2022). High-end estimates suggest that even with a modest 2°C warming scenario, SLR could reach up to 0.9 m by 2100 and 2.5 m by 2300, with more extreme scenarios predicting up to 1.6 m by 2100 and 10.4 m by 2300 under higher emissions scenarios (Fernandez, 2022; van de Wal et al., 2022). These projections underscore the long-term benefits of mitigation, as the differences between scenarios become more pronounced beyond 2100 (Fernandez, 2022; van de Wal et al., 2022). Furthermore, emissions until 2030 could lock in around 1.0 m of global mean SLR by 2300, highlighting the urgency of immediate emissions reductions (Nauels et al., 2024). Regional variations in SLR are also significant, with some areas, such as low-lying Pacific Islands, experiencing higher local sea-level rise than the global average (Nauels et al., 2024). High-end scenarios (HESs) further illustrate the potential for extreme SLR, with projections reaching up to 3.22 m by 2100 under high emissions scenarios, emphasizing the need for risk-averse stakeholders to consider these possibilities in adaptation planning (Dayan et al., 2021). Overall, these findings highlight the critical need for comprehensive mitigation strategies to limit future SLR and its associated risks.

Figure 4.6 from the AR6 framework provides a comprehensive assessment of future greenhouse gas emissions, climate change impacts, risks, and mitigation strategies. The figure integrates socio-economic development and policy factors, emissions pathways (represented by five Shared Socioeconomic Pathways, SSPs), and global surface temperature responses across different scenarios. It categorizes global mean temperature changes into eight levels (C1–C8) and assesses risks through the lens of impacts on socio-economic conditions. The dashed arrow indicates a pathway where impacts and risks influence socio-economic changes, a feature not yet fully incorporated into the assessed scenarios. Emissions include greenhouse gases, aerosols, and ozone precursors, with CO_2 emissions highlighted as an example. Temperature projections for 2100 relative to the pre-industrial period (1850–1900) are depicted, illustrating median values (line) and the combined very likely range (bar) across scenarios. Future risks associated with increasing warming are symbolized by a 'burning ember' figure, reflecting the severity of potential impacts.

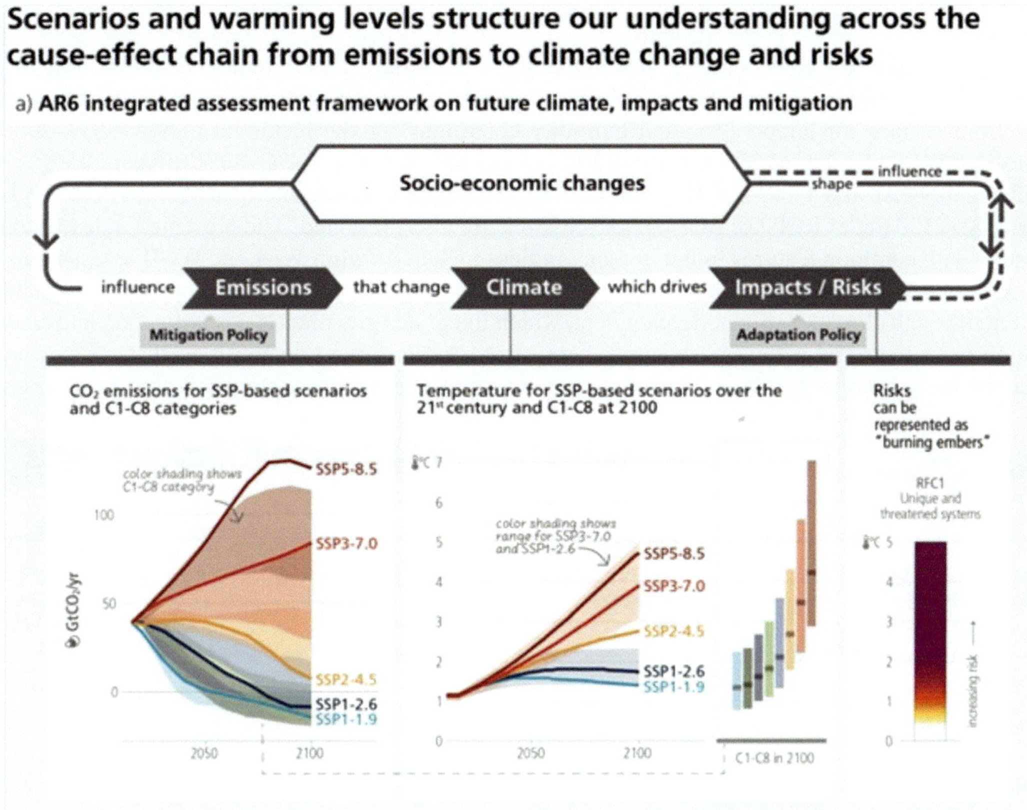

FIGURE 4.6 Schematic of the AR6 framework for assessing future greenhouse gas emissions, climate change, risks, impacts, and mitigation.

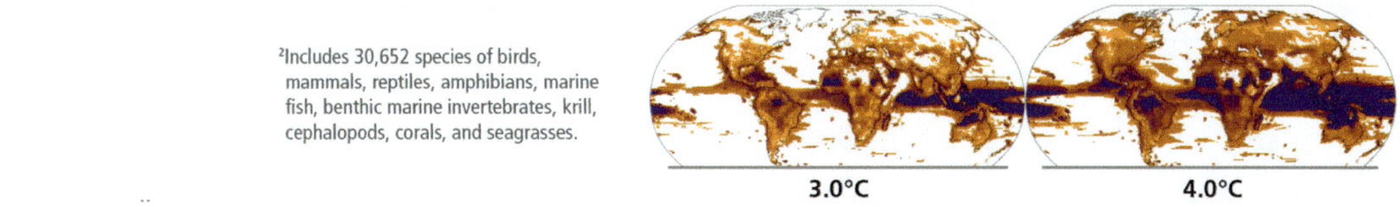

FIGURE 4.7 Risk of species loss.

4.4.2 Species Loss

Figure 4.7 shows the risks of species losses as indicated by the percentage of assessed species exposed to potentially dangerous temperature conditions. These are based on a baseline condition (1850–2005) maximum mean annual temperature experienced by each species, at GWLs of 1.5°C, 2°C, 3°C, and 4°C. Underpinning projections of temperature are from 21 Earth system models and do not consider extreme events impacting ecosystems, such as those in the Arctic.

4.4.3 Environmental Health Risks to Human Health

Figure 4.8 shows the environmental health (heat and humidity) risks to human health as indicated by the days per year of population exposure to hypothermic conditions that pose a risk of mortality from surface air temperature and humidity conditions for the historical period (1991–2005) and at GWLs of 1.7°C–2.3°C (mean = 1.9°C; 13 climate models), 2.4°C–3.1°C (mean = 2.7°C; 16 climate models), and 4.2°C–5.4°C (mean = 4.7°C; 15 climate models). Interquartile ranges of GWLs by 2081–2100 under RCP2.6, RCP4.5, and RCP8.5 are also presented. The presented index is consistent with common features found in many indices included within WGI and WGII assessments.

Figure 4.9 illustrates the critical and narrowing window for achieving a livable and sustainable future through climate-resilient development, which integrates greenhouse gas mitigation and adaptation measures. The pathways shown (ranging from red to green) demonstrate that choices and actions by various stakeholders, including governments, the private sector, and civil society, can

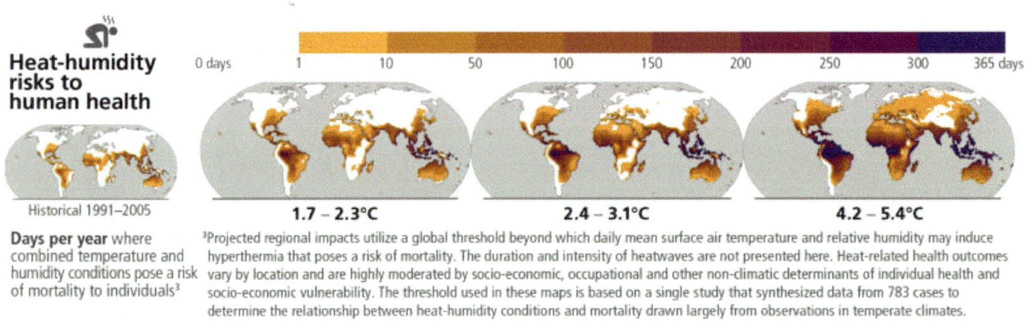

FIGURE 4.8 Risk to human health.

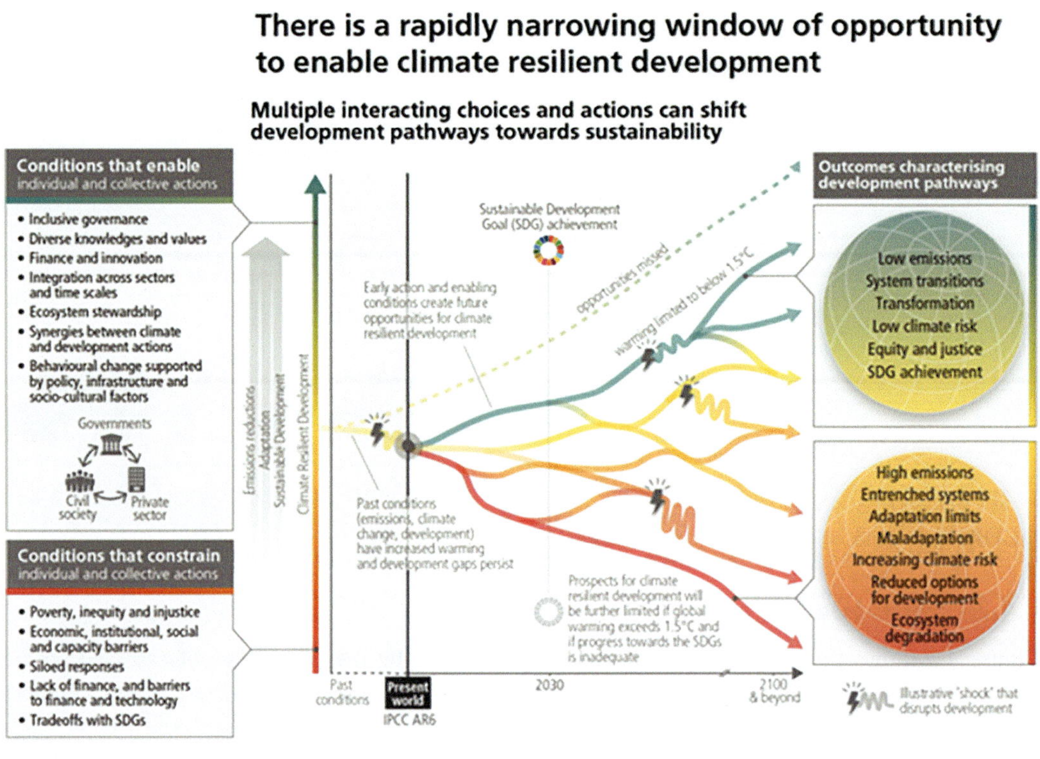

FIGURE 4.9 A rapidly narrowing window of opportunity to secure a livable and sustainable future for all.

influence the trajectory toward sustainability and lower emissions. These pathways are informed by diverse knowledge systems, including cultural values, Indigenous knowledge, local knowledge, and scientific knowledge. Pathways with lower climate resilience (red to yellow) face more severe impacts from climatic and non-climatic events like droughts, floods, and pandemics compared to those with higher resilience (green). Adaptation and adaptive capacity have limits, especially at 1.5°C of global warming, with increasing losses and damages at higher temperatures. The development pathways of countries at all economic stages affect greenhouse gas emissions and shape mitigation challenges and opportunities, influenced by past actions and enabling or constraining conditions. Delaying emissions mitigation efforts reduces the availability of effective adaptation options.

4.5 CONCLUSION

Decades of scientific research have made it clear: human activities—especially fossil fuel use and land-use change—are driving unprecedented changes in Earth's climate system. This chapter detailed how greenhouse gas emissions have pushed global temperatures, sea levels, and extreme weather events beyond natural variability, with the Arctic warming at record rates and extreme events becoming more frequent and severe. Paleoclimate records, advanced climate models, and improved observation systems all point to the same conclusion: today's climate change is rapid, global, and largely anthropogenic.

Climate science has also advanced our ability to project future risks, revealing that the choices we make now will determine the scale of warming, sea rise, and ecosystem disruption for centuries to come. The chapter emphasized that mitigation and adaptation are both urgent and essential—delaying action narrows our options and raises the risks for people and nature worldwide.

The next chapter moves from the global scale of climate science to the practical realities of climate action within organizations. Chapter 5 explores how companies and institutions can measure, manage, and report their greenhouse gas emissions using the Scope 1, 2, and 3 framework. By connecting the science of climate change to the mechanics of emissions accounting, the discussion shifts from understanding the problem to taking responsibility and driving solutions across the value chain.

REVIEW QUESTIONS

1. How has scientific understanding of climate change evolved over recent decades?
2. What is the current consensus on the causes of climate change?
3. How do scientists model and predict future climate change impacts?
4. What is the significance of the IPCC's assessment reports in guiding climate policy?
5. How do human activities disrupt natural greenhouse gas cycles?
6. What role does the U.S. Environmental Protection Agency (EPA) play in climate research?
7. How does atmospheric data inform climate projections?
8. What are the limitations of current climate models?
9. How do environmental feedback loops affect climate predictions?
10. What are some technical innovations that aid in climate research?

REFERENCES

Abram, N. J., McGregor, H., Tierney, J. E., Tierney, J. E., Evans, M. N., McKay, N. P., & Kaufman, D. S. (2016). Early onset of industrial-era warming across the oceans and continents. *Nature*. https://doi.org/10.1038/NATURE19082

Asaadi, A., Schwinger, J., Lee, H., Tjiputra, J., Arora, V. K., Séférian, R., Liddicoat, S., Hajima, T., Santana-Falcón, Y., & Jones, C. (2024). Carbon cycle feedbacks in an idealized simulation and a scenario simulation of negative emissions in CMIP6 Earth system models. *Biogeosciences*. https://doi.org/10.5194/bg-21-411-2024

Ayugi, B., Zhidong, J., Zhu, H., Ngoma, H., Babaousmail, H., Rizwan, K., & Dike, V. N. (2021). Comparison of CMIP6 and CMIP5 models in simulating mean and extreme precipitation over East Africa. *Preprints*. https://doi.org/10.20944/PREPRINTS202102.0111.V1

Batıbeniz, F., Hauser, M., & Seneviratne, S. I. (2023). Countries most exposed to individual and concurrent extremes and near-permanent extreme conditions at different global warming levels. *Earth System Dynamics Discussions*. https://doi.org/10.5194/esd-14-485-2023

Bess, M. D. (2022). *Fossil Fuels and Climate Change*. Cambridge University Press. https://doi.org/10.1017/9781009160315.002

Boy, M., Zhou, P., Kurtén, T., Chen, D., Xavier, C., Clusius, P., Roldin, P., Baykara, M., Pichelstorfer, L., Foreback, B., Bäck, J., Petäjä, T., Makkonen, R., Kerminen, V. M., Pihlatie, M., Aalto, J., & Kulmala, M. (2022). Positive feedback mechanism between biogenic volatile organic compounds and the methane lifetime in future climates. *Npj Climate and Atmospheric Science*. https://doi.org/10.1038/s41612-022-00292-0

Broomfield, J. P. (2022). Ice sheet and climate processes driving the uncertainty in projections of future sea level rise: Findings from a structured expert judgement approach. *Earth's Future*. https://doi.org/10.1029/2022ef002772

Carr, P. H. (2013). Weather extremes from anthropogenic global warming. *Natural Science*. https://doi.org/10.4236/NS.2013.51A020

Chandan, D., & Peltier, W. R. (2018). On the mechanisms of warming the mid-Pliocene and the inference of a hierarchy of climate sensitivities with relevance to the understanding of climate futures. *Climate of the Past*. https://doi.org/10.5194/CP-14-825-2018

Dayan, H., Cozannet, G. L., Speich, S., & Thiéblemont, R. (2021). High-end scenarios of sea-level rise for coastal risk-averse stakeholders. *Frontiers in Marine Science*. https://doi.org/10.3389/FMARS.2021.569992

Diffenbaugh, N. S., & Charland, A. (2016). Probability of emergence of novel temperature regimes at different levels of cumulative carbon emissions. *Frontiers in Ecology and the Environment*. https://doi.org/10.1002/FEE.1320

Durand, G., van den Broeke, M. R., Cozannet, G. L., Edwards, T. L., Holland, P. R., Jourdain, N. C., Marzeion, B., Mottram, R., Nicholls, R. J., Pattyn, F., Paul, F., Slangen, A. B. A., Winkelmann, R., Burgard, C., van Calcar, C., Barré, J.-B., Bataille, A., & Chapuis, A. (2022). Sea-level rise: From global perspectives to local services. *Frontiers in Marine Science*. https://doi.org/10.3389/fmars.2021.709595

Engdaw, M. M., Steiner, A. K., Hegerl, G. C., & Ballinger, A. (2023). Attribution of observed changes in extreme temperatures to anthropogenic forcing using CMIP6 models. *Weather and Climate Extremes*. https://doi.org/10.1016/j.wace.2023.100548

Esper, J., Schulz, P., & Büntgen, U. (2024). Is recent warming exceeding the range of the past 125,000 years? *Atmosphere*. https://doi.org/10.3390/atmos15040405

Fernandez, A. (2022). A high-end estimate of sea level rise for practitioners. *Earth's Future*. https://doi.org/10.1029/2022ef002751

Filonchyk, M., Peterson, M. P., Zhang, L., Hurynovich, V., & He, Y. (2024). Greenhouse gases emissions and global climate change: Examining the influence of CO2, CH4, and N2O. *Science of the Total Environment*. https://doi.org/10.1016/j.scitotenv.2024.173359

Friedlingstein, P., O'Sullivan, M., Jones, M. W., Andrew, R. M., Gregor, L., Hauck, J., ... & Zheng, B. (2022). Global carbon budget 2022. *Earth System Science Data*, 14(11), 4811–4900. https://doi.org/10.5194/essd-14-4811-2022

Gettelman, A., Eidhammer, T., Duffy, M. L., McCoy, D., Song, C. H., & Watson-Parris, D. (2024). The interaction between climate forcing and feedbacks. *ESS Open Archive*. https://doi.org/10.22541/essoar.170612437.79287996/v1

Gorodetskaya, I., Lauer, M., & Kiszler, T. (2022). Polar amplification: Stronger warming in the Arctic and Antarctic. *Frontiers for Young Minds*. https://doi.org/10.3389/frym.2022.703805

Hantemirov, R. M. (2022). Current Siberian heating is unprecedented during the past seven millennia. *Nature Communications*. https://doi.org/10.1038/s41467-022-32629-x

Haszpra, L. (2022). Fejezetek az üvegházhatás-kutatás történetéből Highlights from the History of Greenhouse Effect Research. *Magyar Tudomány*. https://doi.org/10.1556/2065.183.2022.1.5

Hauck, J., Oziel, L., Karakus, O., Zeising, M., Xing, F., Kevonanrsm, K., & Salvi, S. (2023). Ocean biogeochemistry in the coupled ocean-sea ice-biogeochemistry model FESOM2.1-REcoM3. *Geoscientific Model Development*. https://doi.org/10.5194/gmd-2023-2

Hede, T., Leck, C., & Claesson, J. (2015). Amplified feedback mechanism of the forests-aerosols-climate system. *International Journal of Climatology*. https://doi.org/10.1155/2015/262980

Jones, C., Adloff, F., Booth, B., Cox, P. M., Eyring, V., Friedlingstein, P., Frieler, K., Hewitt, H. T., Jeffery, H., Joussaume, S., Koenigk, T., Lawrence, B., O'Rourke, E., Roberts, M., Sanderson, B. M., Séférian, R., Somot, S., Vidale, P. L., Vuuren, D. P. van, … Zaehle, S. (2024). Bringing it all together: science priorities for improved understanding of Earth system change and to support international climate policy. *Earth System Dynamics Discussions*. https://doi.org/10.5194/esd-15-1319-2024

Klingelhöfer, D., Braun, M., Brüggmann, D., & Groneberg, D. (2023). Heatwaves: Does global research reflect the growing threat in the light of climate change? *Globalization and Health*. https://doi.org/10.1186/s12992-023-00955-4

van Kooten, G. C. (2013). Climate science and paleoclimatology. In *Climate Change, Climate Science and Economics* (pp. 59–100). Springer. https://doi.org/10.1007/978-94-007-4988-7_3

Kulmala, M., Kokkonen, T. V., Ezhova, E., Baklanov, A., Mahura, A., Mammarella, I., Bäck, J., Lappalainen, H., Tyuryakov, S. A., Kerminen, V.-M., Zilitinkevich, S., & Petäjä, T. (2023). Aerosols, clusters, greenhouse gases, trace gases and boundary-layer dynamics: On feedbacks and interactions. *Boundary-Layer Meteorology*. https://doi.org/10.1007/s10546-022-00769-8

Kulmala, M., Nieminen, T., Nikandrova, A., Lehtipalo, K., Manninen, H. E., Kajos, M., Kolari, P., Lauri, A., Petäjä, T., Krejci, R., Hansson, H.-C., Swietlicki, E., Lindroth, A., Christensen, T. R., Arneth, A., Hari, P., Bäck, J., Vesala, T., & Kerminen, V.-M. (2014). CO2-induced terrestrial climate feedback mechanism: From carbon sink to aerosol source and back. *Boreal Environment Research*, 19, 122–131.

Linden, A. (2022). *Climate Change Assessment over the Arctic Region*. CRC Press. https://doi.org/10.1201/9781003265177-1

Liu, Y., Zheng, J., Hao, Z., & Zhang, X. (2017). Unprecedented warming revealed from multi-proxy reconstruction of temperature in southern China for the past 160 years. *Advances in Atmospheric Sciences*. https://doi.org/10.1007/S00376-017-6228-X

Ma, N., Jiang, J. H., Hou, K., Lin, Y., Vu, T., Rosen, P. E., Gu, Y., & Fahy, K. A. (2022). 21st Century global and regional surface temperature projections. *ESS Open Archive*. https://doi.org/10.1002/essoar.10512735.1

Marcott, S. A., Shakun, J. D., Clark, P. U., & Mix, A. C. (2013). A reconstruction of regional and global temperature for the past 11,300 years. *Science*. https://doi.org/10.1126/SCIENCE.1228026

Mindlin, J., & van Garderen, L. (2023). Attributing extreme weather events and mean climate change using dynamical and event storylines. *EGU General Assembly 2023*, Vienna, Austria, 23–28 April 2023, EGU23-17183. https://doi.org/10.5194/egusphere-egu23-17183

Nauels, A., Nicholls, Z., Kloenne, U., Hermans, T. H. J., Mengel, M., Smith, C. J., & Palmer, M. D. (2024). Exploring multi-century sea level rise commitments from 21st century cumulative emissions to inform minimum coastal adaptation needs. *EGU General Assembly 2024*, Vienna, Austria, 14–19 April 2024, EGU24-10716. https://doi.org/10.5194/egusphere-egu24-10716

Neofotis, P. (2013). Detection and attribution of anthropogenic climate change impacts. *WIREs Climate Change*, 4, 121–150.

Noy, I., Stone, D. A., & Uher, T. (2024). Extreme event impact attribution: The state of the art. *EGU General Assembly 2024*, Vienna, Austria, 14–19 April 2024, EGU24-1616. https://doi.org/10.5194/egusphere-egu24-1616

Otto, F. E. L. (2023). Attribution of extreme events to climate change. *Annual Review of Environment and Resources*. https://doi.org/10.1146/annurev-environ-112621-083538

Overland, J. E. (2024). Emergence of Arctic extremes. *Climate*. https://doi.org/10.3390/cli12080109

Perron, P., Estrada, F., & Martínez-López, B. (2012). Statistical evidence about human influence on the climate system. *Research Papers in Economics*.

Quilcaille, Y., Gudmundsson, L., Gasser, T., & Seneviratne, S. I. (2024). Collective attribution of historical heatwaves to anthropogenic climate change. *EGU General Assembly 2024*, Vienna, Austria, 14–19 April 2024, EGU24-17768. https://doi.org/10.5194/egusphere-egu24-17768

Rosenzweig, C., & Neofotis, P. (2013). Detection and attribution of anthropogenic climate change impacts. *Wiley Interdisciplinary Reviews: Climate Change*. https://doi.org/10.1002/WCC.209

Samoshkina, I. (2023). Greenhouse effect. In S.M. Scheiner (Ed.), *Encyclopedia of Biodiversity* (pp. 307–323). Elsevier. https://doi.org/10.1016/b978-0-12-822562-2.00132-8

Scambos, T., & Abdalati, W. (2022). How fast is sea level rising? *Arctic, Antarctic, and Alpine Research*. https://doi.org/10.1080/15230430.2022.2047247

Schwartz, S. E. (2018). Resource letter GECC-1: The greenhouse effect and climate change: Earth's natural greenhouse effect. *American Journal of Physics*. https://doi.org/10.1119/1.5045574

Séférian, R., Tjiputra, J., Schwinger, J., Santana-Falcón, Y., Lee, H., Stevelink, S. A. M., Yassue, R. M., Korkolis, Y. P., & Egboluche, Y. G. (2023). Carbon cycle feedbacks in an idealized and a scenario simulation of negative emissions in CMIP6 Earth system models. *EGUsphere*. https://doi.org/10.5194/egusphere-2023-1127

Subedi, S. (2022). Increasing atmospheric extreme events and role of disaster risk management: Dimensions and approaches. *Disaster Resilience and Green Growth*. https://doi.org/10.1007/978-981-16-7727-4_13

Sung, H. M., Kim, J., Shim, S., Ha, J.-C., Byun, Y.-H., & Kim, Y.-H. (2021). Sea level rise drivers and projections from coupled model intercomparison project phase 6 (CMIP6) under the Paris climate targets: Global and around the Korea Peninsula. *Journal of Marine Science and Engineering*. https://doi.org/10.3390/JMSE9101094

Tavassoli, M., & Kamran-Pirzaman, A. (2023). *Comparison of Effective Greenhouse Gases and Global Warming*. IEEE. https://doi.org/10.1109/ICTEM56862.2023.10083954

Taylor, P. T., Langer, P., & Tan, I. (2023). Editorial: Arctic amplification: Feedback process interactions and contributions. *Frontiers in Earth Science*. https://doi.org/10.3389/feart.2023.1140871

Thornhill, G. D., Collins, W. J., Olivié, D. J. L., Skeie, R. B., Archibald, A. T., Bauer, S. E., Checa-Garcia, R., Fiedler, S., Folberth, G. A., Gjermundsen, A., Horowitz, L. W., Lamarque, J.-F., Michou, M., Mulcahy, J., Nabat, P., Naik, V., O'Connor, F. M., Paulot, F., Schulz, M., … Weber, J. (2021). Climate-driven chemistry and aerosol feedbacks in CMIP6 Earth system models. *Atmospheric Chemistry and Physics*. https://doi.org/10.5194/ACP-21-1105-2021

Trok, J. T., Barnes, E. A., Davenport, F. V., & Diffenbaugh, N. S. (2024). Machine learning–based extreme event attribution. *Science Advances*. https://doi.org/10.1126/sciadv.adl3242

Turner, F. E., Edwards, T. L., Nicholls, R. J., O'Neill, J., & Adhikari, M. (2023a). Illustrative multi-centennial projections of global mean sea-level rise and their application. *Earth's Future*. https://doi.org/10.1029/2023ef003550

Turner, F. E., Santos, V. M., Edwards, T. L., Slangen, A. B. A., Nicholls, R. J., Le Cozannet, G., O'Neill, J., & Adhikari, M. (2023b). Multi-centennial projections of global mean sea-level rise and their application. *ESS Open Archive*. https://doi.org/10.22541/essoar.167591057.72675797/v1

Vargo, L., Anderson, B., Dadic, R., Horgan, H. J., Mackintosh, A., King, A. D., & Lorrey, A. (2020). Anthropogenic warming forces extreme annual glacier mass loss. *Nature Climate Change*. https://doi.org/10.1038/S41558-020-0849-2

van de Wal, R. S. W., Nicholls, R. J., Behar, D., Mcinnes, K., Stammer, D., Lowe, J. A., Church, J. A., DeConto, R. M., Fettweis, X., Goelzer, H., Haasnoot, M., Haigh, I. D., Hinkel, J., Horton, B. P., James, T. S., Jenkins, A. A., LeCozannet, G., Levermann, A., Lipscomb, W. H., … White, K. (2022). A high-end estimate of sea level rise for practitioners. *Earth's Future*. https://doi.org/10.1029/2022EF002751

Wehner, M., & Reed, K. A. (2022). Operational extreme weather event attribution can quantify climate change loss and damages. *PLOS Climate*. https://doi.org/10.1371/journal.pclm.0000013

Xu, Y., & Ramanathan, V. (2017). Well below 2 °C: Mitigation strategies for avoiding dangerous to catastrophic climate changes. *Proceedings of the National Academy of Sciences of the United States of America*. https://doi.org/10.1073/PNAS.1618481114

Zhang, B., Song, S., Wang, H., Guo, T., & Ding, Y. (2024). Evaluation of the performance of CMIP6 models in simulating extreme precipitation and its projected changes in global climate regions. *Natural Hazards*. https://doi.org/10.1007/s11069-024-06850-4

5 Unpacking Emissions
Scopes 1, 2, and 3

This chapter provides a comprehensive overview of Scopes 1, 2, and 3 emissions, key metrics used in greenhouse gas (GHG) accounting. It explains the distinctions between these scopes, emphasizing the growing importance of GHG data collection and reporting due to consumer and regulatory pressures. This chapter details the calculation methodologies for each scope, covering stationary and mobile combustion, fugitive emissions, purchased energy, and indirect emissions across the value chain. It also discusses the significance of setting reduction targets, tracking emissions over time, and ensuring the accuracy and credibility of GHG inventories through assurance. Additionally, this chapter explores concepts beyond Scope 3, such as avoided emissions and biogenic emissions, and provides guidance on reporting GHG emissions transparently and comprehensively.

The regulatory environment is evolving to mandate greater transparency and accountability in GHG emissions. In March 2024, the Securities and Exchange Commission (SEC) issued a ruling requiring publicly traded companies to disclose their Scope 1 and 2 emissions and related climate risks starting in 2026. Concurrently, California is developing its own GHG reporting requirements through CA SB 253 and 261, which will require companies operating within the state to report their emissions and disclose climate risks.

Measuring GHG emissions can offer significant benefits:

a. **Operational Efficiency and Cost Savings**: Quantifying GHG emissions often involves assessing energy use, fuel consumption, and other inputs associated with capital expenditures. Reductions frequently lead to lower operating expenses.
b. **Enhanced Capital Expenditure Management**: A comprehensive inventory of equipment, required for accurate Scope emissions reporting, helps organizations manage capital expenditures and address deferred maintenance. This ensures that investments in new equipment are both environmentally and economically justified.
c. **Competitive Advantage**: Companies that can demonstrate robust GHG tracking and reduction initiatives gain a competitive edge with environmentally conscious buyers.
d. **Marketing Potential**: Showcasing a commitment to reducing GHG emissions and addressing climate change can strengthen a company's brand and foster customer loyalty. Consumers are more likely to support brands that actively engage in sustainability efforts.

Essentially, Scope 1 and 2 are those emissions that are owned or controlled by a company, whereas Scope 3 emissions are a consequence of the activities of the company but occur from sources not owned or controlled by it. The three scopes are a way of categorizing the different kinds of emissions a company creates in its own operations and in its wider 'value chain' (its suppliers and customers) (Figure 5.1).

5.1 SCOPE 1 EMISSIONS

Scope 1 emissions are direct GHG emissions from sources that are directly controlled by an organization. This means that the organization has operational control over the sources of these emissions and can directly influence the level of emissions produced. Scope 1 emissions include carbon dioxide (CO_2), methane (CH_4), and nitrous oxide (N_2O) that are emitted from stationary and mobile combustion, and from inadvertent fugitive emissions (Chang, 2023). Fugitive emissions can also

DOI: 10.1201/9781003479666-5

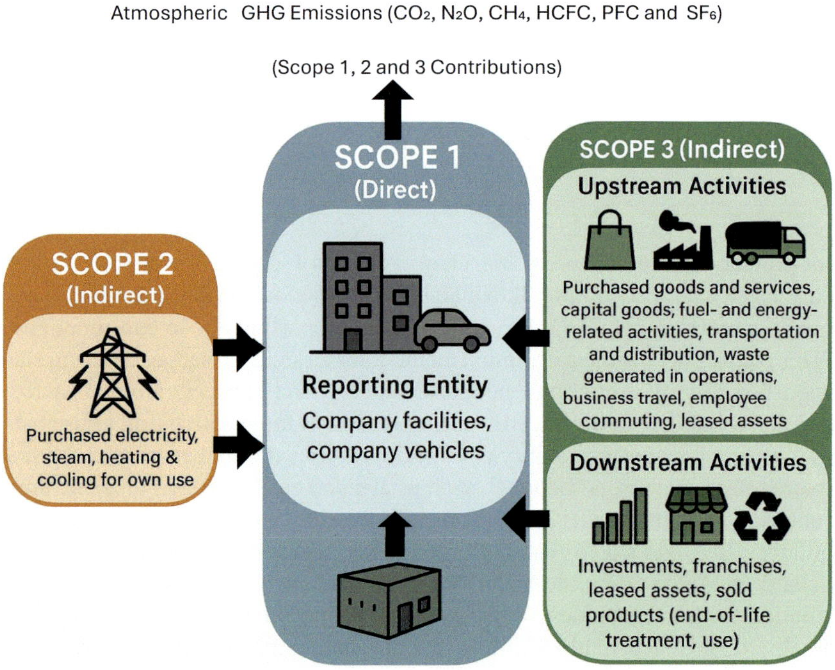

FIGURE 5.1 Scopes 1,2, and 3 diagrammatic expression.

include hydrofluorocarbons (HFCs), perfluorinated compounds (PFCs), sulfur hexafluoride (SF_6), and nitrogen trifluoride (NF_3). Stationary combustion emissions arise from the burning of fuels in fixed installations such as boilers and internal combustion engines, which are significant contributors to nitrogen oxides, unburned hydrocarbons, carbon monoxide, and particulate matter emissions. Mobile combustion emissions, on the other hand, are associated with vehicles and other mobile equipment, where fugitive emissions can occur due to leaks in exhaust systems, often exacerbated by wear and thermal fatigue (Caetano et al., 2023). Fugitive emissions, which are unintentional releases of gases from industrial processes, are particularly challenging to manage due to their sporadic nature and the difficulty in detection and quantification (Shing et al., 2016). These emissions can account for a substantial portion of industrial waste, as seen in the United States, where they contribute significantly to the release of organic compounds from chemical and petrochemical plants (Adkins & Ehlers, 2010).

5.1.1 Stationary Combustion

In the U.S., CO_2 represents over 99% of total CO_2e GHG emissions from commercial, industrial, and electricity generation sources, while CH_4 and N_2O together contribute less than 1%. Although CH_4 and N_2O emissions are often considered immaterial, the GHG Protocol advises estimating these emissions based on available data for inclusion in GHG inventories. Emissions of CH_4 and N_2O vary with fuel characteristics, technology type, combustion conditions, and operational practices. However, the same methods used for calculating CO_2 emissions can accurately estimate CH_4 and N_2O emissions by applying the appropriate emission factors.

5.1.1.1 Continuous Emissions Monitoring Systems Method

Continuous emissions monitoring systems (CEMS) facilitate the continuous measurement of pollutants emitted into the atmosphere from combustion or industrial processes. Various U.S. EPA regulatory programs, including the Acid Rain, New Source Performance Standards, Greenhouse Gas Reporting Program (GHGRP), and Maximum Achievable Control Technology Standards, incorporate provisions for CEMS. Specifically, under Title IV of the U.S. Clean Air Act, electricity generating units are mandated to report CO_2 emissions, as outlined in 40 CFR Part 75. This regulation delineates two methodologies for determining CO_2 emissions using CEMS: (1) employing a monitor to measure the hourly average CO_2 (or O_2) concentration by volume in the flue gas, combined with a flow monitoring system to measure the volumetric flow rate of flue gas, to calculate CO_2 mass emissions, aggregated annually; and (2) using an O_2 concentration monitor along with theoretical CO_2 and flue gas production based on fuel characteristics to determine CO_2 flue gas emissions and mass emissions, calculated based on unit operating time. These methodologies are also specified in 40 CFR Part 98 for GHGRP reporting. Organizations using CEMS to report under Title IV or GHGRP are advised to use the same CO_2 data for their GHG inventory. However, if CEMS do not comply with the requirements of 40 CFR Parts 75 or 98, or if no CEMS are installed, fuel analysis methods should be utilized for CO_2 reporting. Additionally, since CEMS cannot measure CH_4 and N_2O emissions, fuel analysis methods are recommended for these gases. In cases where CEMS do not capture all stationary combustion emissions, such as in facilities with a heat recovery system generator or duct burner, fuel analysis methods should also be used to estimate unmeasured emissions.

5.1.1.2 Fuel Analysis Method

The fuel analysis method for calculating CO_2 emissions involves determining the carbon content of the combusted fuel using either specific information about the fuel or default emission factors and applying this to the amount of fuel burned during the reporting period. The method can be adjusted for unburned carbon and is consistent with regulatory requirements under the Acid Rain Program and the GHGRP. For entities not covered by these programs, the guidance provides a method for GHG emissions calculations using one of the following three equations.

- Equation 5.1 is used when fuel consumption is known only in mass or volume units, without information about fuel heat content or carbon content. It relies on default emission factors based on assumed fuel heat content, making it the least preferred due to higher uncertainty. An "emissions factor" is a factor that converts the activity data into greenhouse gas emissions. The emission factor varies depending on the energy source and the method used (location-based or market-based).
- Equation 5.2 is applied when actual fuel heat content is known or provided in energy units. This equation uses emission factors based on energy units, making it more accurate than Equation 5.1 because it considers the fuel's heat content, which correlates more closely with carbon content.
- Equation 5.3 is recommended for calculating CO_2 emissions when the actual carbon content of the fuel is known. This equation requires fuel use data in mass units and is preferred for CO_2 calculations due to its direct relationship with the fuel's carbon content.

$$\text{Emissions} = \text{Fuel} \times \text{EF}_1 \tag{5.1}$$

where

Emissions = Mass of CO_2, CH_4, or N_2O emitted
Fuel = Mass or volume of fuel combusted
EF_1 = CO_2, CH_4, or N_2O emission factor per mass or volume unit

$$\text{Emissions} = \text{Fuel} \times \text{HHV} \times \text{EF}_2 \tag{5.2}$$

where

Emissions = Mass of CO_2, CH_4, or N_2O emitted
Fuel = Mass or volume of fuel combusted
HHV = Fuel heat content (higher heating value), in units of energy per mass or volume of fuel
EF_2 = CO_2, CH_4, or N_2O emission factor per energy unit

$$\text{Emissions} = \text{Fuel} \times \text{CC} \times 44/12 \tag{5.3}$$

where

Emissions = Mass of CO_2 emitted
Fuel = Mass or volume of fuel combusted
CC = Fuel carbon content, in units of mass of carbon per mass or volume of fuel
44/12 = ratio of molecular weights of CO_2 and carbon.

To determine the quantity of fuel combusted for each fuel type during the reporting year, organizations should first measure the fuel input into each combustion device using systems like fuel flow meters for natural gas and fuel oil, or coal feed belt scales. If direct measurements are unavailable, fuel purchase records, such as utility bills and fuel delivery invoices, can be used. Care must be taken to avoid double counting when using multiple suppliers. Factors like changes in fuel storage, use as feedstock, and fugitive losses should be considered and adjusted for. In cases where direct data collection is impractical, such as with emergency generators, estimates based on operating hours and consumption rates may be used. If exact data is unavailable, fuel consumption can be estimated proportionally based on floor area or using published average consumption values.

5.1.2 Mobile Combustion

Emissions from the combustion of fuels in company-owned or controlled mobile sources, such as cars, trucks, ships, airplanes, and trains, produce GHG emissions like CO_2, CH_4, and N_2O through fuel combustion in vehicles such as cars, trucks, buses, aircraft, ships, locomotives, and construction equipment. It excludes indirect GHG emissions from electric vehicle operation, which fall under Scope 2 emissions.

Mobile combustion emissions, primarily from company-owned vehicles and transport systems, are a significant source of greenhouse gases, including CO_2, CH_4, and N_2O. These emissions are largely derived from the carbon content of combusted fuels, with CO_2 emissions being the most substantial due to the high carbon content in fuels like gasoline and diesel. For instance, in Surabaya, gasoline and diesel contribute 63.36% and 23.64% of CO_2 emissions, respectively, highlighting the dominance of these fuels in mobile combustion emissions (Ismayanti, 2011; Assomadi et al., 2011). The estimation of CH_4 and N_2O emissions typically relies on data related to fuel consumption or the distance traveled by vehicles, as these gases are emitted in smaller quantities than CO_2 but still contribute to the overall greenhouse gas footprint (Cosgrove et al., 1994). In regions like Guangdong Province in China, freight vehicles are identified as major contributors to CO_2 emissions, accounting for 57.10% of the total daily emissions from highway vehicles, with passenger vehicles contributing 42.90% (Li et al., 2022).

To calculate CO_2 emissions from fuel combustion, determine the fuel's carbon content using one of three equations based on available fuel data. Equation 5.4, the least preferred, uses default emission factors when only mass or volume is known, leading to higher uncertainty. Equation 5.5, preferred over Equation 5.1, uses actual fuel heat content provided by the supplier, offering more accuracy. Equation 5.3, the most preferred, uses the fuel's known carbon content expressed as a percentage by mass, directly relating CO_2 emissions to the fuel's carbon content.

$$\text{Emissions} = \text{Fuel} \times EF_4 \tag{5.4}$$

where
Emissions = Mass of CO_2 emitted
Fuel = Mass or volume of fuel combusted
EF_4 = CO_2 emission factor per mass or volume unit

$$\text{Emissions} = \text{Fuel} \times \text{HHV} \times EF_5 \tag{5.5}$$

where
Emissions = Mass of CO_2 emitted
Fuel = Mass or volume of fuel combusted
HHV = Fuel heat content (higher heating value), in units of energy per mass or volume of fuel
EF_5 = CO_2 emission factor per energy unit

$$\text{Emissions} = \text{Fuel} \times \text{CC} \times 44/12 \tag{5.6}$$

where
Emissions = Mass of CO_2 emitted
Fuel = Mass or volume of fuel combusted
CC = Fuel carbon content, in units of mass of carbon per mass or volume of fuel
44/12 = ratio of molecular weights of CO_2 and carbon

One of two equations (Equation 5.7 or 5.8) can be used to calculate CH_4 and N_2O emissions for each type of fuel combusted:

$$\text{Emissions} = \text{Distance} \times EF_7 \tag{5.7}$$

where
Emissions = Mass of CH_4 or N_2O that is emitted
Distance = Distance that the vehicle traveled
EF_7 = CH_4 or N_2O emission factor per distance unit

$$\text{Emissions} = \text{Fuel} \times EF_8 \tag{5.8}$$

where
Emissions = Mass of CH_4 or N_2O that is emitted
Fuel = Volume of fuel that is combusted
EF_8 = CH_4 or N_2O emission factor per volume unit.

5.1.3 Fugitive Emissions

Fugitive emissions are those not emitted through a confined process stream. These can include equipment leaks, refrigerant losses, and emissions from wastewater treatment. The USEPA provides guidance for calculating fugitive greenhouse gas emissions from various sources, including refrigeration and air conditioning equipment, fire suppression systems, and industrial gases. It outlines different calculation methods, each with varying levels of complexity and data requirement.

Fugitive emissions, particularly from industrial processes, refrigeration systems, and fire suppression equipment, are significant concerns due to their contribution to atmospheric pollution and their high global warming potential (GWP), especially when involving hydrofluorocarbons (HFCs) (Shing et al., 2016; Li et al., 2023). These emissions are often unanticipated and result from leaks in industrial sites, which can occur due to discontinuities in containment barriers (Shing et al., 2016). The estimation of fugitive emissions is crucial for both environmental and occupational health reasons, as these emissions contribute to background inhalation exposure that can lead to serious health issues, including cancer (Aziz et al., 2013). Various methods have been developed to estimate these emissions, including manual calculations and computer-aided tools that utilize process flow diagrams (PFDs) and piping and instrumentation diagrams (P&IDs) to provide more comprehensive assessments as more data becomes available (Aziz et al., 2013; Hassim et al., 2012). For instance, in the petrochemical industry, valves, pumps, and flanges are identified as major sources of fugitive emissions, with valves alone accounting for 50% of emissions in some cases (Hassim et al., 2012).

In refrigeration systems, the leakage of HFC-134a is particularly concerning due to its high GWP, with studies showing significant emissions during both the use and disposal phases of equipment like reefer containers and mobile air conditioners (Kim et al., 2012, 2014). Strategies for reducing emissions include substituting high-leakage components with inherently safer designs, such as replacing rising stem valves with ball valves and using hermetic pumps, which can reduce emissions by over 90% in practice (Hassim et al., 2012). These approaches highlight the importance of early design considerations and the adoption of both inherent and add-on prevention strategies to mitigate the impact of fugitive emissions on the environment and human health (Hassim et al., 2012; Ellis & Lackaye, 1989).

The Screening Method is a simplified approach using equipment-specific emission factors and the following equations. Equation 5.9 estimates emissions from equipment installation. Equation 5.10 estimates emissions during equipment operation, considering factors like refrigerant capacity and leak rate. Equation 5.11 estimates emissions from equipment disposal, accounting for remaining refrigerant and recovery efficiency. The Material Balance Method offers a more comprehensive approach, tracking refrigerant flows through inventory changes, purchases, sales, and capacity adjustments. It utilizes Equation 5.12 to represent the material balance and Equation 5.5 to calculate emissions directly. Lastly, the Simplified Material Balance Method is a streamlined version for entities not maintaining refrigerant stocks, focusing on purchases, capacity changes, and recovery using Equation 5.13.

Estimated Emissions from Installation

$$\text{Emissions from installation} = \text{CN} \times (k \div 100) \tag{5.9}$$

where

CN = Amount of refrigerant that was charged into the new piece of equipment

k = Assembly losses as a percentage of the amount that was charged

Estimated Emissions from Operation

$$\text{Emissions from operation} = C \times (x \div 100) \times T \tag{5.10}$$

where

C = Refrigerant capacity of the piece of equipment

x = Annual leak rate as a percentage of the capacity

T = Time in years that the piece of equipment was used during the reporting period (For example, T equals 0.5 if the equipment was used during half the reporting period and then disposed.)

Estimated Emissions from Disposal

$$\text{Emissions from disposal} = \text{CD} \times (y \div 100) \times (1 - z \div 100) \tag{5.11}$$

where

CD = Refrigerant capacity of the piece of equipment that is being disposed of

y = Percentage of the capacity that remains at the time of disposal

z = Percentage of the refrigerant that is recovered

Material Balance of Refrigerant

$$I_B + P + C_B = \mathrm{IE} + S + C_E + \text{Emissions} \tag{5.12}$$

where

I_B=refrigerant in inventory (in storage, not in operating equipment) at the beginning of reporting period

P=refrigerant purchased or otherwise acquired during the reporting period

C_B=total refrigerant capacity of equipment at the beginning of the reporting period

IE=refrigerant in inventory (in storage, not in operating equipment) at the end of reporting period

S=refrigerant sold or otherwise disposed of during the reporting period

C_E=total refrigerant capacity of equipment at the end of the reporting period

Calculating Refrigerant Emissions with the Material Balance Method

$$\text{Emissions} = (I_B - I_E) + P - S + (C_B - C_E) \tag{5.13}$$

5.2 SCOPE 2 EMISSIONS

Scope 2 emissions are indirect greenhouse gas emissions resulting from the generation of purchased energy, such as electricity or heat. They are distinct from Scope 1 emissions directly produced by a company's operations. Scope 2 emissions are measured using invoices for energy procurement and can include generation data from onsite renewables. They result from generating purchased electricity, steam, heat, or cooling that a company consumes. These emissions are considered indirect because they occur at sources owned or controlled by another entity, such as an electricity generator or utility.

Scope 2 emissions are a significant source of greenhouse gas emissions globally. Therefore, it is crucial for companies to account for and report these emissions accurately. The methods used to calculate and report Scope 2 emissions can significantly impact a company's assessment of its environmental performance and its strategies to mitigate emissions.

The guidance outlines two methods for accounting for Scope 2 emissions:

1. **Location-Based Method**: This method reflects the average emissions intensity of the grids from which a company purchases electricity. It emphasizes the connection between the collective demand for electricity and the resulting emissions from local electricity production. It provides insights into the mix of energy sources used in the grids and the average emissions associated with them. This approach is criticized for its potential to misrepresent actual emissions due to its reliance on annual average grid emission factors, which do not account for temporal variations in grid carbon intensity (Miller et al., 2022; Balaras et al., 2023).
2. **Market-Based Method**: This method reflects the emissions from the electricity that companies specifically choose to purchase. It considers the contractual instruments used by companies to procure electricity, such as renewable energy certificates (RECs), guarantees of origin (GOs), or power purchase agreements (PPAs) (Roston et al., 2024). This method aims to capture the emissions associated with the specific energy sources that companies have intentionally selected. However, it has been criticized for potentially supporting misleading emissions reduction claims, as it may not accurately reflect the physical flow of electricity and its associated emissions (Roston et al., 2024; Wald et al., 2021).

Although each method has limitations, reporting emissions using both methods allows companies to demonstrate transparency and provide stakeholders with a better understanding of the emissions associated with their electricity consumption.

Calculating Emissions

To calculate Scope 2 emissions, the guidance recommends using the following equation:

$$\text{Emissions} = \text{Activity Data} * \text{Emission Factor} \tag{5.14}$$

where

Activity Data:

This refers to the amount of electricity, steam, heat, or cooling consumed by the company, typically measured in kilowatt-hours (kWh) or megawatt-hours (MWh).

Example:

- Number of livestock heads (i.e., 250 beef cattle).
- Quantity of nitrogen fertilizer applied to cropland soils).
- The number of days of flooded irrigation of a paddy field.

By multiplying the activity data by the appropriate emission factor, companies can determine the total greenhouse gas emissions associated with their energy consumption.

Example:

- Methane emitted by beef cattle head.
- The percentage of nitrogen applied to cropland emitted as N_2O-N directly.
- Daily emissions from irrigated rice cultivation.

5.2.1 Indirect Emissions from Purchased Electricity

This section presents guidance to calculate GHG emissions from purchased electricity, steam, heat, and cooling. This guidance applies to the calculation of emissions using both the location-based and market-based methods. It is recommended that organizations calculate emissions for each of their facilities separately, rather than aggregating multiple facilities. This increases the accuracy and credibility of the inventory

$$\text{Emissions} = \text{Electricity} \times \text{EF} \tag{5.15}$$

where

Emissions = Mass of CO_2, CH_4, or N_2O emitted
Electricity = Quantity of electricity purchased
EF = CO_2, CH_4, or N_2O emission factor.

Calculation Steps

For both the location-based and market-based methods, emissions are calculated by multiplying the purchased electricity during the reporting year by the appropriate emission factors. The steps involved in calculating emissions from the consumption of purchased electricity are shown below. The same approach applies to steam, heat, and cooling.

- **Step 1**: Determine the amount of electricity purchased.
- **Step 2**: Determine emission factors.
- **Step 3**: Calculate emissions: Use Equation 5.1 to calculate the emissions of CO_2, CH_4, and N_2O. Multiply the emissions of CH_4 and N_2O by their respective GWPs to calculate CO_2 equivalent emissions.

Overall, while both methods have their merits and limitations, the integration of more granular data and innovative accounting models could enhance the reliability and environmental impact of Scope 2 emissions reporting.

5.3 SCOPE 3 EMISSIONS

Scope 3 emissions, which encompass all other indirect emissions not covered by Scope 1 and Scope 2, are indeed the most extensive and complex to measure, as they include a wide range of activities both upstream and downstream in the supply chain. These emissions are divided into 15 categories, covering upstream activities such as purchased goods, transportation, and employee commuting, as well as downstream activities like product use and end-of-life disposal (Shrimali, 2022; Huang et al., 2009). The complexity in measuring Scope 3 emissions arises from the need for detailed data sharing across supply chains, which is often hindered by legal, regulatory, and data privacy challenges (Stenzel & Waichman, 2023). In the automotive industry, for instance, an interdisciplinary approach integrating cloud computing and machine learning has been proposed to enhance the accuracy of Scope 3 emission quantification and optimize supply chains (Wang et al., 2024). Similarly, the water sector faces significant challenges in addressing Scope 3 emissions, particularly those associated with residential hot water use, which dominate the sector's carbon footprint (Kenway et al., 2023).

Regulatory frameworks, such as those in Australia, are increasingly focusing on Scope 3 emissions, requiring industries to implement measures like fugitive methane emissions reduction and collaborative efforts with customers to manage these emissions (Holding, 2023). The use of economic input-output analysis and hybrid life cycle assessment models is common for estimating these emissions. However, they come with uncertainties and require improvements to minimize errors (Schmidt, 2022; Lee & Ma, 2013). Furthermore, selecting emission factors is crucial, as discrepancies in these factors can lead to significant variations in reported emissions, affecting the quality of carbon management strategies (Downie & Stubbs, 2012).

Scope 3, therefore, pertains to all the indirect emissions within a company's value chain. The 15 categories under Scope 3 serve as a structured model for companies to quantify, oversee, and curtail emissions across their corporate value chain (Hertwich & Wood, 2018; Huang et al., 2009). These categories are exclusively formulated to prevent any possibility of the company accounting for emissions twice under multiple categories.

5.3.1 Steps to Calculate Scope 3 Emissions

5.3.1.1 Step One: Identify Scope 3 Activities

The first step is to identify the relevant Scope 3 activities for the company. The Scope 3 Standard categorizes Scope 3 emissions into 15 categories, including purchased goods and services, business travel, waste generated in operations, and use of sold products.

5.3.1.1.1 Upstream Activities

Upstream emissions are generated during the production of a company's goods or services

Category 1: Purchased Goods & Services

The procurement, manufacturing, and delivery of goods and services obtained by your organization during the reporting year that do not fall under any of categories 2–8. This includes the extraction, production, and transportation of such goods and services that were acquired or purchased by your organization.

Category 2: Capital Goods

The emissions related to the manufacturing, extraction, or transportation of a product that come from capital goods, which are physical assets, such as equipment, machinery, and buildings, that organizations use to produce goods or provide services, within the organization's reporting year.

Category 3: Fuel- and Energy-Related Activities (Not Included in Scope 1 or Scope 2)

The indirect GHG emissions resulting from fuel and energy-related activities that are not covered in Scope 1 or Scope 2. This category includes GHG emissions that occur outside your organization's own operations but are still associated with its activities or products. For instance, if a company's employees travel to work using personal vehicles, the GHG emissions associated with their transportation would be included in Category 3 emissions.

Category 4: Upstream Transportation and Distribution

The emissions associated with the movement and delivery of goods bought by your organization during the reporting year, between your organization's primary suppliers and its own operations, using transportation and facilities not owned or managed by your organization. This includes logistics, outbound logistics for sold products, as well as transportation and distribution between your organization's own facilities. These transportation and distribution services were procured by your organization during the reporting year.

Category 5: Waste Generated in Operations

Includes the GHG emissions associated with the use of a company's products or services, including emissions resulting from the energy used to operate or maintain those products. This can include emissions from the use of consumer goods, such as appliances or electronics, as well as emissions from the use of industrial equipment or machinery. For instance, if a company produces and sells air conditioners, the GHG emissions associated with the energy used to operate those air conditioners would be included in Category 5 emissions.

Category 6: Business Travel

During the reporting year, your organization facilitated the transportation of its employees for business-related activities using vehicles that were neither owned nor operated by the organization.

Category 7: Employee Commuting

These emissions are those emitted during the reporting year when your organization facilitated the transportation of its employees between their homes and worksites using vehicles that were neither owned nor operated by your organization.

Category 8: Upstream Leased Assets

The lessee reported the emissions associated with the operation of assets leased by your organization during the reporting year, which were not included in their Scope 1 and Scope 2 emissions.

5.3.1.1.2 Downstream Activities

Downstream emissions occur during their usage and disposal.

Category 9: Downstream Transportation and Distribution

During the reporting year, your organization transported and distributed its products to end consumers, including retail and storage, using vehicles and facilities that were neither owned nor controlled by your organization and were paid for by the organization.

Category 10: Processing of Sold Products

This category involves the processing of intermediate products sold to downstream organizations, such as manufacturers, during the reporting year.

Category 11: Use of Sold Products

The consumption of products and services sold by your organization during the reporting year.

Category 12: End-of-Life Treatment of Sold Products

The disposal and treatment of products sold by your organization at the end of their useful life during the reporting year.

Category 13: Downstream Leased Assets

The lessor reported on the operation of assets owned by your organization and leased to other entities during the reporting year, which were not included in their Scope 1 and Scope 2 emissions.

Category 14: Franchises

The franchisor reported on the operation of franchises that were not included in their Scope 1 and Scope 2 emissions during the reporting year.

Category 15: Investments

The operation of investments, such as equity and debt investments and project finance, that were not included in their Scope 1 or Scope 2 emissions during the reporting year.

5.3.1.2 Step Two: Set the Scope 3 Boundary

The next step is to set the boundary for the Scope 3 inventory. This involves determining which activities will be included in the inventory and which will be excluded. This is crucial, as Scope 3 emissions often represent the largest portion of a company's carbon footprint, and accurate boundary setting is necessary to avoid underestimation. Setting the Scope 3 boundary is a critical step in greenhouse gas (GHG) accounting. It involves determining which indirect emissions along a company's value chain will be included in the inventory. The GHG Protocol Corporate Standard offers flexibility in choosing Scope 3 activities, while the Scope 3 Standard aims for greater completeness and consistency.

1. **Mapping the Value Chain**:
   - Companies should begin by mapping their entire value chain, identifying all activities that could potentially generate GHG emissions.
   - This includes listing all Scope 3 categories and activities, purchased and sold goods and services, and suppliers or value chain partners.
   - The goal is to create a comprehensive overview, although 100% completeness may not always be feasible .
2. **Boundary Requirements**:
   - The standard mandates accounting for all Scope 3 emissions, with any exclusions disclosed and justified.
   - Emissions from each of the 15 Scope 3 categories should be accounted for according to the minimum boundaries defined in the standard.
   - Companies can include optional activities within each category and exclude activities within the minimum boundary if justified .
   - The standard covers emissions of six main GHGs: CO_2, CH_4, N_2O, HFCs, PFCs, and SF_6.
   - Biogenic CO_2 emissions (from biomass) are excluded from the scopes but should be separately reported .
3. **Disclosing and Justifying Exclusions**:
   - While completeness is encouraged, it is understood that accounting for all Scope 3 emissions might not be possible.
   - If certain categories are not applicable (e.g., no leased assets), companies should report zero emissions or "not applicable."
   - If emissions cannot be estimated due to lack of data, companies may exclude activities, but this must be disclosed and justified in the report.
   - Exclusions should not compromise the relevance of the inventory, and any activity expected to contribute significantly to emissions should not be excluded.
4. **Accounting for Downstream Emissions**:
   - The applicability of downstream Scope 3 categories (emissions from sold goods and services) depends on whether the products are final or intermediate.
   - If the end-use of intermediate products is unknown, downstream emissions can be excluded, provided that justification is given in the report.
5. **Accounting for Biogenic Emissions and Removals**:
   - Biogenic CO_2 emissions (from biomass combustion or biodegradation) are excluded from the scopes but must be reported separately.
   - GHG removals (e.g., carbon sequestration) are not included in Scope 3 but can be reported separately.

5.3.1.3 Step Three: Collect Data

After a company has identified the activities to include in its Scope 3 boundary, the next step is to collect the necessary data to calculate its Scope 3 emissions. The priority should be to gather high-quality primary data for significant activities, supplemented by secondary data for less critical processes. Input-output life cycle assessment (EIO-LCA) models can help identify significant upstream emission sources, allowing companies to focus on gathering detailed emissions data from key suppliers (Lee & Ma, 2013; Huang et al., 2009). Allocation is crucial in this process, as it involves distributing emissions from shared facilities or systems to individual products or services using physical or economic criteria. This step ensures accurate attribution of emissions to the correct products or services, which can be particularly challenging in sectors with complex supply chains, such as electronics manufacturing (Huang et al., 2009). The process of calculating Scope 3 emissions is further complicated by the need for methodological harmonization across different sectors and regions, as current methods often yield non-comparable results due to varying starting points and approaches (Auvinen et al., 2014).

This process is likely to require engagement from several internal departments and external partners.

1. Guidance for Prioritizing Data Collection Efforts

Companies should prioritize collecting data for Scope 3 activities that are expected to have the most significant GHG emissions, offer the most significant GHG reduction opportunities, and are most relevant to the company's business goals. There are several approaches to identifying these priority activities:

- Prioritizing based on the magnitude of GHG emissions:
  - Use initial GHG estimation methods to estimate emissions from each Scope 3 activity.
  - Rank all Scope 3 activities from largest to smallest according to their estimated GHG emissions.
- Prioritizing based on financial spend or revenue:
  - Rank upstream activities by their contribution to the company's total spend.
  - Rank downstream activities by their contribution to the company's total revenue.
- Prioritizing based on other criteria:
  - Consider activities the company has influence over.
  - Consider activities that contribute to the company's risk exposure.
  - Consider activities that stakeholders deem critical.
  - Consider activities identified as significant by sector-specific guidance.

2. Overview of Quantification Methods and Data Types

There are two main methods to quantify emissions:

- **Direct Measurement**: Directly measuring GHG emissions from a source.
- **Calculation**: Multiplying activity data by an emission factor.

Calculation is the most common method for quantifying Scope 3 emissions. It requires two types of data:

- **Activity Data**: A quantitative measure of an activity that results in GHG emissions (e.g., liters of fuel consumed, kilowatt-hours of electricity consumed).
- **Emission Factors**: A factor that converts activity data into GHG emissions data (e.g., kg CO_2e emitted per liter of fuel consumed).

3. Guidance for Selecting Data

The quality of the Scope 3 inventory depends on the quality of the data used. Companies should collect data of sufficient quality to ensure that the inventory appropriately reflects the GHG emissions of the company and serves the decision-making needs of users.

There are two main types of data used to calculate Scope 3 emissions:

Primary Data: Data from specific activities within a company's value chain.

Guidance for Collecting Primary Data

Primary activity data may be obtained through meter readings, purchase records, utility bills, engineering models, direct monitoring, or other methods. Companies should collect energy or emissions data from suppliers and other value chain partners to obtain site-specific data for priority Scope 3 categories and activities. To collect primary data, companies should:

- Identify relevant suppliers.
- Determine the type and level of data to request from suppliers.
- Assess the quality of supplier data.

Secondary Data: Data that is not from specific activities within a company's value chain.

Companies should prioritize collecting high-quality primary data for high-priority activities. However, secondary data can be used when primary data is not available or not of sufficient quality.

Guidance for Collecting Secondary Data and Filling Data Gaps

When using secondary data, companies should prioritize databases and publications that are internationally recognized, provided by national governments, or peer-reviewed. If data of sufficient quality is not available, companies may use proxy data to fill data gaps. Proxy data is data from a similar activity that is used as a stand-in for the given activity.

4. Data quality indicators

When selecting data sources, companies should use the following data quality indicators as a guide:

- Technological representativeness
- Temporal representativeness
- Geographical representativeness
- Completeness
- Reliability

5. Improving Data Quality Over Time

Collecting, assessing, and improving data quality is an iterative process. Companies should seek to improve the data quality of the inventory over time by replacing lower-quality data with higher-quality data as it becomes available.

5.3.1.4 Step Four: Allocating Emissions

Allocation is the process of assigning GHG emissions from a single facility or system to its various outputs. It becomes necessary when a facility or system produces multiple outputs, and emissions are quantified for the entire facility or system, rather than for each individual output. In such cases, the emissions need to be divided among the different outputs.

Allocation is required when a single facility or system produces multiple outputs and emissions are quantified for the entire facility or system, rather than individual outputs. For instance, a production facility might manufacture various products, but the energy use and emissions data are collected for the entire plant. In this situation, allocation is needed to assign the emissions to each product.

Allocation is not necessary when a facility or system produces only one output and emissions from producing each output are quantified separately. Allocation is also typically not needed when using secondary data, as the activity data and emission factors usually refer to a single product. It's best to avoid or minimize allocation when possible, as it introduces uncertainty into emissions estimates. This is particularly true when a facility produces diverse products with varying GHG contributions.

To avoid or minimize allocation, companies can obtain product-level GHG data from value chain partners following the GHG Protocol Product Standard. Separately sub-metering energy use and other activity data (e.g., at the production line level) and using engineering models to estimate emissions for each product separately further help minimize allocations.

If allocation is unavoidable, companies should first determine the total facility or system emissions and then choose the most appropriate allocation method and factor. The choice of method depends on the specific circumstances and should reflect the causal relationship between production and emissions, ensuring accurate and credible estimates.

The main allocation methods include:

- **Physical Allocation**: This method allocates emissions based on a physical relationship between inputs/outputs and emissions. It can be based on mass, volume, energy content, chemical composition, or the number of units.
- **Economic Allocation**: This method allocates emissions based on the market value of each output/product.

Examples of Allocating Emissions

- **Manufacturing**: If a facility produces multiple products with similar energy and material inputs but different market values, physical allocation is more appropriate.
- **Transportation**: Emissions from transporting multiple products in a single vehicle can be allocated based on weight, volume, or a combination, depending on the limiting factor of the vehicle's capacity.
- **Commercial Buildings**: Emissions from buildings like retail stores or warehouses can be allocated based on volume or area occupied by each product.

Different allocation methods can yield different results, so companies should evaluate the range of possible outcomes before selecting a method. For each facility or system, a single, consistent allocation factor should be used to avoid over- or under-counting emissions. Emissions from waste generated during production should not be allocated to the waste itself but to the facility's other outputs. By understanding and applying these allocation principles, companies can ensure accurate and transparent accounting of their Scope 3 emissions.

5.3.1.5 Step Five: Setting a GHG Reduction Target and Tracking Emissions Over Time

Greenhouse gas (GHG) accounting and reporting enable companies to monitor and disclose their emissions performance over time. Companies may track Scope 3 emissions to demonstrate progress toward GHG reduction targets, manage risks and opportunities, and meet stakeholder expectations.

The process involves:

1. Choosing a Base Year and Determining Base Year Emissions
   - A base year is a reference point for tracking emissions performance.
   - Companies should ideally use a single base year for all scopes (1, 2, and 3) but can choose a more recent year for Scope 3 if needed.
   - Base year emissions are calculated according to the standard's requirements.
2. Setting Scope 3 Reduction Goals
   - While not mandatory, setting reduction targets is a key part of GHG management.
   - Companies can set targets for total emissions (all scopes), total Scope 3 emissions, individual Scope 3 categories, or a combination.
   - Targets can be absolute (reducing the quantity of emissions) or intensity-based (reducing emissions per unit of output).
   - Offsets can be used to meet targets, but internal reductions are preferred.

3. Recalculating Base Year Emissions (If Necessary)
    - Recalculation is required if there are significant changes in company structure (mergers, acquisitions) or inventory methodology.
    - This ensures consistency and meaningful comparisons over time.
    - Recalculation is also needed for significant outsourcing or insourcing of activities or changes in included Scope 3 categories.
    - Changes in calculation methods or data accuracy may also trigger recalculation.
4. Accounting for Scope 3 Emissions and Reductions Over Time
    - The inventory method is used to track changes in emissions compared to the base year.
    - This method reflects the aggregate effect of a company's activities on emissions.
    - The project method can be used to assess reductions from specific Scope 3 projects, reported separately from the inventory.
5. Accounting for Avoided Emissions
    - Avoided emissions are reductions achieved outside the company's Scopes 1, 2, and 3.
    - These require project-based accounting and are reported separately from the inventory.
6. Addressing Double Counting of Scope 3 Reductions
    - Double counting occurs when multiple entities claim the same GHG reduction.
    - While acceptable for reporting and tracking, it should be avoided if reductions have monetary value or receive credit in a program.

5.3.1.6 Step Six: Assurance

Assurance is the degree of confidence that a company's greenhouse gas (GHG) inventory is complete, accurate, consistent, transparent, relevant, and free from material misstatements. While not mandatory under the GHG Protocol Scope 3 Standard, assurance is valuable for both the reporting company and stakeholders when making decisions based on the inventory results.

1. Benefits of Assurance
    - **Increased Confidence for Decision-Making**: It gives senior management greater confidence in the reported information, which is crucial for setting reduction targets and making informed decisions.
    - **Enhanced Internal Practices**: Assurance helps improve internal accounting and reporting practices, including data collection, calculation, and reporting systems. It also facilitates learning and knowledge transfer within the company.
    - **Improved Efficiency**: The assurance process can streamline subsequent inventory updates, making them more efficient.
    - **Greater Stakeholder Confidence**: Assurance enhances the credibility of the reported information for external stakeholders, increasing their trust in the company's GHG management efforts.
2. Types of Assurance
    - **First-Party Assurance**: Conducted internally by individuals within the reporting company who are independent of the GHG inventory process.
    - **Third-Party Assurance**: Conducted by an external organization independent of the inventory process. This type is generally considered more credible due to its objectivity and independence.
3. Competencies of Assurers

    A competent Scope 3 GHG inventory assurer should possess:
    - Expertise in assurance frameworks.
    - Knowledge and experience in corporate GHG accounting and/or life cycle assessment.

- Familiarity with the key steps in the Scope 3 inventory process.
- Knowledge of the company's activities and industry sector.
- Ability to assess emission sources and potential errors or misrepresentations.
- Credibility, independence, and professional skepticism.

4. Assurance Process
   - **Planning and Scoping**: Determining the scope of the assurance engagement and identifying potential risks and material misstatements.
   - **Identifying Emission Sources**: Identifying the emission sources included in the Scope 3 inventory.
   - **Performing the Assurance Process**: Gathering evidence, performing analytical procedures, and evaluating the company's GHG accounting processes and controls.
   - **Evaluating Results**: Assessing the evidence gathered and evaluating the company's GHG assertions against the chosen criteria (e.g., GHG Protocol standards).
   - **Determining and Reporting Conclusions**: Forming an opinion on the reliability of the inventory and communicating the findings in an assurance statement.
5. Levels of Assurance

   There are two levels of assurance:
   - **Limited Assurance**: Provides a moderate level of assurance based on a less rigorous review of the evidence.
   - **Reasonable Assurance**: Provides a high level of assurance based on a more comprehensive and rigorous review of the evidence.
6. Key Concepts in Assurance
   - **Assertion**: A statement by the company on its Scope 3 emissions, presented to the assurer.
   - **Subject Matter**: The Scope 3 emissions data and supporting information in the inventory report.
   - **Criteria**: The benchmarks used to evaluate the subject matter, including standards, methodologies, and data quality.
   - **Evidence**: The data and documentation used to calculate emissions and support the assertion.
   - **Assurance Standards**: Standards guiding the assurance process (e.g., ISO 14064-3).
   - **Assurance Opinion**: The assurer's conclusion on the inventory's reliability.
7. Preparing for Assurance
   - Preparing a written assertion (inventory results).
   - Having a complete data management plan.
   - Providing access to sufficient and appropriate evidence.

Assurance plays a crucial role in enhancing the credibility and reliability of Scope 3 GHG inventories. By understanding the assurance process and its key concepts, companies can better prepare for and benefit from assurance, ultimately improving their GHG management efforts and stakeholder communication. Assuring Scope 3 inventories can be challenging due to the reliance on various data sources and assumptions, as well as the inherent uncertainty in Scope 3 emissions. The lack of control over emission sources outside the company's direct control can also make it difficult to obtain sufficient evidence. The assurance statement should include:

1. An introduction describing the reporting company and its assertion.
2. A description of the assurance process, including the assurers' competencies, a summary of the work performed, and the assurance criteria used.
3. A conclusion stating the level of assurance achieved, the materiality threshold (if any), and any exceptions or issues encountered during the assurance process.

5.3.1.7 Step Seven: Reporting

A credible greenhouse gas (GHG) emissions report should present information based on relevance, accuracy, completeness, consistency, and transparency. It should be based on the best available data and transparently communicate its limitations.

Companies must publicly report the following information:

1. A Scope 1 and Scope 2 emissions report in conformance with the GHG Protocol Corporate Standard.
2. Total Scope 3 emissions reported separately by each of the 15 Scope 3 categories.
3. For each Scope 3 category, total GHG emissions in metric tons of CO_2 equivalent, excluding biogenic CO_2 emissions and independent of GHG trades (e.g., offsets or allowances).
4. A list of Scope 3 categories and activities included in the inventory.
5. A list of Scope 3 categories or activities excluded from the inventory, with justifications for their exclusion.
6. Once a base year has been established:
   - The year chosen as the Scope 3 base year.
   - The rationale for choosing the base year.
   - The base year emissions recalculation policy.
   - Scope 3 emissions by category in the base year, consistent with the base year emissions recalculation policy.
   - Appropriate context for any significant emissions changes that triggered base year emissions recalculations.
7. For each Scope 3 category, any biogenic CO_2 emissions reported separately.
8. For each Scope 3 category:
   - A description of the types and sources of data used to calculate emissions (including activity data, emission factors, and GWP values).
   - A description of the data quality of reported emissions data.
   - A description of the methodologies, allocation methods, and assumptions used to calculate Scope 3 emissions.
9. For each Scope 3 category, the percentage of emissions calculated using data obtained from suppliers or other value chain partners.

5.3.1.7.1 Reporting Guidance

Companies should aim for a report that is relevant, transparent, accurate, consistent, and complete. It's important to differentiate between a summary report and a full public report containing all the required data. If a summary report is used, it should reference a publicly available full report. Specific guidance is provided for:

- **Reporting GHG Emissions**: Companies can report aggregated emissions in CO_2e and are not required to report emissions by individual gas.
- **Describing Methodologies, Allocation Methods, and Assumptions**: Companies should report assumptions for each Scope 3 category, including details on data sources, time boundaries, and calculation methods.
- **Reporting on Supplier Engagement and Performance**: This includes metrics on the number of suppliers providing data, their reporting practices, and their GHG performance over time.
- **Reporting on Product Performance**: This includes product performance indicators, intensity metrics, annual emissions from product use, product lifetime/durability, and explanations for changes in emissions over time.

- **Reporting Historic and Future Emissions Separately**: This is recommended for Categories 5, 11, and 12 to avoid misinterpreting the timing of emissions.
- **Reporting on Uncertainty**: Companies should describe the level of uncertainty in the data, either qualitatively or quantitatively, and outline efforts to address it.

5.4 BEYOND SCOPE 3: OTHER EMISSIONS CATEGORIES

Emerging concepts such as Scope 4 emissions and biogenic emissions are expanding the understanding of a company's environmental impact beyond the traditional Scopes 1, 2, and 3 emissions. Scope 4 emissions, also known as avoided emissions, refer to the reductions achieved outside a company's value chain, such as using energy-efficient products. These emissions are not yet standardized within the existing emissions accounting framework, and their legitimacy is debated, as they can distract from the critical task of reducing absolute emissions (Young-Ferris et al., 2024).

The concept of Scope 4 is seen as an extension of the existing framework, but it lacks the formal recognition and standardization that Scopes 1, 2, and 3 have achieved (Young-Ferris et al., 2024). On the other hand, biogenic emissions, which are derived from natural sources, highlight the interaction between anthropogenic and natural carbon cycles. For instance, in the forestry sector, biogenic carbon fluxes from forest land and harvested wood products can significantly offset value chain emissions, as demonstrated by Stora Enso's net removal of 11.5 million Mg CO_2-eq per year when considering these factors (Hammar et al., 2020). This highlights the potential of biogenic emissions to contribute to climate change mitigation by counteracting a substantial portion of anthropogenic emissions. Furthermore, the concept of beyond Scope 3 emissions, which includes impacts such as crowding out, market expansion, and innovation, suggests that companies can have indirect impacts on competitors and other market agents outside their supply chain (Csutora & Dobák, 2019; Csutora & Harangozó, 2019).

These emerging categories of emissions accounting underscore the complexity of accurately assessing a company's environmental impact and the need for comprehensive methodologies that consider both direct and indirect environmental effects. As the understanding of these concepts evolves, companies must integrate them into their climate strategies to achieve deeper decarbonization and contribute effectively to global climate goals (Johnson et al., 2023).

5.5 THE STRATEGIC IMPORTANCE OF GHG ACCOUNTING

Robust GHG accounting offers several significant benefits, including operational efficiency, regulatory compliance, market competitiveness, and stakeholder engagement. By identifying emissions hotspots, companies can enhance operational efficiency through targeted cost-saving improvements, as highlighted by using environmental management accounting approaches that provide more accurate emissions data (Almihoub et al., 2013). This precision in accounting not only aids in internal management but also aligns with established protocols like the Greenhouse Gas Protocol, which serves as a comprehensive guide for developing credible GHG inventories. Regulatory compliance is another critical benefit, as accurate GHG accounting helps businesses meet government regulations and reporting standards, mitigating legal risks (Singh & Mahapatra, 2013). This compliance is increasingly important as frameworks like the European Union's Corporate Sustainability Reporting Directive and the U.S. SEC's proposed climate disclosure rules demand transparency in emissions reporting, particularly for Scope 3 emissions (Emborg et al., 2023). Furthermore, demonstrating environmental stewardship through robust GHG accounting can enhance market competitiveness by building consumer trust and strengthening brand reputation. This is supported by the growing acceptance of the Greenhouse Gas Protocol, which signals a commitment to sustainability and can provide competitive advantages (Hileman, 2001). Finally, transparent GHG reporting fosters stakeholder engagement by facilitating collaboration with investors, customers, and suppliers. Reliable data from GHG reporting programs inform mitigation policies and communicate emissions-related information to stakeholders, enhancing corporate accountability and sustainability practices (Singh & Mahapatra, 2013; Gulluscio et al., 2020).

Overall, the integration of GHG accounting into corporate strategies is essential for achieving both operational and sustainability goals, as it provides a structured approach to managing and reducing emissions across various levels of business operations (Csutora & Harangozó, 2017).

5.6 CONCLUSION

Understanding and managing greenhouse gas emissions requires more than just measuring totals—it demands clarity on where emissions originate and how they move through an organization's operations and value chain. This chapter broke down the critical distinctions between Scope 1 (direct), Scope 2 (indirect from purchased energy), and Scope 3 (all other indirect) emissions, providing a practical framework for companies to map their carbon footprint. We explored the methodologies for calculating emissions across stationary and mobile sources, the challenges of tracking fugitive and value chain emissions, and the growing regulatory and market pressures for transparent, comprehensive reporting.

Robust GHG accounting isn't just about compliance. It drives operational efficiency, supports risk management, and builds trust with customers and stakeholders. As regulatory requirements tighten and expectations for climate action rise, companies that master emissions tracking and reporting gain a clear competitive edge. The chapter also highlighted the importance of data quality, independent assurance, and stakeholder engagement in building credible inventories—and introduced emerging concepts like avoided and biogenic emissions that broaden the sustainability conversation.

As we move to the next chapter, the focus shifts from corporate accounting frameworks to the broader policy landscape. Chapter 6 examines the pivotal role of the U.S. Environmental Protection Agency (EPA) in shaping national climate action—covering regulatory measures, voluntary programs, and partnerships that drive emissions reductions across sectors. By linking organizational strategies with public policy, the next chapter bridges internal climate accountability with the collective efforts needed for large-scale, systemic change.

REVIEW QUESTIONS

1. What defines Scope 1 emissions, and why are they essential in GHG accounting?
2. How do Scope 2 emissions differ from Scope 1, and what role does purchased electricity play?
3. What makes Scope 3 emissions challenging to measure, and why are they important?
4. How can companies accurately calculate and report their Scope 3 emissions?
5. What role does the National Grid play in managing emissions across all three scopes?
6. How can companies reduce their Scope 1 emissions from stationary and mobile combustion sources?
7. What are fugitive emissions, and how can they be minimized in industrial operations?
8. Why is it essential for companies to track emissions across their supply chain?
9. How does emissions data transparency affect stakeholder trust and environmental accountability
10. What technologies are available to help organizations monitor and reduce emissions in all scopes?

REFERENCES

Adkins, M., & Ehlers, P. (2010). Preventing fugitive emissions. *Manufacturing Chemist,* 81(11), 43–44.

Almihoub, A. A. A., Mula, J. M., & Rahman, M. M. (2013). Are There effective accounting ways to determining accurate accounting tools and methods to reporting emissions reduction. *Journal of Sustainable Development*. https://doi.org/10.5539/JSD.V6N4P118

Assomadi, A. F. (2011). *Kajian emisi co2 menggunakan persamaan mobile 6 dan mobile combustion dari sektor transportasi di kota surabaya*. http://digilib.its.ac.id/ITS-Undergraduate-3100011043613/16090

Auvinen, H., Clausen, U., Davydenko, I., Diekmann, D., Ehrler, V. C., & Lewis, A. (2014). Calculating emissions along supply chains — Towards the global methodological harmonisation. *Research in Transportation Business and Management*. https://doi.org/10.1016/J.RTBM.2014.06.008

Aziz, N. N. N. A., Hassim, M. H., & Hurme, M. (2013). Computer aided estimation of fugitive emission rates and occupational air concentration in process design. *Chemical Engineering Research & Design*, 91(8), 1373–1382. https://doi.org/10.1016/J.CHERD.2013.04.010

Balaras, C. A., Dascalaki, E. G., Patsioti, M., Droutsa, K. G., Kontoyiannidis, S., & Cholewa, T. (2023). Carbon and Greenhouse Gas Emissions from Electricity Consumption in European Union Buildings. *Buildings*. https://doi.org/10.3390/buildings14010071

Caetano, A. C., da Costa, A. M. S., Janeiro, V., Soares, P. H., Cancino, L. R., & Andrade, C. M. G. (2023). Fugitive emissions from mobile sources—Experimental analysis in buses regulated by the Euro 5 standard. *Atmosphere*, 14(4), 613. https://doi.org/10.3390/atmos14040613

Chang, Y. W. (2023). Understanding greenhouse gas management: scopes and boundaries for carbon footprinting. *Springer Environmental Series*. https://doi.org/10.4324/9781003274049-14

Cosgrove, D., Evill, B., & Subasic, K. (1994). The estimation of transport emissions. *Transport Research Forum*. https://trid.trb.org/view/415358

Csutora, M., & Dobák, K. (2019). Accounting for beyond Scope 3 GHG emissions. In *Proceedings of the 23rd Conference of the Environmental and Sustainability Management Accounting Network (EMAN)*, Prague, Czech Republic.

Csutora, M., & Harangozó, G. (2017). Twenty years of carbon accounting and auditing – a review and outlook. *Society and Economy*. https://doi.org/10.1556/204.2017.39.4.1

Csutora, M., & Harangozó, G. (2019). Four stages of carbon accounting: A small country perspective. In *Proceedings of the 23rd Conference of the Environmental and Sustainability Management Accounting Network (EMAN)*, Prague, Czech Republic.

Downie, J., & Stubbs, W. (2012). Corporate Carbon Strategies and Greenhouse Gas Emission Assessments: The Implications of Scope 3 Emission Factor Selection. *Business Strategy and The Environment*, 21(6), 412–422. https://doi.org/10.1002/BSE.1734

Ellis, H. M., & Lackaye, R. (1989). *Estimating fugitive emissions of volatile compounds from equipment leaks. Journal of the Air & Waste Management Association*, 39(12), 1619–1622. https://doi.org/10.1080/08940630.1989.10466656

Emborg, M., Lloyd, S. M., & Olsen, S. I. (2023). Why process-level Scope 3 accounting is needed for delivering supply chain greenhouse gas emission reduction. *Integrated Environmental Assessment and Management*. https://doi.org/10.1002/ieam.4816

Gulluscio, C., Puntillo, P., Luciani, V., & Huisingh, D. (2020). Climate change accounting and reporting: a systematic literature review. *Sustainability*. https://doi.org/10.3390/SU12135455

Hammar, T., Hansson, P.-A., Seleborg, M., Stendahl, J., & Jonsson, R. (2022). *Climate effects of a forestry company – including biogenic carbon fluxes and substitution effects (2021 update)* (Report No. 120). Swedish University of Agricultural Sciences, Department of Energy and Technology. https://doi.org/978-91-576-9948-0

Hassim, M. H., Hassim, M. H., Hurme, M., Amyotte, P., & Khan, F. (2012). Fugitive emissions in chemical processes.The assessment and prevention based on inherent and add-on approaches. *Journal of Loss Prevention in The Process Industries*, 25(5), 820–829. https://doi.org/10.1016/J.JLP.2012.03.004

Hertwich, E. G., & Wood, R. (2018). The growing importance of scope 3 greenhouse gas emissions from industry. *Environmental Research Letters*, 13(10), 104013. https://doi.org/10.1088/1748-9326/AAE19A

Hileman, B. (2001). Accounting for greenhouse gases: New reports show that what's good for the environment is good for the bottom line. *Chemical & Engineering News*, 79(46), 21–25. https://doi.org/10.1021/cen-v079n046.p021

Holding, M. (2023). Australia's regulation of scope 3 emissions. *The APPEA Journal*, 63(2), S391–S394. https://doi.org/10.1071/aj22132

Huang, Y. A., Weber, C. L., & Matthews, H. S. (2009). Categorization of Scope 3 emissions for streamlined enterprise carbon footprinting. *Environmental Science & Technology*, 43(22), 8509–8515. https://doi.org/10.1021/ES901643A

Ismayanti, R. I. (2011). *Kajian emisi co2 menggunakan persamaan mobile 6 dan mobile combustion dari sektor transportasi di kota surabaya.*

Johnson, M. P., Rötzel, T. S., & Frank, B. (2023). Beyond conventional corporate responses to climate change towards deep decarbonization: a systematic literature review. *Management Review Quarterly*. https://doi.org/10.1007/s11301-023-00318-8

Kenway, S., Pamminger, F., Yan, G., Hall, R., Lam, K. L., Skinner, R., Olsson, G., Satur, P., & Allan, J. (2023). Opportunities and challenges of tackling Scope 3 "Indirect" emissions from residential hot water. *Water Research X*, 21, 100192. https://doi.org/10.1016/j.wroa.2023.100192

Kim, E.-K., Kim, S., Lee, Y. P., Byun, S., & Kim, H. (2014). Fugitive emission characteristics of HFC-134a from reefer container. *Journal of Korean Society for Atmospheric Environment*, 30(2), 110–118. https://doi.org/10.5572/KOSAE.2014.30.2.110

Lee, C.-H., & Ma, H.-W. (2013). Improving the integrated hybrid LCA in the upstream scope 3 emissions inventory analysis. *International Journal of Life Cycle Assessment*, 18(4), 541–553. https://doi.org/10.1007/s11367-012-0469-9

Li, Y., Wu, Q., Zhang, Y., Huang, G., Jin, S., & Fang, S. (2022). Mapping highway mobile carbon source emissions using traffic flow big data: A case study of Guangdong Province, China. *Frontiers in Energy Research*, 10, 891742. https://doi.org/10.3389/fenrg.2022.891742

Li, Y., Yang, J., Wu, X., Liu, Y., Zhang, Y., Zhou, P., & Chen, G. (2023). Leakage, diffusion and distribution characteristics of refrigerant in a limited space: A comprehensive review. *Thermal Science and Engineering Progress*, 40, 101731. https://doi.org/10.1016/j.tsep.2023.101731

Miller, G. J., Novan, K., & Jenn, A. (2022). Hourly accounting of carbon emissions from electricity consumption. *Environmental Research Letters*, 17(4), 044073. https://doi.org/10.1088/1748-9326/ac6147

Roston, M., Seiger, A., & Mathieson, A. (2024). What's Scope 2 good for? *Oxford Open Climate Change,* 4(1), kgae011. https://doi.org/10.1093/oxfclm/kgae011

Schmidt, M. (2022). Determining the Scope 3 Emissions of Companies. *Chemical Engineering & Technology*, *45*(7), 1218–1230. https://doi.org/10.1002/ceat.202200181

Shrimali, G. (2022). Scope 3 emissions: Measurement and management. *Journal of Impact and ESG Investing*, 1(4), 45–62. https://doi.org/10.3905/jesg.2022.1.051

Shing, S. Y., Foo, C. Y. D., & Hassim, H. M. (2016). Considering fugitive emissions during the conceptual design stage. *Chemical Engineering*, 123(11), 64. http://eprints.utm.my/id/eprint/69079/

Singh, N., & Mahapatra, A. (2013). *Designing greenhouse gas reporting systems: Learning from existing programs* (Working Paper). World Resources Institute. https://www.wri.org/publication/designing-greenhouse-gas-reporting-systems-learning-fromexisting-programs

Stenzel, A., & Waichman, I. K. (2023). *Supply-chain data sharing for scope 3 emissions. Climate Action*, 2(1). https://doi.org/10.1038/s44168-023-00032-x

Von Wald, G., Cullenward, D., Mastrandrea, M. D., & Weyant, J. P. (2021). Accounting for the greenhouse gas emission intensity of regional electricity transfers. *Environmental Science & Technology*, 55(10), 6571–6579. https://doi.org/10.1021/ACS.EST.0C08096

Wang, Y., Hao, Y., Hou, Y., Quan, Q., & Li, Y. (2024). Optimizing Scope 3 emissions in the automotive manufacturing industry: a multidisciplinary approach. *Carbon Research.* https://doi.org/10.1007/s44246-024-00131-2

Young-Ferris, A., Malik, A., Calderbank, V., & Jacob-John, J. (2024). Making things (that don't exist) count: a study of Scope 4 emissions accounting claims. *Accounting, Auditing & Accountability*. https://doi.org/10.1108/aaaj-04-2023-6406

6 The U.S. EPA's Role in Climate Action

This chapter delves into the critical role of the United States Environmental Protection Agency (EPA) in combating climate change. It explores the agency's comprehensive approach, encompassing regulatory measures, voluntary programs, partnerships, and educational initiatives. This chapter details the EPA's efforts to reduce greenhouse gas emissions, promote renewable energy, improve energy efficiency, and enhance climate resilience across various sectors, including energy, transportation, agriculture, and waste management. It also highlights the agency's commitment to environmental justice, ensuring climate action benefits all communities—particularly those disproportionately affected by pollution and climate change.

The EPA plays a crucial role in climate action by developing and enforcing regulations to reduce greenhouse gas emissions, promoting renewable energy, and improving energy efficiency. The EPA collaborates with federal, state, local, and tribal governments, businesses, and non-profit organizations to implement policies and initiatives that address climate change. This includes setting national standards for air quality, providing grants for environmental projects, conducting scientific research on climate impacts, and fostering partnerships to enhance sustainability efforts. The EPA ensures that communities and individuals are informed and engaged in mitigating climate-related risks and promoting environmental stewardship through education and public outreach.

6.1 EPA INVENTORY OF GHG EMISSIONS AND SINKS

EPA prepares an annual report called the *Inventory of U.S. Greenhouse Gas Emissions and Sinks* (Inventory), which tracks U.S. greenhouse gas emissions and sinks by source, economic sector, and greenhouse gas, going back to 1990. This annual report provides a comprehensive accounting of total greenhouse gas emissions for all man-made sources in the United States, including carbon dioxide removal from the atmosphere by "sinks" (e.g., through the uptake of carbon and storage in forests, vegetation, and soils) from the management of lands in their current use or as lands are converted to other uses. The gases covered by the Inventory include carbon dioxide, methane, nitrous oxide, hydrofluorocarbons, perfluorocarbons, sulfur hexafluoride, and nitrogen trifluoride. The importance of such inventories is underscored by their role in policy development and implementation, as they provide the necessary data to evaluate and establish effective climate change mitigation strategies (Mingshan, 2013).

The Greenhouse Gas Equivalencies Calculator complements this effort by translating complex emissions data into relatable terms, thereby enhancing public understanding and transparency (Burtraw & Woerman, 2012). The Greenhouse Gas Equivalencies Calculator uses the Avoided Emissions and Generation Tool (AVERT) U.S. national weighted average CO_2 marginal emission rate to convert reductions of kilowatt-hours into avoided units of carbon dioxide emissions. The Greenhouse Gas Equivalencies Calculator allows the conversion of emissions or energy data to the equivalent amount of carbon dioxide (CO_2) emissions from using that amount. The calculator helps translate abstract measurements into concrete terms, such as the annual emissions from cars, households, or power plants. Steps of the calculations are outlined in Table 6.1 below. This calculator is useful in communicating your greenhouse gas reduction strategy, reduction targets, or other initiatives aimed at reducing greenhouse gas emissions.

DOI: 10.1201/9781003479666-6

TABLE 6.1
GHG Emission Calculations for Electricity, Gasoline, Fuel Oil and Coal Consumption

	Calculation
Electricity consumed (kilowatt-hours)	852.3 lbs CO_2/MWh × 1 metric ton/2,204.6 lbs × 1/(1 − 0.073) MWh delivered/MWh generated × 1 MWh/1,000 kWh = 4.17×10^{-4} metric tons CO_2/kWh (eGRID, U.S. annual CO_2 total output emission rate [lb/MWh], year 2021 data)
Gallons of gasoline consumed	8,887 g of CO_2/gallon of gasoline = 8.887×10^{-3} metric tons CO_2/gallon of gasoline In the preamble to the joint EPA/Department of Transportation rulemaking on May 7, 2010 that established the initial National Program fuel economy standards for model years 2012–2016, the agencies stated that they had agreed to use a common conversion factor of 8,887 g of CO_2 emissions per gallon of gasoline consumed (Federal Register 2010)
Gallons of diesel consumed	10,180 g of CO_2/gallon of diesel = 10.180×10^{-3} metric tons CO_2/gallon of diesel In the preamble to the joint EPA/Department of Transportation rulemaking on May 7, 2010 that established the initial National Program fuel economy standards for model years 2012–2016, the agencies stated that they had agreed to use a common conversion factor of 10,180 g of CO_2 emissions per gallon of diesel consumed (Federal Register 2010)
Barrels of oil consumed	5.80 mmbtu/barrel × 20.31 kg C/mmbtu × 44 kg CO_2/12 kg C × 1 metric ton/1,000 kg = 0.43 metric tons CO_2/barrel Carbon dioxide emissions per barrel of crude oil are determined by multiplying heat content times the carbon coefficient times the fraction oxidized times the ratio of the molecular weight of carbon dioxide to that of carbon (44/12). The average heat content of crude oil is 5.80 mmbtu per barrel (EPA, 2023). The average carbon coefficient of crude oil is 20.31 kg carbon per mmbtu (EPA, 2023). The fraction oxidized is assumed to be 100% (IPCC, 2006)
Home electricity use	12,154 kWh per home × 852.3 lbs CO_2/MWh generated × 1/(1–0.073) MWh delivered/MWh generated × 1 MWh/1,000 kWh × 1 metric ton/2,204.6 lb = 5.067 metric tons CO_2/home. In 2021, 125.0 million homes in the United States consumed 1,519 billion kilowatt-hours (kWh) of electricity (EIA 2022a). On average, each home consumed 12,154 kWh of delivered electricity (EIA 2022a). The national average carbon dioxide output rate for electricity generated in 2021 was 852.3 lbs CO_2 per megawatt-hour (EPA 2023a), which translates to about 919.1 lbs CO_2 per megawatt-hour for delivered electricity, assuming transmission and distribution losses of 7.3% (EIA 2022b; EPA 2023b)
Coal-fired power plant emissions for one year	805,305,916.8 metric tons of CO_2 × 1/207 power plants = 3,890,366.75 metric tons CO_2/power plant In 2021, a total of 207 power plants used coal to generate at least 95% of their electricity (EPA, 2023). These plants emitted 805,305,916.8 metric tons of CO_2 in 2021

6.2 BUILDING ENERGY CODES

Building energy codes require new construction and major renovations in existing buildings to meet minimum energy efficiency requirements. Well-designed and enforced codes can institute construction practices that reduce building life cycle costs and occupants' total housing or commercial costs. Building energy code requirements can also help reduce peak energy demand, greenhouse gas (GHG) emissions, and other air pollutants. Recognizing these benefits, most states have adopted building energy codes for residential and commercial buildings.

6.2.1 Leadership in Energy and Environmental Design

Leadership in Energy and Environmental Design (LEED) is the world's most widely used green building rating system. LEED certification provides a framework for healthy, highly efficient, and cost-saving green buildings that offer environmental, social, and governance benefits. LEED certification is a globally recognized symbol of sustainability achievement, and it is backed by an entire industry of committed organizations and individuals, paving the way for market transformation.

LEED-certified buildings are critical to addressing climate change and meeting ESG goals, enhancing resilience, and supporting more equitable communities. LEED is a holistic system that doesn't simply focus on one building element, such as energy, water, or health. Instead, it looks at the big picture, factoring in all critical elements that work together to create the best building possible. The goal of LEED is to create better buildings that:

1. Reduce contribution to global climate change
2. Enhance individual human health
3. Protect and restore water resources
4. Protect and enhance biodiversity and ecosystem services
5. Promote sustainable and regenerative material cycles
6. Enhance community quality of life.

LEED v5 is the newest version of LEED. It marks a transformative milestone in the built environment's alignment with a low-carbon future and addresses critical imperatives such as equity, health, ecosystems, and resilience. LEED buildings use less energy and water, utilize renewable energy and fewer resources, create less waste, and preserve land and habitat. LEED certification is a global solution for cities, communities, and neighborhoods. Through sustainable design, construction, and operations, LEED can help new and existing buildings reduce carbon emissions, energy use, and waste, conserve water, prioritize safer materials, and lower our exposure to toxins.

LEED-certified structures prioritize energy efficiency by integrating renewable energy technologies and smart building automation systems, significantly reducing energy consumption and improving operational efficiency (Morshed et al., 2024). However, the effectiveness of LEED in ensuring energy efficiency is sometimes debated, as some studies indicate that not all LEED-certified buildings outperform non-certified ones in energy efficiency due to variations in design orientation, occupancy, and construction periods (Al-Zubaidy, 2015). Despite these challenges, LEED remains a vital tool in achieving sustainability goals, as it encourages the adoption of diverse and dynamic energy conservation technologies and design approaches (Liao & Tam, 2021).

6.3 ENERGY EFFICIENCY

6.3.1 Energy Star

The EPA tackles climate change by collaborating with various industries on energy efficiency. One key program is ENERGY STAR, which incentivizes businesses and consumers to choose energy-saving products and practices. This includes recognizing energy-efficient options across various categories, like appliances (ENERGY STAR Products), new homes (ENERGY STAR New Homes), commercial buildings (ENERGY STAR Commercial Buildings), and even industrial facilities (ENERGY STAR Industrial Plants). By promoting these energy-conscious choices, the EPA and its partners work together to reduce greenhouse gas emissions and combat climate change. ENERGY STAR is the government-backed symbol for energy efficiency, providing simple, credible, and unbiased information that consumers and businesses rely on to make well-informed decisions. Thousands of industrial, commercial, utility, state, and local organizations—including nearly 40% of the Fortune 500—rely on their partnership with the EPA to deliver cost-saving energy efficiency solutions. This success aligns with broader international efforts to enhance energy efficiency, underscoring the importance of innovative policies in industrialized and emerging economies to overcome market barriers and adapt to technological advances (Brown, 2015).

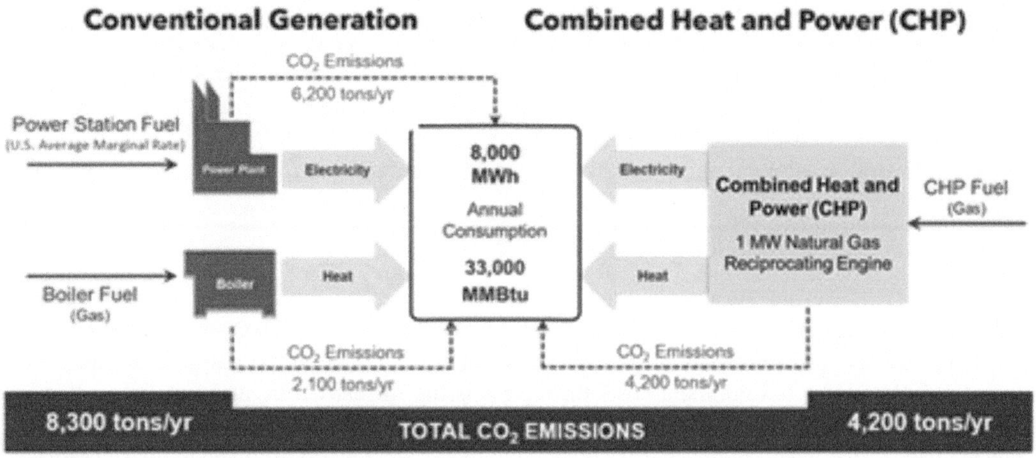

FIGURE 6.1 Conventional generation vs. CHP: CO_2 emissions. (Source: USEPA https://www.epa.gov/chp/chp-benefits.)

To earn the EPA's ENERGY STAR, a commercial building must earn an ENERGY STAR score of 75 or higher on the EPA's 1–100 scale, indicating that it is more energy efficient than 75% of similar buildings nationwide. A building's ENERGY STAR score is calculated based on several factors, including energy use, hours of operation, and other operating characteristics. Since 1992, ENERGY STAR and its partners have helped American families and businesses save 5 trillion kilowatt-hours of electricity, avoid more than $500 billion in energy costs, and achieve 4 billion metric tons of greenhouse gas reductions—all through voluntary action. By the end of 2023, more than 43,000 buildings across America had earned EPA's ENERGY STAR certification. Together, these buildings have saved nearly $6 billion on energy bills and prevented more than 23 million metric tons of greenhouse gas emissions—equal to the annual emissions of more than 3 million homes.

6.3.2 Combined Heat and Power Partnership

The Combined Heat and Power Partnership (CHP) is an EPA program that promotes CHP systems to reduce greenhouse gas emissions and improve energy efficiency. CHP systems generate electricity and heat from a single fuel source, which can be more efficient than separate heat and power generation. CHP can also be used with renewable fuels like biogas and biomass to further reduce emissions. CHP systems can also provide backup power and heat during outages, making them more resilient than traditional grid-based power, as shown in Figure 6.1.

6.4 RENEWABLE ENERGY

6.4.1 Green Power Partnership

The Green Power Partnership (GPP), established by the EPA in 2001, is a voluntary program that encourages organizations to use renewable energy sources (green power) to reduce environmental impact and promote the development of the American green power market. The GPP provides guidance, resources, and recognition to organizations that transition to green power, which includes electricity produced from solar, wind, geothermal, biogas, biomass, and low-impact small hydroelectric sources. The program has been instrumental in promoting the voluntary green power market, which allows consumers and institutions to purchase renewable energy to meet their electricity needs. This market not only provides a revenue stream for renewable energy projects but also raises consumer awareness about the benefits of renewable energy. The GPP has also seen significant participation from higher education institutions, which are often leaders in sustainability initiatives and are characterized by various institutional, geographical, and financial factors. Despite the challenges in large-scale renewable energy adoption and implementation—as noted in broader studies on renewable energy efficiency and effectiveness—programs like the GPP are crucial in supporting the transition from traditional energy sources to more sustainable alternatives.

6.4.2 RE-Powering America's Land Initiative

The EPA's RE-Powering America's Land Initiative encourages the development of renewable energy sources on previously contaminated lands (like landfills and mines) suitable for such projects. This initiative aims not only to generate clean energy but also to revitalize abandoned areas and potentially reduce development costs through existing infrastructure and streamlined permitting. For instance, feasibility studies have been conducted at various sites, such as the former Asarco smelter in Montana, where biomass was chosen due to the local wood products industry (Moriarty, 2013b), and the Kaiser Aluminum Landfill in Louisiana, where anaerobic digestion of food waste was explored (Moriarty, 2013). Similarly, the Kolthoff Landfill in Ohio was assessed for solar photovoltaic installations, highlighting the potential for solar energy on contaminated lands (Salasovich et al., 2013). These projects demonstrate the potential of using marginal and contaminated lands for renewable energy, which could meet a significant portion of energy demands, as seen in the U.S., where such lands could contribute up to 39% of the 2013 energy demand for biofuel and electricity (Niblick & Landis, 2016). The initiative also addresses common barriers in brownfield-to-renewable energy projects, such as cleanup costs and liability risks, by fostering strong partnerships and public support, which are crucial for overcoming these challenges and facilitating project success (Jensen, 2010). Moreover, as discussed in the context of sustainable land restoration, the circular economy approach emphasizes the importance of policy frameworks that integrate resource circularity with land restoration, enhancing ecological vitality and social welfare (Priyadarshini & Abhilash, 2020). Overall, these efforts not only contribute to renewable energy production but also promote sustainable land use and community development, turning environmental liabilities into opportunities for clean energy and economic growth (Brown & Ghag, 2023; EPA, 2014). Figure 6.2 illustrates the potential advantage of re-powering.

6.5 METHANE REDUCTION, RECOVERY, AND REUSE

Methane, a potent greenhouse gas with a global warming potential 25 times greater than carbon dioxide, plays a significant role in climate change, contributing to over one-third of anthropogenic warming despite its lower atmospheric concentration than CO_2 (Thakur & Solanki, 2022). The EPA has implemented programs such as AgSTAR and the Coalbed Methane Outreach Program (CMOP) to mitigate methane emissions from agriculture and coal mining by promoting technologies that capture and convert methane into energy resources (Melvin et al., 2016). These initiatives have been effective, with EPA programs reducing 127.9 teragrams of methane emissions from 1993 to 2013, significantly impacting atmospheric methane concentrations and global temperatures (Melvin et al., 2016).

6.5.1 AgSTAR

AgSTAR is a collaborative program sponsored by EPA and USDA that promotes the use of biogas recovery systems to reduce methane emissions from livestock waste. AgSTAR promotes the use of biogas recovery systems to reduce methane emissions from livestock waste. Anaerobic digester outputs are shown in Figure 6.3. Biogas recovery also helps achieve other social, environmental, agricultural, and economic benefits. AgSTAR assists those who enable, purchase, or implement

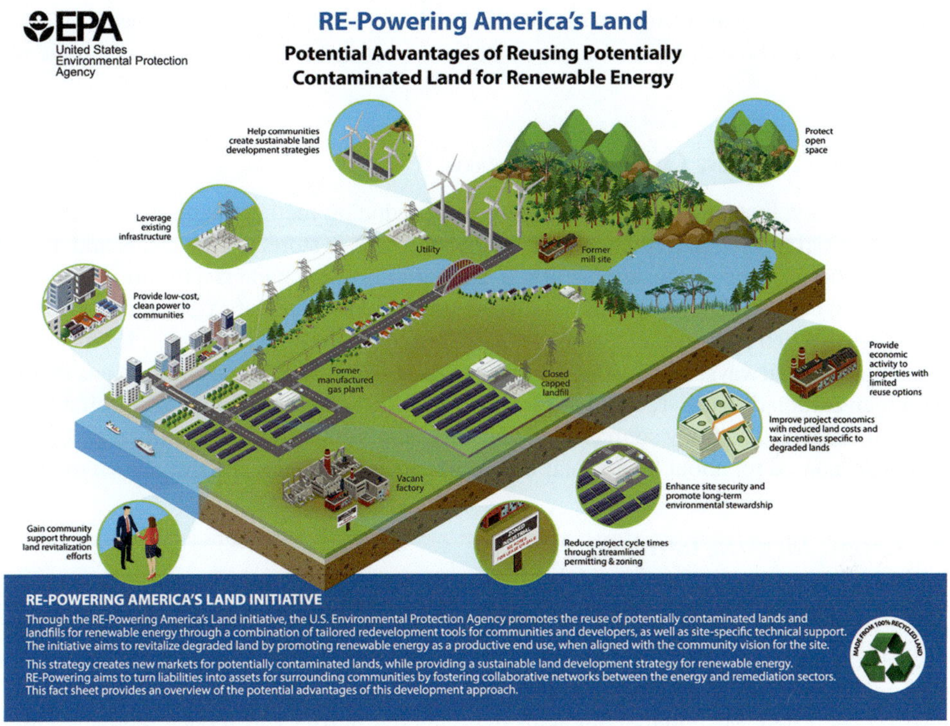

FIGURE 6.2 Advantages of RE-powering. (Source: USEPA https://www.epa.gov/re-powering/what-re-powering.)

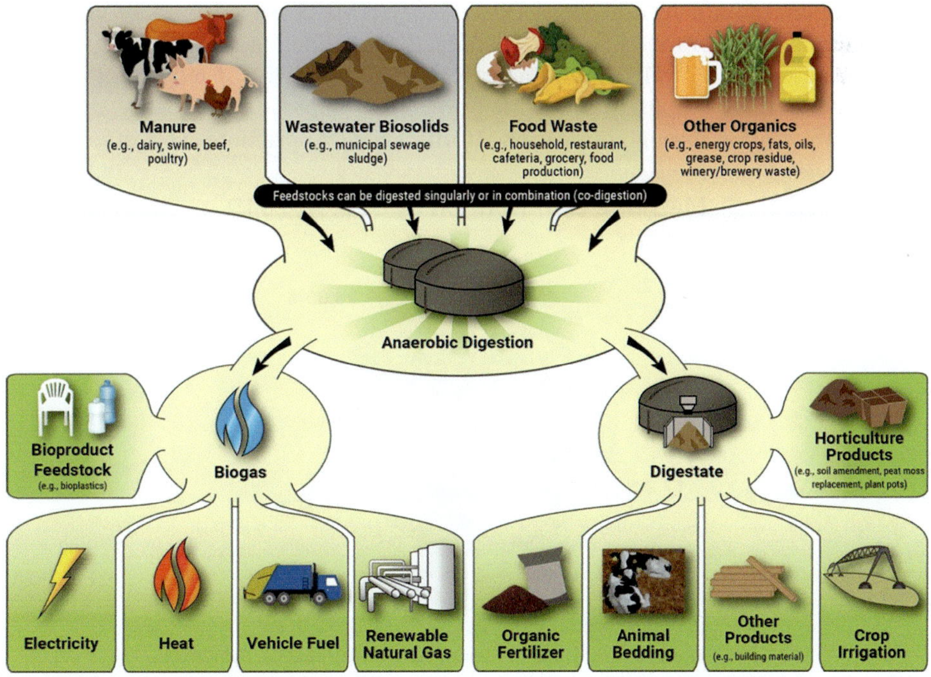

FIGURE 6.3 Anaerobic digester outputs. (Source: USEPA https://www.epa.gov/agstar/how-does-anaerobic-digestion-work.)

anaerobic digesters by identifying project benefits, risks, options, and opportunities. AgSTAR provides information and organizes events to create a supporting environment for anaerobic digester implementation.

6.5.2 Coalbed Methane Outreach Program

EPA's CMOP is a voluntary program with the goal of reducing methane emissions from coal mining activities. Coal mining activities release methane, which is stored in the cleats and micropores of coal when coal-bearing seams are disturbed. In coal mines, methane is a mining hazard because it is explosive at certain concentrations, and its accumulation can result in a mine atmosphere deficient in oxygen. When released, methane is a precursor to ground-level ozone, a harmful pollutant that can irritate the airways. In the atmosphere, methane is a powerful greenhouse gas that is more powerful than carbon dioxide. Technologies can help capture methane from coal mining before it is released and use it as an energy source. Methane (CH_4) emissions from coal mining and abandoned coal mines accounted for about 8% of total U.S. methane emissions in 2019, as shown in Figure 6.4.

6.5.3 Global Methane Initiative

The Global Methane Initiative (GMI) is an international public–private partnership focused on reducing barriers to recovering and using methane as a valuable energy source. GMI provides technical support to deploy methane-to-energy projects around the world, enabling Partner Countries to launch methane recovery and use projects. GMI focuses on three key sectors: oil and gas, biogas, and coal mines. The GMI collaborates with international partners to reduce methane emissions, emphasizing the importance of cutting human-caused methane emissions by 45% this decade to prevent dangerous climate tipping points and improve air quality (Coalition, 2021). This target aligns with the Global Methane Pledge (GMP), which aims for a 30% reduction by 2030 compared to 2020 levels, although the GMP does not specify contributions from individual countries or sectors.

The International Methane Emissions Observatory (IMEO) supports these efforts by providing reliable, policy-relevant data to facilitate methane reduction actions, integrating diverse data streams to create a centralized emissions record. The benefits of reducing methane emissions are substantial, including rapid warming reduction, improved air quality, enhanced food security, and job creation through mitigation efforts. The GMI promotes cost-effective methane

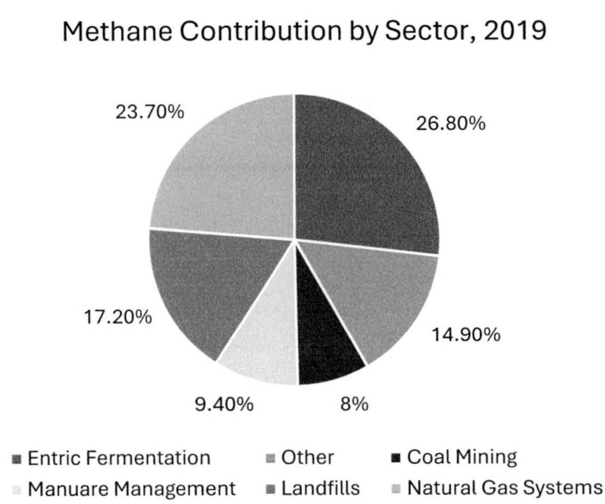

FIGURE 6.4 Methane contribution by sector, 2019. (Source: USEPA https://www.epa.gov/cmop/about-coal-mine-methane.)

recovery through partnerships across various sectors, including the private sector and NGOs. Furthermore, a zero anthropogenic methane emissions scenario demonstrates the profound impacts on atmospheric composition and climate, highlighting the potential of methane emissions reductions and their air-quality co-benefits. Collectively, these initiatives and findings underscore the critical role of methane mitigation in achieving climate goals and preventing adverse environmental impacts.

6.5.4 Landfill Methane Outreach Program

LMOP is a voluntary program that works cooperatively with industry stakeholders and waste officials to reduce or avoid methane emissions from landfills. LMOP encourages the recovery and beneficial use of biogas generated from organic municipal solid waste (MSW). Landfill gas (LFG) and other biogas generated from MSW (collectively called biogas) contain methane that can be captured and used as a renewable fuel for many end uses, including electricity generation, industrial heat applications, and vehicle fuel. Capturing and using biogas reduces local air pollution, creates health benefits, generates revenue and jobs in the community, and may offset the use of non-renewable resources.

6.5.5 Voluntary Methane Programs for the Oil and Natural Gas Industry

EPA collaborates with oil and natural gas operators and industry stakeholders to encourage identifying and implementing technologies and practices to reduce methane emissions from the oil and gas sector. Reducing methane emissions reduces operational risk, increases efficiency, and demonstrates company concern for the environment, with benefits spanning from climate change to air quality improvements to conservation of a non-renewable energy resource. EPA facilitates technology transfer and implementation workshops and webinars through the Natural Gas STAR Program, shares technical resources, and collaborates with oil and gas operators.

6.6 STATE, TRIBAL, AND LOCAL PARTNERSHIPS

6.6.1 State and Local Climate and Energy Program

EPA's State and Local Climate and Energy Program offers free tools, data, and technical expertise about energy strategies, including energy efficiency, renewable energy, and other emerging technologies, to help state, local, and tribal governments achieve their environmental, energy, and economic objectives. While many of these resources were developed for state and local governments, tribes may also find them useful.

6.6.2 Heat Island Reduction Program

Heat islands are urbanized areas that experience higher temperatures than outlying areas. Structures such as buildings, roads, and other infrastructure absorb and re-emit the sun's heat more than natural landscapes such as forests and water bodies. Urban areas, where these structures are highly concentrated and greenery is limited, become "islands" of higher temperatures relative to outlying areas. Daytime temperatures in urban areas are about 1°F–7°F higher than temperatures in outlying areas, and nighttime temperatures are about 2°F–5°F higher.

6.6.3 Advance Program

The EPA's Advance Program is a voluntary initiative established in 2012 to help areas that meet air quality standards maintain good air quality. It works with states, tribes, and local governments to reduce ozone and particle pollution through various programs and collaborations. The program offers technical assistance, grants, and opportunities to share successful strategies between participating communities.

6.7 TRANSPORTATION

The EPA addresses GHG emissions from transportation and waste management through targeted programs such as Clean School Bus USA and the Responsible Appliance Disposal (RAD) program. Transportation is a significant contributor to GHG emissions, and strategies to mitigate these emissions include improving vehicle efficiency, adopting low-carbon fuels, and reducing travel demand, all of which offer substantial co-benefits like energy savings and pollution reduction (Lutsey & Sperling, 2008).

Transportation is a major source of air pollution, harming human health and contributing to climate change. Pollutants come from vehicles like cars, trucks, and airplanes. Stricter fuel standards, cleaner-burning gasoline, and advanced engine technologies can significantly reduce emissions. Electric vehicles and improved transportation planning offer even cleaner solutions for the future. Sources of transportation-related air pollution and solutions for reducing these are shown in Figure 6.5.

6.7.1 Clean School Bus USA

School buses travel about four billion miles annually, providing the safest transportation to and from school for more than 25 million American children daily. The Clean School Bus USA program specifically targets reducing diesel exhaust exposure, which is crucial for protecting vulnerable populations, including children, from harmful pollutants (Hodges & Potter, 2010). While new buses must meet EPA's tougher emission standards, many older school buses continue to emit harmful diesel exhaust. EPA's Clean School Bus is a national program designed to help communities reduce emissions from older diesel school buses. School districts, fleet owners and operators, bus drivers, parents, and students all have a role in helping to reduce diesel emissions from school buses.

6.7.2 National Clean Diesel Campaign

The Diesel Emissions Reduction Act (DERA) Program funds grants and rebates that protect human health and improve air quality by reducing harmful emissions from diesel engines.

6.7.3 Ports Initiative

EPA's Ports Initiative works in collaboration with the port industry, communities, and all levels of government to improve environmental performance and increase economic prosperity. This effort helps people living and working near ports across the country breathe cleaner air and live better lives. The long-term Ports Initiative goal is for U.S. ports to become global leaders in clean, efficient freight and passenger transportation. Ongoing actions outlined in the initiative's elements are laying the groundwork for achieving both the long-term goal and the vision.

6.7.4 SmartWay Transport Partnership

The EPA's SmartWay program is a collaborative effort between the public and private sectors to address sustainability in the freight transportation industry. Launched in 2004, this voluntary program helps companies track and reduce their freight emissions by offering resources for identifying efficient transportation methods and technologies. By promoting environmentally conscious practices, SmartWay aims to bolster global energy security and minimize environmental risks.

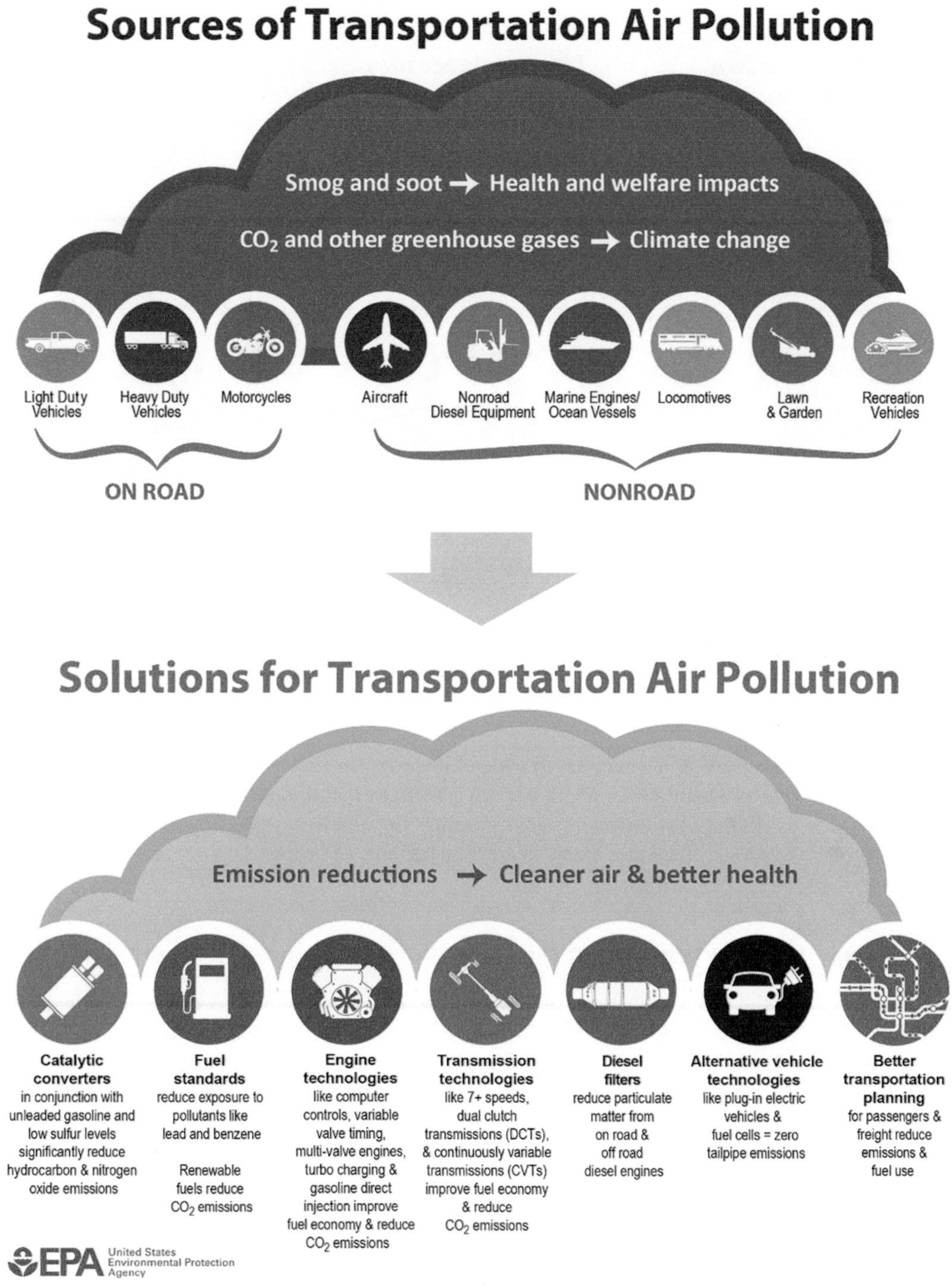

FIGURE 6.5 Source for and of transportation air pollution. (Source: USEPA https://www.epa.gov/transportation-air-pollution-and-climate-change/overview-air-pollution-transportation.)

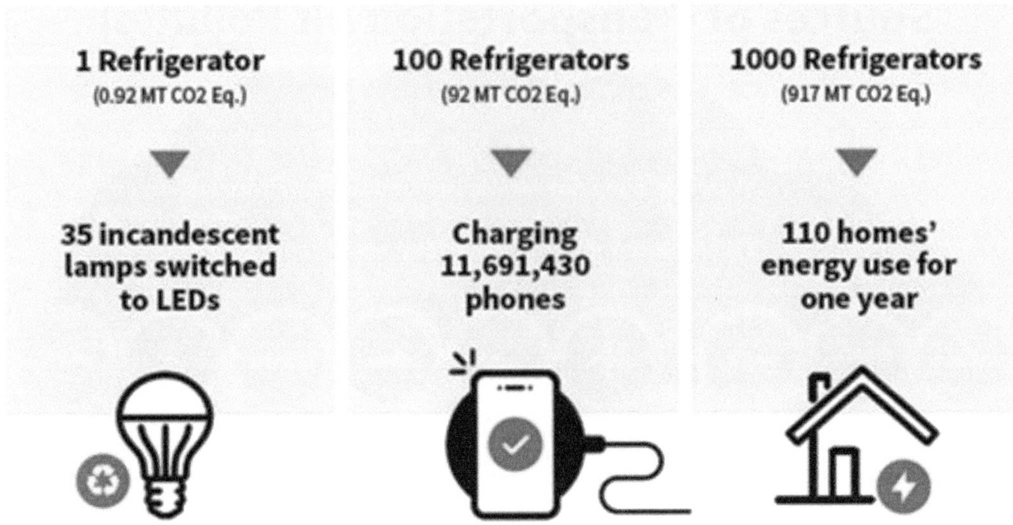

FIGURE 6.6 RAD benefits from properly disposing of old refrigerators (20+ years old). (Source: USEPA https://www.epa.gov/rad/program-benefits#Related_Pages_2.)

6.8 RECYCLING AND WASTE

6.8.1 RAD Program

The EPA's RAD program is a voluntary partnership program that works with utilities, retailers, manufacturers, state and local government agencies, affiliates, and others to dispose of old refrigerated appliances using the best environmental practices available—going beyond federal requirements to protect Earth's climate and ozone layer. The benefits of proper disposal of old refrigerators are further explained in Figure 6.6.

6.8.2 Food Loss and Waste 2030 Champions

In November 2016, the U.S. Department of Agriculture (USDA) and the EPA announced the formation of the U.S. Food Loss and Waste 2030 Champions group and presented the first set of 2030 Champions. U.S. Food Loss and Waste 2030 Champions are businesses and organizations that have made a public commitment to reduce food loss and waste in their own operations in the United States by 50% by the year 2030.

6.9 WATER

6.9.1 WaterSense

WaterSense, a voluntary partnership program sponsored by the U.S. Environmental Protection Agency (EPA), is both a label for water-efficient products and a resource for helping you save water. The WaterSense label makes it simple to find water-efficient products, new homes, and programs that meet EPA's criteria for efficiency and performance. WaterSense-labeled products and services are certified to use at least 20% less water, save energy, and perform as well as or better than regular models. WaterSense partners with manufacturers, retailers, and distributors, homebuilders, irrigation professionals, and utilities to bring WaterSense to your community. Our partnerships encourage innovation in manufacturing and support sustainable jobs for American workers.

6.9.2 Climate Ready Estuaries

The Climate Ready Estuaries program works with the coastal management community to assess climate change vulnerabilities, develop and implement adaptation strategies, and engage and educate stakeholders. CRE provides technical guidance and assistance about climate change adaptation. The Climate Ready Estuaries program works to help the National Estuary Programs and all environmental managers address climate change in watersheds and coastal areas. This effort, initiated in 2008, brings together EPA's oceans and coastal protection programs and Climate Change Programs to build additional capacity in the NEPs and coastal communities as they prepare to adapt to the effects of climate change. The EPA Office of Research and Development and Creating Resilient Water Utilities are supporting this effort as well. Climate Ready Estuaries assists NEPs and coastal communities in becoming "climate ready" by providing tools and assistance to assess climate change vulnerability and plan for adaptation.

6.9.3 Creating Resilient Water Utilities

EPA's CRWU initiative provides drinking water, wastewater, and stormwater (water sector) utilities with practical tools, training, and technical assistance to increase system resilience to climate change impacts. CRWU helps promote a clear understanding of climate data and helps water sector utilities identify potential adaptation strategies, implementation options, and infrastructure financing.

6.9.4 Sustainable Water Infrastructure

The U.S. has invested billions of dollars over the years to build an extensive network of drinking water, wastewater, and stormwater infrastructure to provide the public with safe and clean water. Much of the network of water treatment plants, distribution lines, sewer lines, and storage facilities was built after World War II. Some of that infrastructure is now over 100 years old.

Historically, investment has not been enough to meet the ongoing need to maintain and renew these systems. Over the coming decades, this pattern of underinvestment needs to change, and practices must be put in place to sustain the water services provided by water infrastructure and utilities. Doing so is vital to public, economic, and environmental health.

6.10 HIGH-POTENCY EMISSIONS REDUCTION

6.10.1 SF6 Emission Reduction Partnership for Electric Power Systems

Established in 1999, the SF6 Emission Reduction Partnership for Electric Power Systems is a collaborative effort between the EPA and the electric power industry to reduce sulfur hexafluoride (SF_6) emissions through voluntary, cost-effective technical and management solutions. SF6 is a highly potent greenhouse gas that is used in the electric power systems industry for insulation and current interruption in electric transmission and distribution equipment.

6.10.2 GreenChill Partnership

Under the GreenChill Certification Levels program, a food retail store can achieve Platinum, Gold, or Silver Level Certification. To achieve Platinum-Level Store Certification, a store must exclusively use non-ozone-depleting refrigerants approved by the EPA's SNAP program and meet either an average HFC refrigerant charge of 0.5 pounds per 1,000 BTUH total evaporator cooling load with an annual HFC emissions rate of 5% or less, or use refrigerants with GWPs below 150. For Gold-Level Certification, the store must also use SNAP-approved non-ozone-depleting refrigerants, maintain an HFC charge of 1.25 pounds per 1,000 BTUH, and have an annual HFC emissions rate of 15% or less. Silver-Level Certification requires the use of SNAP-approved non-ozone-depleting refrigerants, an HFC charge of no more than 1.75 pounds per 1,000 BTUH, and a 15% or lower annual HFC emissions rate.

6.11 INITIATIVES AND RESOURCES

6.11.1 Climate Change Adaptation Resource Center

EPA's Adaptation Resource Center (ARC-X) is an interactive resource that helps local governments effectively deliver services to their communities despite climate changes. Decision-makers can create an integrated package of information tailored specifically to their needs. The ARC-X focuses on helping communities adapt to the changes documented in the U.S. National Climate Assessment.

6.11.2 Center for Corporate Climate Leadership

EPA's Center for Corporate Climate Leadership serves as a resource center to help organizations of all sizes measure and manage greenhouse gas (GHG) emissions. Organizations can use the resource center's technical tools, ground-tested guidance, and educational resources to inventory GHG emissions, set and achieve GHG emission reduction targets, understand climate-related financial risks and opportunities, and implement climate transition plans. The resource center was launched in 2012, and its resources have evolved since then to align with best practices around measuring and managing corporate GHG emissions. The resource center offers tools, guidance, and information that support organizations' efforts toward addressing the four GHG elements that typically make up climate-related disclosures:

1. Developing a GHG inventory to account for an organization's scope 1, scope 2, and scope 3 emissions.
2. Setting and achieving a GHG reduction target.
3. Understanding and disclosing climate-related financial risks and opportunities.
4. Developing a climate transition plan.

6.11.3 Green Vehicle Guide

EPA developed the Green Vehicle Guide to help you find information on vehicles that are more efficient and less polluting. Vehicles that operate primarily on gasoline or diesel have historically accounted for over 99% of cars and passenger truck sales. However, sales of cars that operate on other fuels—particularly electricity—are growing. New models of both electric vehicles and plug-in hybrid electric vehicles are entering the market in increasing numbers each year. Other alternative fuel vehicles include those that run on compressed natural gas (CNG) or E85 (a mixture of about 85% ethanol and 15% gasoline).

Many alternative fuels "burn" cleaner than gasoline or diesel, so there are fewer tailpipe emissions. The amount of greenhouse gases (GHG) emitted when the fuel is produced depends on the source of the fuel—see GHG Emissions for more information. These fuels can also be produced domestically, reducing our dependence on imported petroleum. In addition, reducing vehicle emissions and increasing your fuel economy helps the environment and your wallet.

6.11.4 Smart Growth

EPA works with local, state, and national agencies; nonprofits and community organizations; academia; and the development sector to encourage development strategies that support thriving local economies, provide attractive and affordable neighborhoods for people of all income levels, and mitigate and adapt to climate change. EPA's smart growth efforts help create healthier and more environmentally sustainable communities.

In the United States, development decisions are predominantly under state and local jurisdictions, and policies and regulations vary by community and state. The federal government generally does not directly regulate development, although many federal policies, particularly those related to the environment, transportation, and housing, do affect how communities develop. The federal government can also help municipalities, states, and the development sector better understand the impacts of development patterns.

6.11.5 America Recycles: Circular economy

A circular economy keeps materials and products in circulation for as long as possible. Circularity is embraced within the sustainable materials management (SMM) approach that EPA and other federal agencies have pursued since 2009. A circular economy approach under the SMM umbrella demonstrates continuity in our emphasis on reducing negative lifecycle impacts of materials, including climate impacts, reducing the use of harmful materials, and decoupling material use from economic growth and meeting society's needs. EPA has a broad vision to help the nation address the full impacts of materials on our communities and has set out a transformative vision for our waste management system—one that is inclusive, more equitable, and reflects the urgency of the climate crisis—by releasing a series of strategies that will be dedicated to building a circular economy for all.

6.11.6 Sector Snapshots

The sector snapshots are interactive tools that provide environmental and economic information about industries participating in EPA's Smart Sectors program. The sector snapshots employ a novel approach by assembling a range of environmental and economic data from different publicly available sources to provide an integrated, easy-to-understand overview of each sector over the last 25 years.

6.11.7 Environmentally Preferable Purchasing

The Environmentally Preferable Purchasing (EPP) program harnesses the power of the over 690-billion-dollar federal pocketbook to catalyze a more sustainable marketplace for all—reducing climate impacts, improving the health of frontline communities, preventing pollution, and increasing U.S. industry competitiveness. The program helps U.S. federal government purchasers utilize private sector standards and ecolabels to identify and procure environmentally preferable products and services, providing a convenient and streamlined way to make sense of the often complex sustainable products marketplace.

6.11.8 Comprehensive Procurement Guideline Program

The Comprehensive Procurement Guideline (CPG) program is part of EPA's SMM initiative that promotes a systems approach to reducing materials use, associated greenhouse gas emissions that contribute to climate change, and other environmental impacts over the materials' life cycle. CPG continues to promote using materials recovered from the MSW stream. Buying products made with recovered materials ensures that the materials collected in recycling programs will be used again to manufacture new products.

6.12 CLIMATE CHANGE RESOURCES FOR EDUCATORS AND STUDENTS

The EPA's Climate Change Resources for Educators and Students page offers a comprehensive collection of federal and non-governmental resources to support teaching and learning about climate change. It includes:

EPA Resources

- **Climate Change Indicators**: Data-rich information on environmental changes.
- **Climate Change Science**: Basics and causes of climate change.
- **Climate Change Impacts**: Effects on communities, ecosystems, and industries.
- **What You Can Do about Climate Change**: Strategies for individual action.
- **Generate! Board Game**: An interactive game exploring energy choices.
- **ENERGY STAR Kids**: Resources on energy efficiency.

Federal Resources for Educators

- **NOAA**: Educational content on climate and earth sciences.
- **National Park Service**: Lesson plans on climate change impacts on parks.
- **Smithsonian National Museum of Natural History**: Resources on global climate change.
- **U.S. Department of Agriculture Forest Service**: Information on forests and climate change.
- **U.S. Department of Energy**: Materials on energy efficiency and renewable energy.

Federal Resources for Students

- **NASA Climate Kids**: Interactive content on climate change for younger students.
- **U.S. Energy Information Administration Energy Kids**: Information on energy sources and conservation.
- **NASA Eyes on the Earth and SciJinks**: Tools and games for middle to high school students.

Other Educational Resources

- **The "Ology" site hosted by American Museum of Natural History**: Games and activities about climate change.
- **MIT Environmental Solutions Initiative**: Flexible activities and podcasts for high school teachers.
- **National Geographic**: Lesson plans and activities on earth science and conservation.

6.13 INVESTING IN CLIMATE CHANGE FUTURE

6.13.1 Federal Sustainability Plan

President Joe Biden's administration released the "Federal Sustainability Plan" in December of 2021, which outlined an ambitious path to achieve a net-zero emissions building goal by 2045 (The White House, 2021). A plan is set to use past data to drive annual water and energy reduction goals for buildings. These goals are based on leading performance benchmarks for different building-type categories. The plan's immediate-term priorities involve improving energy efficiency and eliminating onsite fossil fuel use. Improving energy efficiency in buildings has been shown to lead to massive cost savings and is great for business. For example, between 2008 and 2015, the General Services Administration (GSA) saved $340 million in energy and water costs from building efficiency improvements implemented since 2003. As such, another action of the Federal Sustainability

Plan involves the development of the first-ever Federal Building Performance Standard, which requires agencies to cut energy use and electrify equipment and appliances. This requirement is set to be implemented in 30% of the building spaces owned by the federal government by 2030. The Federal Building Performance Standard is only specifically designed to be implemented in federally owned buildings; however, the general standards that will be created can be used as references for any building and renovation projects in the future.

6.13.2 Bipartisan Infrastructure Law

President Biden's Bipartisan Infrastructure Law, enacted in 2021, marks a historic investment in the United States' infrastructure, climate resilience, and environmental protection. The law allocates over $60 billion to the EPA over five years, targeting various environmental initiatives. This unprecedented funding supports the modernization of water infrastructure, extensive environmental cleanups, and air quality enhancement, with a strong emphasis on advancing environmental justice and combating climate change.

Key accomplishments under this law include awarding more than $11 billion to state and local governments for water infrastructure projects and significant allocations for drinking and clean water projects. A top priority is replacing lead service lines, with $6.5 billion already designated for this effort. Additionally, the Clean School Bus Program has funded nearly $1 billion for electrified and low-emission school buses, improving air quality around schools. The law also addresses legacy pollution by providing funds for cleaning up 152 Superfund sites and making significant investments in brownfield sites. More than $410 million was allocated for various cleanup and redevelopment programs. Another notable achievement is the investment in solid waste management and recycling, with more than $73 million granted to 25 communities and $32 million provided to states for improving waste management practices.

6.13.3 Inflation Reduction Act Community Change Grants Program

EPA's new Environmental and Climate Justice Community Change Grants program (Community Change Grants) has announced a Notice of Funding Opportunity (NOFO) for approximately 2 billion dollars in Inflation Reduction Act (IRA) funds for environmental and climate justice activities to benefit disadvantaged communities through projects that reduce pollution, increase community climate resilience, and build community capacity to address environmental and climate justice challenges. These place-based investments will be focused on community-driven initiatives that are responsive to community and stakeholder input. They are designed to deliver on the transformative potential of the IRA for communities most adversely and disproportionately impacted by climate change, legacy pollution, and historical disinvestments.

The EPA is committed to ensuring equitable climate solutions, particularly through initiatives like the IRA, Community Change Grants Program, and the Bipartisan Infrastructure Law. These efforts empower disadvantaged communities by addressing historical inequities and building climate resilience. The Justice40 Initiative, a key component of these efforts, aims to direct 40% of federal investments in clean energy and climate to disadvantaged communities, such as those suffering from structural disinvestment and marginalization. However, challenges remain in integrating equity into infrastructure planning and design, as frameworks cannot guide equitable decision-making in these areas. The proposed three-tiered framework for equitable infrastructure emphasizes restorative justice and community engagement to address and mitigate risk disparities. Furthermore, advancing health equity in community-based climate actions involves building community agency and power, although many projects still fail to transform community power relationships. Addressing environmental racism is also crucial, as historical policies have disproportionately affected marginalized communities' health and welfare. By incorporating these insights, the EPA's initiatives can better ensure that climate solutions are effective, just, and inclusive, ultimately fostering a more equitable distribution of resources and benefits across all communities.

6.14 CONCLUSION

The U.S. Environmental Protection Agency (EPA) stands at the center of America's climate response, using a blend of regulation, voluntary programs, partnerships, and education to drive down greenhouse gas emissions across the economy. This chapter highlights how EPA initiatives—such as ENERGY STAR, AgSTAR, the Green Power Partnership, and LEED—are reducing emissions from buildings, energy, agriculture, transportation, and waste. The agency's annual greenhouse gas inventory, along with tools like the Greenhouse Gas Equivalencies Calculator, makes emissions data transparent and actionable for policymakers and the public alike. The EPA's focus on environmental justice ensures that climate action benefits all communities, particularly those most vulnerable to pollution and climate-related impacts.

The EPA's approach is comprehensive: it sets standards, incentivizes innovation, supports state and local action, and fosters collaboration across sectors. Programs targeting methane, high-potency gases, and circular economy practices show how technical solutions and stakeholder engagement can deliver real climate results. Federal investments—through laws and grants—are scaling up clean infrastructure, resilience, and community-led solutions nationwide.

As the climate challenge intensifies, the EPA's evolving toolkit remains crucial for achieving national and global climate objectives. However, success depends on accurate reporting, ongoing innovation, and equitable access to the benefits of climate action.

The next chapter shifts the lens from U.S. policy and programs to the global stage. Chapter 7 examines the carbon hierarchy—ranking countries by their emissions, analyzing the drivers behind national and per capita footprints, and assessing the challenges of balancing economic growth with sustainability. By comparing emissions across nations and sectors, the discussion moves from local action to global accountability, setting the stage for coordinated solutions to the world's most pressing climate problems.

REVIEW QUESTIONS

1. What are the main objectives of the U.S. Environmental Protection Agency (EPA) in addressing climate change?
2. How does the ENERGY STAR program contribute to emissions reduction in homes and businesses?
3. What is the purpose of the EPA's Climate Change Adaptation Resource Center (ARC-X)?
4. How does LEED certification encourage energy-efficient building practices?
5. What are the roles of voluntary programs like the GPP in climate action?
6. How does the EPA collaborate with states, tribes, and local governments on climate initiatives?
7. What are methane reduction programs, and how do they target high-emission sectors?
8. How does the EPA's SmartWay Transport Partnership help reduce emissions in transportation?
9. What role does the RAD Program play in reducing waste-related emissions?
10. How does the IRA influence the EPA's climate change goals?

REFERENCES

Al-Zubaidy, M. S. K. (2015). A literature evaluation of the energy efficiency of Leadership in Energy and Environmental Design (LEED)-certified buildings. *American Journal of Civil Engineering and Architecture*, 3(1), 1–8. https://doi.org/10.12691/ajcea-3-1-1

Brown, L. A., & Ghag, J. (2023). Contaminated land, green energy, and the energy crisis: Should contaminated land sites be targeted and used to produce renewable energy in the United Kingdom? *Environmental Law Review*, 25(1), 23–39. https://doi.org/10.1177/14614529231162320

Brown, M. A. (2015). Innovative energy-efficiency policies: An international review. *Wiley Interdisciplinary Reviews: Energy and Environment*, 4(1), 1–20. https://doi.org/10.1002/wene.125

Burtraw, D., & Woerman, M. (2012). An assessment of U.S. progress towards its pledge on climate change mitigation (ECP Paper No. 12). European Climate Platform (ECP), Centre for European Policy Studies (CEPS) and Mistra's Indigo Programme. https://www.ceps.eu.

Coalition, C. C. A. (2021). *Global Methane Assessment: Benefits and Costs of Mitigating Methane Emissions*. United Nations Environment Programme.

EPA. (2014). *RE-Powering America's Land*. United States Environmental Protection Agency.

Hodges, T., & Potter, J. (2010). *Transportation's Role in Reducing U.S. Greenhouse Gas Emissions: Volume 1: Synthesis Report and Volume 2: Technical Report*. U.S. Department of Transportation.

Jensen, B. B. (2010). *Brownfields to Green Energy: Redeveloping Contaminated Lands with Large-Scale Renewable Energy Facilities*. Environmental Law Institute.

LEED Rating System. (n.d.). LEED Rating System | U.S. Green Building Council, https://www.usgbc.org/leed

Liao, Z., & Tam, C. (2021). A review of housing certification standards with a focus on energy efficiency. *Earth and Environmental Science*, 787(1), 1–10. https://doi.org/10.1088/1755-1315/787/1/012168

Lutsey, N., & Sperling, D. (2008). Transportation and greenhouse gas mitigation. (Working Paper No. qt6fz1z05g). Institute of Transportation Studies, University of California, Davis. https://escholarship.org/uc/item/6fz1z05g.

Melvin, A. M., Sarofim, M. C., & Crimmins, A. (2016). Climate benefits of U.S. EPA programs and policies that reduced methane emissions 1993–2013. *Environmental Science & Technology*, 50(9), 4523–4530. https://doi.org/10.1021/acs.est.6b00367

Mingshan, S. (2013). The research progress on greenhouse gas inventory methods. *Progress in Geography*, 32(5), 12–24.

Moriarty, K. (2013a). *Feasibility Study of Anaerobic Digestion of Food Waste in St. Bernard, Louisiana. A Study Prepared in Partnership with the Environmental Protection Agency for the RE-Powering America's Land Initiative: Siting Renewable Energy on Potentially Contaminated Land and Mine Sites.*

Moriarty, K. (2013b). *Feasibility Study of Biopower in East Helena, Montana. A Study Prepared in Partnership with the Environmental Protection Agency for the RE-Powering America's Land Initiative: Siting Renewable Energy on Potentially Contaminated Land and Mine Sites*. National Renewable Energy Laboratory (U.S.). https://digital.library.unt.edu/ark:/67531/metadc839818/m2/1/high_res_d/1064534.pdf

Morshed, A., Manjur, K. A., Shahjalal, M., & Yahia, A. K. M. (2024). Optimizing energy efficiency: A comprehensive analysis of building design parameters. *American Journal of Science and Technology Engineering Mathematics Education*, 4(4), 120–130.

Niblick, B., & Landis, A. E. (2016). Assessing renewable energy potential on United States marginal and contaminated sites. *Renewable & Sustainable Energy Reviews*, 52, 19–30. https://doi.org/10.1016/j.rser.2015.12.045

Priyadarshini, P., & Abhilash, P. C. (2020). Fostering sustainable land restoration through circular economy-governed transitions. *Restoration Ecology*, 28(4), 703–712. https://doi.org/10.1111/rec.13181

Salasovich, J., Geiger, J., Mosey, G., & Healey, V. (2013). *Feasibility Study of Economics and Performance of Solar Photovoltaics at the Kolthoff Landfill in Cleveland, Ohio. A Study Prepared in Partnership with the Environmental Protection Agency for the RE-Powering America's Land Initiative: Siting Renewable Energy on Potentially Contaminated Land and Mine Sites.*

Thakur, S., & Solanki, H. (2022). Role of methane in climate change and options for mitigation—a brief review. *International Journal of Applied Science and Environment*, 1(2), 80–90. https://doi.org/10.56588/iabcd.v1i2.80

The White House. (2021). *Federal Sustainability Plan Catalyzing America's Clean Energy Industries and Jobs.*

U.S. Environmental Protection Agency, Washington, DC. U.S. EPA (2023) #430-R-24-004 https://www.epa.gov/energy/greenhouse-gas-equivalencies-calculator-calculations-and-references

7 The Carbon Hierarchy
Countries from Highest to Lowest Emissions

This chapter delves into the complexities of carbon emissions, examining the hierarchy of countries from highest to lowest emitters. It explores the sources of carbon emissions, their environmental impact, and the urgency of addressing them to mitigate climate change. This chapter discusses the concept of total carbon emissions, per capita emissions, and lifecycle emissions, providing a comprehensive understanding of carbon footprints. It also highlights the importance of accurate data on carbon emissions for tracking progress and informing policy decisions. This chapter then focuses on the top emitters, analyzing their emission trends, sectoral contributions, and the drivers behind their emissions. It also examines the per capita emissions of these countries, highlighting the inequalities in carbon footprints across the globe. Additionally, this chapter explores the challenges developing nations face with lower total emissions but high growth rates in balancing economic growth with environmental sustainability. It also discusses the role of cities in carbon emissions and the efforts of initiatives like Carbon Disclosure Project (CDP) to promote environmental disclosure and action. Finally, this chapter presents a case study of the *U.S. National Climate Assessment*, detailing the climate change impacts, risks, and responses in the United States.

The primary sources of carbon emissions are burning fossil fuels (coal, oil, and natural gas) for energy and transportation, deforestation, and various industrial processes. As the global economy has grown, so too have CO_2 emissions, contributing to a rise in average global temperatures. This increase in temperature has led to a range of adverse effects, including more frequent and severe weather events, rising sea levels, and disruptions to ecosystems and agriculture.

Addressing carbon emissions is urgent because the window of opportunity to prevent the most catastrophic impacts of climate change is closing rapidly. The Intergovernmental Panel on Climate Change (IPCC) has warned that significant reductions in CO_2 emissions are necessary by mid-century to keep global warming below 1.5°C compared to pre-industrial levels, a threshold beyond which the risks of severe climate impacts significantly increase. Therefore, understanding and mitigating carbon emissions is critical to global efforts to combat climate change and ensure a sustainable future for all.

7.1 CARBON EMISSIONS METRICS

Carbon emissions are measured through various metrics to provide a comprehensive understanding of their impact, with total emissions, per capita emissions, and lifecycle emissions being key indicators. Total emissions quantify the aggregate greenhouse gases emitted by a country or entity, serving as a primary indicator of global contributions to climate change (Muthu, 2015). Total carbon emissions are the aggregate amount of CO_2 emitted by a country or entity over a specific period, usually measured in metric tons. This measure is crucial for understanding the overall impact on the atmosphere and for comparing the contributions of different countries to global emissions. Per capita emissions, on the other hand, reflect individual carbon footprints and highlight disparities between high-emission developed nations and lower-emission developing ones. For instance, national average per capita footprints can vary significantly, from 1 tCO_2e/y in African countries to approximately 30 tCO_2e/y in Luxembourg and the United States, illustrating the uneven distribution

DOI: 10.1201/9781003479666-7

of emissions across the globe (Hertwich & Peters, 2009). Per capita emissions, therefore, provide insight into energy use efficiency and equity and highlight emission level disparities between developed and developing countries.

A comprehensive understanding of carbon footprints also involves examining the lifecycle emissions of products and services, from production to disposal. Lifecycle emissions provide a comprehensive view of the hidden environmental costs of products and services, encompassing all stages from raw material extraction to disposal (Lützkendorf, 2020). These metrics are crucial for supporting built-environment professionals in quantifying, assessing, and controlling GHG emissions across different scales, such as construction products, buildings, and cities (Lützkendorf, 2020). As an example, the lifecycle of smartphones, encompassing stages from mining and manufacturing to transportation and end-of-life treatment, significantly contributes to greenhouse gas emissions, which are crucial metrics for policymakers aiming to prioritize interventions for climate action. Life cycle assessments (LCA) of smartphones, such as those conducted on the Sony Mobile Z5 and Apple iPhone, reveal that integrated circuit production and display manufacturing are major contributors to the overall environmental impact, with the global warming potential (GWP) for a smartphone like the Sony Mobile Z5 calculated at 57 kg CO_2e over a 3-year lifespan, excluding network usage (Ercan et al., 2016; Arndt & Ewe, 2017). These assessments are vital for identifying key stages in the product lifecycle that contribute most to emissions, thereby guiding strategic interventions to reduce these impacts (Reijnders, 2015). Moreover, understanding these differences is essential for equitable climate action, as they highlight the disproportionate environmental burdens borne by certain countries or populations, often due to the global distribution of manufacturing and resource extraction activities (Anenberg et al., 2022). By focusing on these metrics, policymakers can develop targeted strategies that not only aim to reduce emissions but also address global inequalities, ensuring that climate action is both effective and just (Cordella et al., 2021; Anenberg et al., 2022). This comprehensive approach, supported by robust data and LCA methodologies, identifies major reduction options and supports the development of policies that mitigate climate change while considering environmental justice (Cordella et al., 2021; Anenberg et al., 2022). Additionally, emission metrics like the GWP and forcing equivalent index (FEI) are used to compare the climate impacts of different greenhouse gases, considering their radiative efficiencies and lifetimes.

Reducing carbon footprints requires concerted efforts at multiple levels, including individual behavior changes, corporate responsibility, and robust policy frameworks. By measuring and understanding carbon footprints, we can prioritize actions that offer the greatest potential for emission reductions and move toward a more sustainable and low-carbon future.

7.2 DATA SOURCES FOR EMISSIONS ESTIMATIONS

Reliable data sources are essential for tracking progress and implementing effective climate policies, as they provide the necessary information to identify trends and allocate resources efficiently. The Emissions Database for Global Atmospheric Research (EDGAR) is a key source, offering comprehensive datasets on global, regional, and national greenhouse gas emissions by sector from 1970 to 2019. This dataset includes CO_2 emissions from fossil fuel combustion and industry, as well as CH_4, N_2O, and fluorinated gases, and is complemented by data on net CO_2 emissions from land use, land-use change, and forestry (LULUCF) (Minx et al., 2022). The Climate Actions and Policies Measurement Framework (CAPMF) further supports climate policy assessment by providing a harmonized database of 130 policy variables across 50 countries and the EU-27, covering the period from 1990 to 2022. This framework highlights the relationship between stronger climate actions and significant emission reductions, particularly through decreased energy intensity.

The United Nations Framework Convention on Climate Change (UNFCCC) also requires countries to submit standardized greenhouse gas inventories, fostering transparency and accountability. These submissions, updated as of December 2022, are crucial for maintaining a consistent and reliable record of emissions data (Pflüger & Gütschow, 2021, 2022). These data sources form a robust foundation for monitoring and guiding global climate efforts.

These data repositories reflect production-based emissions and highlight consumption-based emissions embedded in trade. Developed nations often outsource carbon-intensive industries to developing countries, shifting their carbon burden. Addressing this requires global cooperation to track and reduce emissions along supply chains. Data-driven insights can enhance climate negotiations by exposing disparities and driving consensus on shared responsibilities. Robust data collection and reporting mechanisms are thus critical for informed decision-making at both national and international levels. The issue of carbon emissions embedded in international trade highlights the complex dynamics between production-based and consumption-based emissions, particularly as developed nations outsource carbon-intensive industries to developing countries, effectively shifting their carbon burden. This phenomenon is evident in the historical increase of emission transfers from developing to developed countries, which peaked around 2006 (Grubb et al., 2022). Consumption-based accounting (CBA) has emerged as a crucial tool to address these disparities by attributing emissions to the final consumer rather than the producer, thus revealing the hidden carbon footprint of imported goods (Fezzigna et al., 2019).

The European Union, for instance, is a net importer of emissions, with a significant portion of its consumption-based emissions originating from countries like China and Russia (Fezzigna et al., 2019). This shift in responsibility underscores the need for global cooperation to track and reduce emissions along supply chains, as traditional production-based measures often underestimate the true environmental impact of consumption in developed countries (Haga & Zhou, 2021). Data-driven insights, such as those derived from the OECD Inter-Country Input-Output Database, are essential for understanding these divergences and informing climate negotiations (Yamano & Guilhoto, 2020). Moreover, while consumption-based policies could potentially reduce carbon leakage and promote cleaner technologies, they must be carefully evaluated for their distributional effects to ensure fairness and effectiveness in global climate policy (Lininger, 2015). Robust data collection and reporting mechanisms are thus critical for informed decision-making at both national and international levels, facilitating a more equitable distribution of emission responsibilities (Grubb et al., 2022; Lininger, 2015).

The World Bank compiles and publishes data on greenhouse gas emissions as part of its broader efforts to support sustainable development. This data includes country-level emissions and sectoral breakdowns, offering valuable insights into the sources and trends of emissions across different regions. The EDGAR database, maintained by the European Commission, provides detailed emissions data for various pollutants, including CO_2. Researchers and policymakers use EDGAR's comprehensive datasets to analyze emissions patterns and develop mitigation strategies.

National greenhouse gas inventories, submitted annually by countries to the UNFCCC, are another critical source of emissions data. These inventories follow standardized methodologies for estimating and reporting emissions, ensuring consistency and comparability across countries. They provide detailed information on emissions from various sectors, including energy, industrial processes, agriculture, and waste management.

Furthermore, there is the issue of "hidden" emissions embedded in trade. Many developed countries import goods from developing nations, effectively outsourcing a portion of their emissions. This means that consumption-based emissions, which account for the carbon footprint of goods and services consumed within a country regardless of where they were produced, can be much higher than production-based emissions reported in national inventories. Addressing this discrepancy requires a more holistic approach to emissions accounting that considers the global supply chain.

7.3 THE TOP EMITTERS BY TOTAL CARBON EMISSIONS

In 2022, the top six GHG emitters were China, the United States, India, the EU27, Russia, and Brazil, collectively accounting for 61.6% of global emissions. Among these nations, China, the United States, and India saw increases in emissions compared to 2021, with India experiencing the largest relative increase at 5%. Conversely, the EU27 reduced emissions by 0.8% from 2021, contributing 6.7% to global emissions.

Examining countries with over 1% of global GHG emissions, Australia uniquely managed to reduce its emissions in both 2021 (by 1.9%) and 2022 (by 0.3%) relative to 2020. Although Australia's emission intensity has steadily decreased over the past decade, its per capita emissions remain higher than those of the US or Russia. On the other hand, Indonesia exhibited the most significant increase, with a 10% rise in GHG emissions from 2021 to 2022. In the United States, GHG emissions rose by 1.6% in 2022 compared to 2021, reaching approximately 6.0 Gt CO_2eq.

Over the past three decades, the EU27 has generally reduced its GHG emissions, achieving a 27% reduction from 1990 by 2022, equating to 3.6 Gt CO_2eq. India's emissions grew by 5.0% (or 0.19 Gt CO_2eq) in 2022 compared to 2021, marking a 5.7% increase over pre-pandemic 2019 levels. India's emissions have nearly tripled since 1990, reflecting a steady upward trend over the past 30 years. Meanwhile, Russia's GHG emissions decreased by 1% in 2022 compared to 2021 but showed a 2% increase compared to 2019 levels. Figures 7.1 and 7.2 illustrate these statistics.

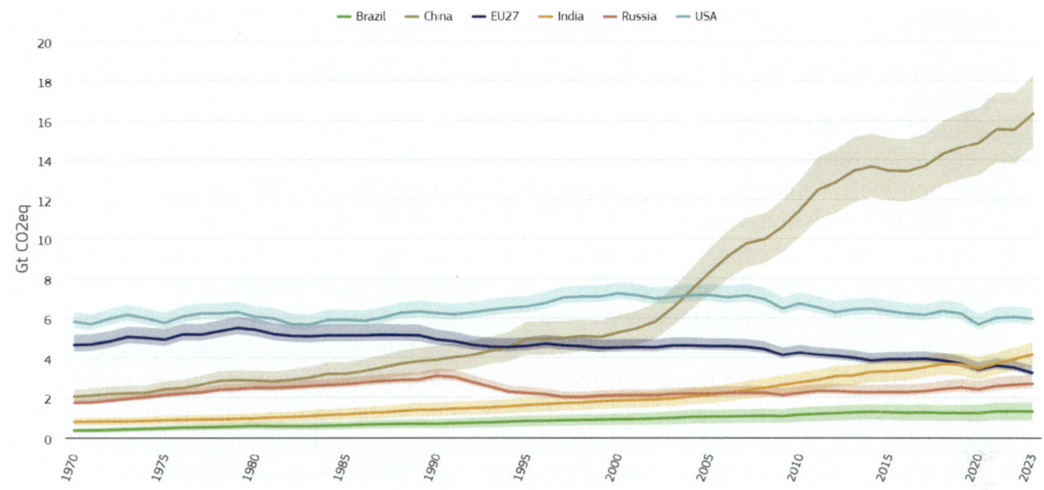

FIGURE 7.1 GHG emission in top emitting economies 1970–2022.

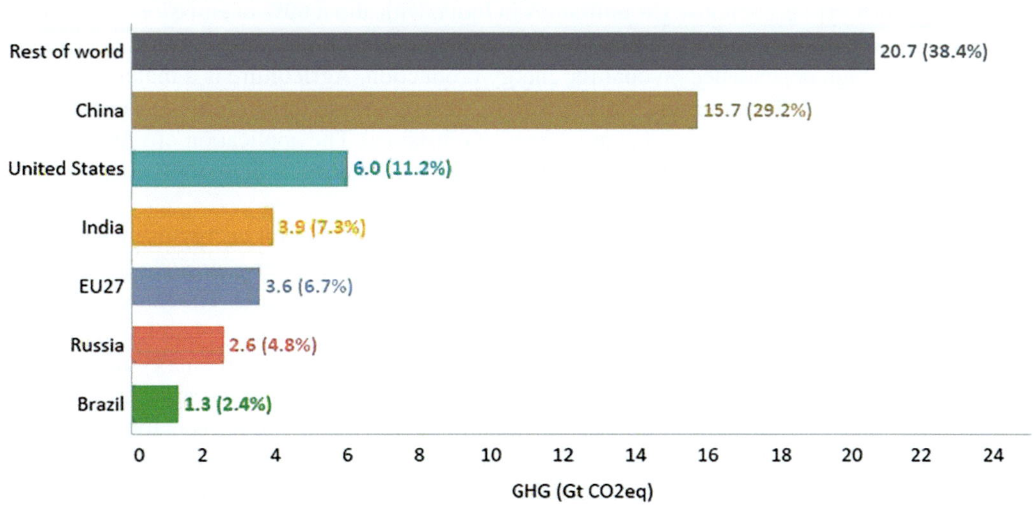

FIGURE 7.2 GHG emission contribution 2022.

	Share in global	Change 2019-2020	Change 2020-2021	Change 2021-2022	Change 2019-2022	Change 2020-2022	CAGR (1990-2022)
China	29.2%	1.9%	5.1%	0.3%	7.4%	5.4%	4.3%
United States	11.2%	-8.7%	5.5%	1.6%	-2.2%	7.2%	-0.1%
India	7.3%	-5.7%	6.7%	5.0%	5.7%	12.1%	3.2%
EU27	6.7%	-7.7%	5.6%	-0.8%	-3.4%	4.7%	-1.0%
Russia	4.8%	-3.9%	7.2%	-1.0%	2.0%	6.1%	-0.5%
Brazil	2.4%	-0.3%	5.1%	-2.4%	2.3%	2.6%	2.0%
Indonesia	2.3%	-4.9%	2.1%	10.0%	6.8%	12.3%	3.4%
Japan	2.2%	-5.3%	1.2%	0.6%	-3.6%	1.8%	-0.3%
Iran	1.8%	-1.6%	3.9%	1.6%	3.9%	5.6%	3.3%
Mexico	1.5%	-6.5%	3.5%	7.1%	3.7%	10.9%	1.8%
Saudi Arabia	1.5%	-0.8%	3.3%	3.9%	6.4%	7.3%	3.9%
Canada	1.4%	-8.2%	3.0%	3.2%	-2.4%	6.4%	0.8%
South Korea	1.3%	-4.3%	4.5%	-0.7%	-0.8%	3.7%	2.5%
Türkiye	1.3%	3.5%	8.5%	3.1%	15.8%	11.9%	3.5%
Australia	1.1%	-3.9%	-2.0%	1.7%	-4.1%	-0.3%	0.7%
South Africa	0.99%	-9.8%	-0.5%	-2.5%	-12.5%	-3.1%	0.8%
Global		-3.7%	4.8%	1.4%	2.3%	6.2%	1.5%
International Aviation	0.8%	-52.3%	15.4%	23.3%	-32.1%	42.3%	1.5%
International Shipping	1.4%	-8.5%	5.7%	5.7%	2.2%	11.7%	2.0%

FIGURE 7.3 Share of global emission.

In 2022, greenhouse gas (GHG) emissions were predominantly composed of CO_2, primarily due to fossil fuel combustion, accounting for 71.6% of the total emissions. Methane (CH_4) contributed 21%, while nitrous oxide (N_2O) and fluorinated gases (F-gases) made up 4.8% and 2.6%, respectively. Figure 7.3 illustrates the emission trends since 1990, showcasing data for both the global scale and the top emitters. Globally, fossil CO_2 emissions have surged by over 70% since 1990. During the same period, CH_4 emissions increased by 32.4%, N_2O by 36.5%, and F-gases saw a fourfold rise.

In the EU27, there has been a consistent decline in emissions of CO_2, CH_4, and N_2O, with CH_4 showing the most significant reduction. The USA also displays a downward trend for these three gases, with the reduction in N_2O emissions paralleling that of CO_2. Conversely, in China and India, fossil CO_2 emissions have skyrocketed—a factor of 5.0 in China and 4.0 in India—while CH_4 emissions increased by factors of 1.6 and 1.3, respectively, and N_2O emissions by factors of 1.4 and 2. Following a sharp decline from 1990 to 1998 in Russia due to the USSR's collapse and the ensuing economic crisis, CO_2, CH_4, and N_2O emissions have risen, with CH_4 increasing faster than CO_2 and N_2O. Brazil has seen a general increase in CO_2, CH_4, and N_2O emissions since 1990. Figure 7.4 provides data for the six (6) emitting world economies.

India's greenhouse gas emissions are largely driven by economic growth and rising energy use, with the energy sector being a significant contributor. The electricity generation sector accounts for a substantial portion of greenhouse gas emissions in India, with about 69% of emissions attributed to this sector. Coal combustion is a major source of carbon dioxide emissions in India, contributing to the country's high dependence on coal for energy production. Agriculture is a major contributor to greenhouse gas emissions in India, with practices such as enteric fermentation, rice cultivation, and crop residue burning being significant sources of emissions. The application of nitrogen-based fertilizers and manures in agriculture leads to the release of nitrous oxide emissions, contributing to the overall greenhouse gas emissions from the sector.

Energy use is the major source of GHG emissions in Brazil, with total energy use experiencing an annual average growth rate close to 3%. In 2015, the industries accounting for 37% of total energy use in Brazil were transport, food & beverages, and electricity, gas, and water. The share of domestic use of energy produced abroad was higher than 20% in textiles and wearing apparel, electrical and machinery, transport equipment, and construction, extending the problem of energy-related emissions mitigation from the national to the global level. The agricultural and livestock sector caused approximately 24% of total GHG emissions in Brazil in 2017, with enteric fermentation of animals and the use of agricultural soil being the main causes of emissions, accounting for almost 86% of emissions within agricultural activities. Additionally, transportation is one of the largest sectors of society in CO_2 emissions, responsible for 8.6% of GHG emissions in Brazil, where heavy-duty vehicles are responsible for emitting 49.9% of CO_2 in the transportation sector.

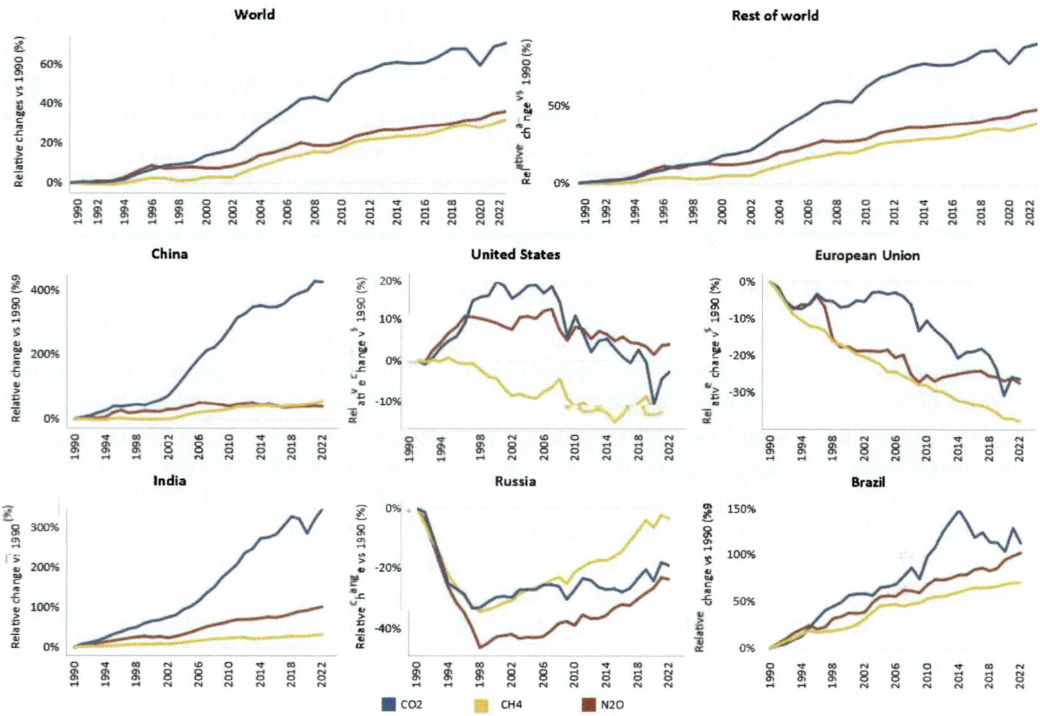

FIGURE 7.4 GHG emission trends of top six emitting economies.

In China, the agricultural sector contributes 39% of methane and 1% of carbon dioxide emissions globally, with rice production being a significant factor. The chemical industry is also identified as a major contributor to GHG emissions, with crude oil and coal being the main sources of direct emissions. Coal mining, metal melting, electric and heat power, and petroleum refining are central causes of carbon emissions, with the energy structure in China being coal-dominated. The aluminum production sector is power-intensive, with power generation significantly impacting overall GHG emissions.

The main sources of GHG emissions in the USA are primarily from the energy and transportation sectors. Oil and natural gas production contributes to about 40% of US greenhouse gas emissions, with the production segment being the dominant contributor to methane emissions. The petrochemical industry contributes about 4.5% of US greenhouse gas emissions, with refining accounting for 88% of emissions from petroleum products. The shipping sector contributed 2.89% to global anthropogenic emissions in 2018, indicating a rising share in greenhouse gas emissions. The industrial sector is responsible for about 35% of global energy use, with the chemicals, iron and steel, cement, and oil refineries sectors being the biggest energy consumers.

One of the top GHG-emitting industries in the European Union is the metals industry, particularly in countries like Germany, Poland, Austria, and the Czech Republic, which has been identified as a significant contributor to GHG emissions, responsible for half of all GHG emissions in the industry. Carbon-intensive materials industries such as iron and steel, cement, and aluminum are highlighted as posing challenges to the EU's commitment to significantly cut emissions by 2030. Industrial sector installations included in the European Emission Trading System (EU ETS) emitted a substantial amount of GHG emissions, with the production of clinker, lime, ammonia, blast furnace operations, and refineries being notable contributors. Energy-intensive industries located in Central and Eastern European countries, both inside and outside the EU, face significant energy intensity and associated GHG emissions, mainly from energy production, particularly from coal-fired power stations.

The main anthropogenic sources of greenhouse gases in Russia include the combustion of fossil fuels in boilers and furnaces, industrial activities, energy consumption, and the oil and gas sector. The energy sector, particularly the combustion of coal, is a major contributor to GHG emissions in Russia. Russia's extensive economic development pattern, heavily reliant on natural resources and fossil fuels, leads to large quantities of GHG emissions. The oil and gas sector, while not the main supplier of greenhouse gases, plays a significant role in the extraction and use of carbon fuels, contributing to emissions.

7.4 THE TOP EMITTERS BY PER CAPITA EMISSIONS

Per capita emissions represent the emissions of an average person in a country or region—they are calculated as the total emissions divided by population. There are large inequalities in the carbon footprint of people across the world. Research has shown mid-tier countries exhibit higher emissions per unit of GDP than low- and high–per capita income countries, with total emissions of mid-tier countries continuing to grow. Southeast Asia and the Pacific have relatively high emission levels, while Latin America, the Caribbean, South Asia, and sub-Saharan Africa have emissions below the global average. Previous studies found an inverse correlation between urban population density and per capita emissions from land transport, but a positive relationship was found in a dataset of over 200 cities from 28 countries, highlighting the distinctiveness of North American cities. China's carbon emission per capita is steadily increasing across the country, with varying effects of socioeconomic factors on carbon emissions at the provincial level.

In 2022, the EU27's greenhouse gas (GHG) emissions were 8.09 metric tons of CO_2 equivalent (CO_2eq) per capita, reflecting a 0.8% decline from the 2021 figure of 8.15 metric tons CO_2eq per capita (Figure 7.5). Additionally, GHG emissions per unit of GDP in terms of purchasing power parity (PPP) decreased by 4.3% from 2021, reaching 0.175 metric tons CO_2eq per thousand USD in 2022. The USA per capita GHG emissions in 2022, recorded at 17.90 metric tons CO_2eq, were 4.3% lower than the pre-pandemic levels in 2019, which stood at 18.70 metric tons CO_2eq per capita. This figure is comparable to Russia's emissions, which are significantly higher than those of other major emitters. Furthermore, the per capita emissions for 2022, at 17.98 metric tons CO_2eq, were on par with those of the United States but notably higher than those of China by 64% and the EU27 by 122%.

India's per capita carbon dioxide emissions are relatively low compared to its aggregate emissions, which can be attributed to societal organization, lifestyle, and poverty, as represented in Figure 7.5. The rural-urban divide, spatial divide in energy-based emissions, poverty, lifestyle,

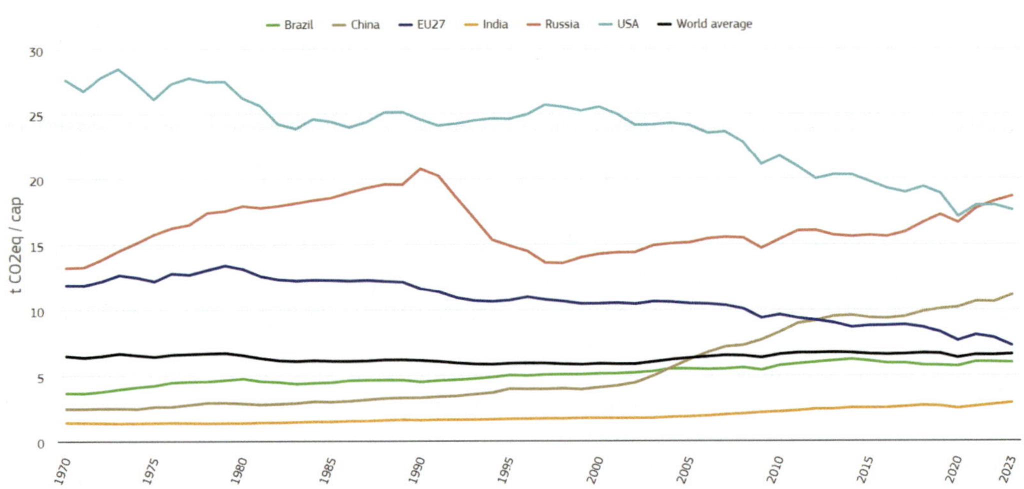

FIGURE 7.5 GHG emission per capita.

societal organization, and dependence on biomass contribute to India's lower per capita emissions. Projections indicate that India's per capita emissions may increase by 40% by 2030 due to continued economic growth and rising energy use. India's consumption-based emissions increased significantly between 2000 and 2016, with per capita consumption having a greater impact than population growth. However, improvements in energy efficiency and the use of renewable energy have helped reduce domestic emissions in India.

Brazil's low per-capita energy-related greenhouse gas (GHG) emissions of 2.4 tons CO_2 in 2014 are attributed to abundant clean energy sources. The major sources of energy use contributing to emissions in Brazil include transport, food & beverages, and electricity, gas, and water industries. Brazil's unique emissions profile is driven by increasing individual standards of living, exports, and population size, unchallenged by technological and structural improvements toward lower emissions intensities and more efficient production structures. Wealthier areas in Brazil, known for diverse and value-added production, have lower emission-intensive industries, but economically prosperous regions drive emissions due to demand and supply factors.

The main drivers of national per capita emissions in the USA are attributed to population size, wealth, and consumption patterns. Household carbon emissions in the USA are mainly affected by income, with the highest income group having a per capita carbon footprint about 2.6 times that of the lowest income group. Previous research has shown that carbon dioxide emissions in the USA are primarily driven by the consumption of carbon-intensive products such as gasoline, electricity, and animal-based food products, where the top quintile income households were identified as the main contributors to emission increases before the peak, while the third- and fourth-income quintiles became emission mitigation leaders after 2010.

EU countries are converging on stable per capita emissions, indicating a long-term decline, not just a temporary dip due to economic factors. This progress is linked to effective climate policies, with some countries like Germany and France on track to meet their emission targets without needing even stricter measures. A study on the environmental consequences of economic growth in the EU found that the relationship between per capita income and emissions of NO_x and NH_3 falls into the Environmental Kuznets Curve pattern, with the growth of primary energy consumption increasing the emissions of both gases. Another analysis revealed an important inequality in the distribution of per capita CO_2 emissions across EU countries, with potential mitigation targets to decrease this inequality.

The dynamic of Russian greenhouse gas (GHG) emissions is mainly determined by changes in the volume and type of fossil fuel combustion. Population size and the level of economic development, particularly gross regional product (GRP) per capita, have a decisive influence on the increase in air emissions in Russian regions. Economic growth in Russia greatly relies on natural resources, especially fossil fuels, leading to large quantities of GHG emissions. Therefore, the relationship between GDP per capita and economy-related GHG emissions per capita in Russia follows an inverted U-shaped pattern, supporting the Environmental Kuznets Curve (EKC) hypothesis.

7.5 THE LOWER EMITTERS

Inequalities in average per-capita emissions between world regions remain large, as shown **at the** top of Figure 7.6. On top of these gaps, important inequalities in carbon footprints are observed within regions (i.e., carbon footprints of the bottom 50% of emitters, the middle 40%, and the top 10% of the population within regions according to my benchmark results). Emission levels and shares for other groups are presented in the Supplementary Information.

In East Asia, the poorest 50% emit on average 2.9 tCO_2e per annum, while the middle 40% emit nearly 8 t, and the top 10% almost 40 t. This contrasts sharply with North America, where the bottom 50% emit fewer than 10 t, the middle 40% around 22 t, and the top 10% around 69 tCO_2e. This, in turn, can be contrasted with the emissions in Europe, where the bottom 50% emit 5 t, the middle 40% around 10.5 tCO_2e, and the top 10% around 30 tCO_2e. Emission levels in South and

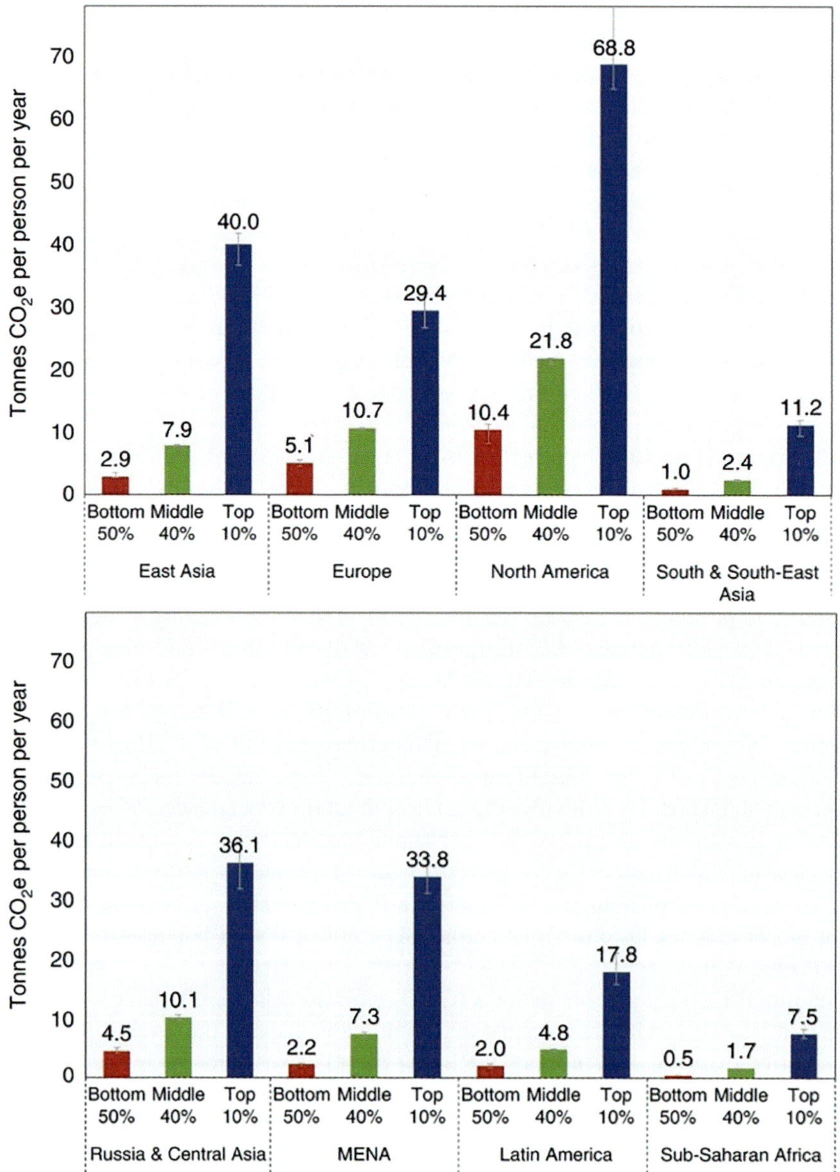

FIGURE 7.6 Per-capita emissions by group in 2019 (tCO_2e per capita).

Southeast Asia are notably lower than in the other regions, from around 1 tCO_2e for the bottom 50% to 11 t on average for the top 10%.

It is striking that the poorest half of the population in the United States has emission levels comparable with the European middle 40%, despite being almost twice as poor as this group in PPP terms[17]. Conversely, the top 10% of the population in East Asia emits notably more than its European counterpart (40 tCO_2e vs 29 tCO_2e, respectively). It also appears that Russia and Central Asia have an emissions distribution broadly similar to that of Europe, but with higher top 10% emissions (mainly due to higher income and wealth inequalities in Russia and Central Asia) and lower bottom 50% emissions. Sub-Saharan Africa lags behind, with the bottom 50% emitting around 0.5 t per capita per year, and the top 10% emitting around 7.5 t.

Developing nations with lower total emissions but high growth rates face the challenge of balancing economic growth with environmental sustainability. This is due to the curvilinear relationship between economic development and greenhouse gas emissions, where middle-income countries tend to have high emissions. Efforts to avoid the acceleration of global warming have historically focused on countries with high CO_2 emissions levels and large populations, often neglecting African countries, which are more vulnerable to climate change due to their limited infrastructure and technological capacity to cope with its impacts. The long-term economic growth rates projected for Africa's rapid development period will make Africa a potential emission hotspot in the near future.

Developing nations need to balance economic growth with environmental sustainability. It is crucial for these nations to disentangle economic growth from energy-related emissions to reduce emissions and thrive. However, the required energy-system transformation imposes considerable costs on developing nations, making it challenging to drastically reorient development paths toward low-carbon growth. Mitigation measures could meet further resistance due to adverse distributional impacts and political economy reasons. Therefore, efforts should focus on feasible mitigation actions such as fossil-fuel subsidy reform, decentralized modern energy, and fuel switching in the power sector.

The implications of high growth rates in developing nations for global emissions reduction efforts are significant. A study analyzing the trends and drivers of emissions in countries where emissions grew faster than the global average found that these "emerging emitters" would require faster and deeper reductions in emissions from other countries to meet international climate goals. The evolution of energy intensity and carbon intensity globally indicates a steady decline in these intensities for all groups of countries, with economic development being "decoupled" from total primary energy supply and CO_2 emissions. However, carbon emissions are growing faster than total primary energy supply in the world as a whole, posing challenges for global emissions reduction efforts.

Developing nations can adopt strategies to address the complex situation of lower total emissions and high growth rates. These strategies include scaling financial grant provisions for investing in technological innovation programs to reduce environmental challenges and mitigate emission growth rates. Additionally, under carbon reduction emission constraints, countries can focus on developing sectors with growth potential and lower carbon intensity, such as the services sector, agriculture, and the food manufacturing sector.

7.6 TOP URBAN EMITTERS

The global gridded model of carbon footprints (GGMCF) is fundamentally based on the EU's GHS-SMOD gridded population model. This model provides a globally consistent, spatially resolved (250 m) estimate of carbon footprints (also called Scope 3 emissions) per capita and in absolute terms across 189 countries. Findings highlight that a few large or affluent cities contribute a significant portion of the national total emissions in many countries. Consequently, focused actions by a small number of local mayors and governments can substantially reduce the national carbon footprint. Globally, carbon footprints are predominantly concentrated in a limited number of dense, high-income cities and wealthy suburbs. Specifically, 100 cities are responsible for 18% of global emissions (Jhariya et al., 2024; Gurney, 2022). In most countries (98 out of 187 assessed), the top three urban areas account for more than a quarter of national emissions. Figures 7.7 and 7.8 illustrate CO_2 and city populations.

Among the top 500 cities, 41 are located in countries with low total and per capita emissions, such as Dhaka, Cairo, and Lima. In these cities, a combination of high population density and affluence results in carbon footprints similar to those of the highest-income cities. For large and wealthy cities, the total Scope 3 footprint—indirect emissions associated with the city's consumption of goods and services—is significantly larger than their direct emissions.

Implementing radical decarbonization measures, such as limiting non-electric vehicles and requiring 100% renewable electricity, can lead to substantial emissions reductions beyond city

Urban Cluster	Country	Footprint/cap (t CO2)	Population	Footprint (Mt CO2)	Global ranking	Domestic ranking
Seoul	South Korea	13.0 ±2.4	21,254,000	276.1 ±51.8	1	1
Guangzhou	China	6.1 ±1.0	44,309,000	272.0 ±46.2	2	1
New York	USA	17.1 ±5.5	13,648,000	233.5 ±75.4	3	1
Hong Kong SAR	China	34.6 ±6.3	6,029,000	208.5 ±37.8	4	1
Los Angeles	USA	14.6 ±3.2	13,482,000	196.4 ±43.7	5	2
Shanghai	China	7.6 ±1.9	23,804,000	181.0 ±44.6	6	2
Country of Singapore	Singapore	30.8 ±6.5	5,235,000	161.1 ±34.1	7	1
Chicago	USA	21.1 ±5.1	7,260,000	152.9 ±37.2	8	3
Tokyo/Yokohama	Japan	4.0 ±0.6	32,999,000	132.8 ±21.4	9	1
Riyadh	Saudi Arabia	20.7 ±4.6	5,747,000	118.8 ±26.4	10	1
Dubai	UAE	22.3 ±6.2	4,971,000	110.8 ±31.0	11	1
Wuxi	China	9.2 ±2.1	12,010,000	110.4 ±25.8	12	3

FIGURE 7.7 Among top 500 cities, populations and Mt CO_2 ranking.

Urban Cluster	Country	Population	Footprint (Mt CO2)	Footprint per Capita (t CO2/cap)	Global ranking
Hong Kong SAR	China	6,029,000	208.5 ±37.8	34.6 ±6.3	1
Mohammed Bin Zayed City	UAE	188,000	6.2 ±5.3	32.9 ±27.9	2
Abu Dhabi	UAE	911,000	29.9 ±15.6	32.9 ±17.1	3
Country of Singapore	Singapore	5,235,000	161.1 ±34.1	30.8 ±6.5	4
Hulun Buir	China	198,000	6.0 ±6.4	30.0 ±32.3	5
Al-Ahmadi	Kuwait	2,697,000	80.6 ±19.3	29.9 ±7.2	6
Doha	Qatar	1,128,000	32.4 ±11.5	28.7 ±10.2	7
Hinggan	China	242,000	6.9 ±7.1	28.6 ±29.3	8
Chifeng	China	677,000	19.0 ±11.0	28.0 ±16.2	9
Al-Jahrah	Kuwait	178,000	4.8 ±3.3	27.2 ±18.4	10
Litong	China	270,000	7.1 ±6.6	26.3 ±24.4	11
New Orleans	USA	596,000	15.6 ±7.8	26.1 ±13.0	12

FIGURE 7.8 Among the top 500 cities per capita.

boundaries. In affluent, high-consumption areas, such measures may require only a small financial investment relative to the median income, yet they can significantly reduce total footprint emissions. Therefore, local actions at the city and state levels can impact both national and global emissions.

7.7 CARBON DISCLOSURE PROJECT

The CDP is crucial in promoting environmental transparency and accountability by encouraging corporations, cities, and regions to disclose their environmental impacts, fostering informed decision-making and continuous improvement in environmental practices. The CDP's scoring system, which ranges from D- to A, incentivizes participants to enhance their environmental performance. This system evaluates various aspects, such as the measurement and verification of emissions, alignment with Science-Based Targets (SBT), and the financial implications of climate risks and opportunities. The importance of the CDP is further highlighted in the context of cities, where factors like population and GDP significantly influence carbon emissions, underscoring the need for urban areas to participate in voluntary disclosure to manage their environmental impact effectively.

In the U.S. utilities sector, the reliability and comparability of GHG disclosures to the CDP are critical for assessing the environmental performance of major emitters (Stanny, 2018). The CDP has encouraged listed companies to disclose climate change information in China, contributing to sustainable development amid rapid economic growth. Moreover, the CDP acts as an information intermediary, helping to bridge the gap between firms and stakeholders by providing standardized environmental performance data, which is increasingly valued by investors and customers (Fabrizio & Kim, 2016). Overall, the CDP's framework incentivizes better environmental practices and enhances the transparency and comparability of environmental data across different sectors and regions. The CDP's framework encourages transparency and aligns with regulatory requirements, enhancing corporate reputation and competitiveness.

The CDP is crucial in assisting stakeholders by providing insights into risks, opportunities, and pathways for reducing emissions, as evidenced by its influence on over 23,000 companies globally. The presence of emission trading systems and board risk management committees significantly boosts firms' carbon disclosure scores, indicating a proactive approach to climate change risk management (Hossain & Farooque, 2019). This transparency is vital for investors, as it allows for a comprehensive analysis of a company's climate strategy, which is increasingly important due to the direct and indirect risks posed by climate change, such as extreme weather events and regulatory changes (Kopp, 2008). The CDP's impact extends to financial markets, where its reporting influences stock prices and online search trends, highlighting the importance of timely and detailed disclosures (Rodríguez et al., 2021). In Brazil, companies view the CDP as a tool for gaining stakeholder acceptance and legitimacy, employing strategies to respond to stakeholder demands (Farias et al., 2017).

Despite the benefits, challenges such as inconsistent reporting standards and the risk of greenwashing persist, underscoring the need for standardized reporting frameworks to enhance corporate transparency and environmental reporting practices. As climate challenges intensify, initiatives like the CDP become indispensable tools for fostering collaboration and driving impactful climate actions.

7.7.1 CDP A-List Companies

CDP provides detailed guidance for full corporate and SME questionnaires, with additional resources available on the Guidance for Companies page and the disclosure hub. The importance of environmental disclosure is underscored by the high demand from over 700 financial institutions managing more than US$142 trillion in assets, requesting disclosures from over 33,000 companies in 2024. In 2023, more than 23,000 companies, representing two-thirds of global market capitalization, disclosed through CDP. The benefits of disclosure include improved reputation through transparency, competitive advantage in the market, progress tracking with benchmarking, and the identification of risks and opportunities for strategic planning. Additionally, CDP helps companies stay ahead of regulatory requirements by aligning with global environmental disclosure standards. Case studies from companies like Stanley Black & Decker, WPP, and Telstra highlight how CDP has facilitated the advancement of their sustainability strategies. Beyond disclosure, companies can engage further by becoming CDP Supply Chain members, joining ambitious environmental campaigns, or exploring the CDP A List, which features leaders in environmental transparency and performance. CDP scores from D- to A, guiding them from initial disclosure to understanding what to measure and finally to taking tangible, measurable actions. Even companies that achieve an A – although among the most transparent regarding climate change, deforestation, or water security – are not at the end of their environmental journey. A List companies are recognized for their leadership in environmental transparency and performance. They meet evolving requirements in line with emerging science, stakeholder feedback, and market demands for greater environmental transparency. These companies keep up with increasingly stringent standards, demonstrating continuous environmental improvement. Figure 7.9 shows the scores of companies.

Company	Region	Climate	Forests	Water
Beiersdorf AG	Europe	A	A	A
Danone	Europe	A	A	A
HP Inc	North America	A	A	A
KAO Corporation	Asia	A	A	A
Kering	Europe	A	A	A
Klabin S/A	South America	A	A	A
Lenzing AG	Europe	A	A	A
L'Oréal	Europe	A	A	A
Mayr-Melnhof Karton Aktiengesellschaft	Europe	A	A	A
Philip Morris International	North America	A	A	A
Sekisui House, Ltd.	Asia	A	A	A

FIGURE 7.9 CDP scores companies from D- to A.

7.7.2 CDP A-List Cities

Disclosing through the CDP-ICLEI Track offers cities numerous strategic advantages, making it an essential practice for those striving to achieve climate leadership. By participating in this globally recognized reporting platform, cities can comprehensively understand their environmental impact, identify gaps, and benchmark their performance against peers, thereby uncovering opportunities for improvement. The process of transparent annual reporting aligns with global best practices and ensures accountability, which is critical in the fight against climate change. Additionally, cities gain unparalleled insights into climate actions and progress, which can be communicated to their communities, businesses, and the broader world. Reporting through CDP-ICLEI Track also streamlines the process of participating in multiple key climate initiatives, such as ICLEI, C40, WWF One Planet City Challenge, and the Global Covenant of Mayors for Climate & Energy, by allowing cities to report via a single questionnaire. This saves time and enhances the efficiency of tracking progress against UN-backed climate campaigns like Race to Zero and Race to Resilience. Furthermore, cities that disclose through CDP-ICLEI Track unlock access to tailored support and services, which can bolster their climate action efforts and attract necessary financing. Ultimately, the data provided by these cities contributes to global climate policy and initiatives by informing major entities such as the United Nations, IPCC, WHO, and the World Bank, thus driving impact far beyond their own jurisdictions and helping to build a sustainable future for the planet.

The A List, which was started by CDP in 2018, recognizes cities that have received the highest score for their transparency and bold climate action and celebrates their achievements, vision, and commitment in the fight against climate change. Many cities are building on the momentum in their environmental action, making it mainstream to their operations. This is reflected in the growing number of cities that are consistently receiving an 'A', despite the tightening criteria to achieve this. This is particularly noticeable in the Global South, with cities in Ecuador, Peru, and Jordan receiving their second-ever 'A'; cities in Chile and Malaysia receiving their third 'A'; and cities in Argentina and South Africa receiving their sixth 'A'. Figure 7.10 below shows the cities with the highest scores for their transparency in taking bold climate actions.

City	Region
Auckland Council	Oceania
Belfast City Council	UK
Birmingham City Council	UK
Blackburn with Darwen Borough Council	UK
Bogotá District Capital	Latin America
Broward County	North America
Cambridge City Council	UK
Can Tho City	Southeast Asia
Canberra	Oceania
City Government of Makati	Southeast Asia
City of Adelaide	Oceania
City of Ann Arbor	North America
City of Aspen	North America
City of Austin	North America
City of Barcelona	Europe

FIGURE 7.10 Cities with the highest score for their transparency and bold climate action.

7.8 CASE STUDY: U.S. NATIONAL CLIMATE ASSESSMENT

The Fifth National Climate Assessment (NCA) is the United States government's preeminent report on climate change impacts, risks, and responses. It is a congressionally mandated interagency effort that provides the scientific foundation to support informed decision-making across the United States. Figure 7.11 illustrates the observed changes in various climate aspects across every U.S. region. The five maps highlight the following trends:

1. Warming is evident in all regions, based on changes in annual average temperatures from 2002–2021 compared to the 1901–1960 average for the contiguous U.S., Hawai‘i, and Puerto Rico, and from 1925–1960 for Alaska.
2. The frequency of warm nights (days with minimum temperatures of at least 70°F) has increased everywhere except the Northern Great Plains, where it has decreased, and Alaska, where such warm nights are rare.
3. Most regions have seen an increase in average annual precipitation, except the Northwest, Southwest, and Hawai‘i, where it has decreased.
4. Heavy precipitation events have increased nationwide, except in Hawai‘i and the U.S. Caribbean, where they have decreased (from 1958 to 2021).
5. Relative sea levels are rising along much of the U.S. coast, with a mix of increases and decreases in Oregon, Washington, and Alaska (from 1990 to 2020).

Additionally, every fraction of a degree of warming exacerbates risks across multiple sectors in the U.S., and without significant reductions in greenhouse gases, these risks are expected to rise.

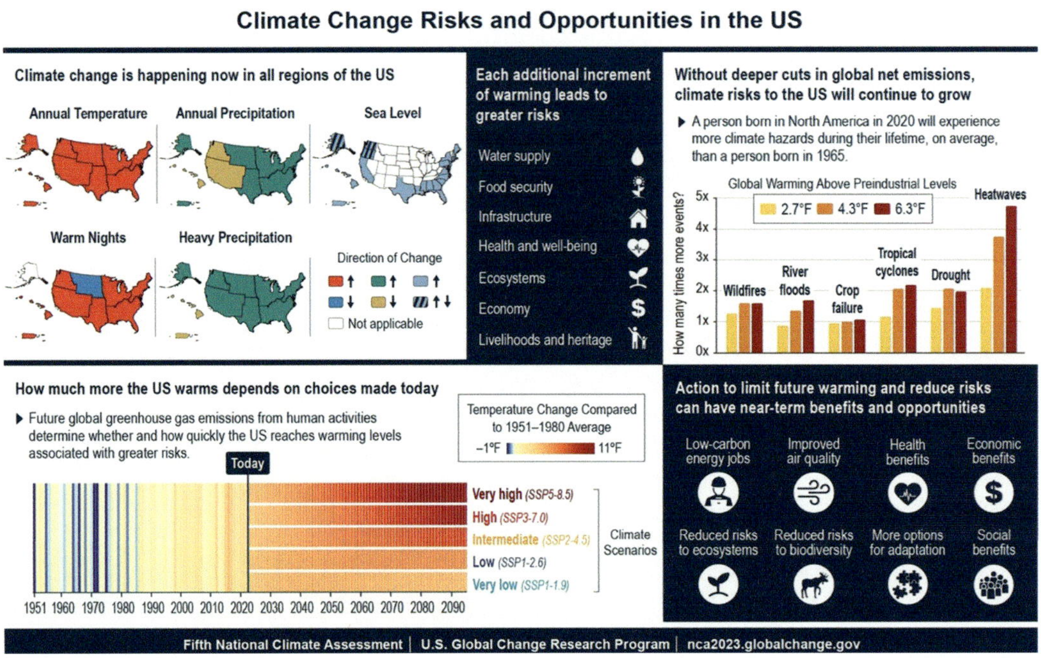

FIGURE 7.11 Climate change risk and opportunities in United States.

People born in North America in 2020 will face more climate-related hazards compared to those born in 1965, with the number of extreme climate events they experience depending on future warming levels. The climate stripes chart shows observed changes in U.S. annual average surface temperature from 1951 to 2022, as well as projected changes for 2023–2095 across five scenarios, from high emissions to low emissions, highlighting the impact of different levels of greenhouse gas reductions. Although significant climate benefits from aggressive emissions cuts may not be visible before mid-century, there are numerous near-term benefits and opportunities from reducing greenhouse gas emissions.

In the 1980s, the country experienced, on average, one (inflation-adjusted) billion-dollar disaster every 4 months. Now, there is one every 3 weeks, on average. Between 2018 and 2022, the country experienced 89 billion-dollar events, as shown in Figure 7.12. Extreme events cost the United States nearly $150 billion annually—a conservative estimate that does not account for loss of life, healthcare-related costs, or damages to ecosystem services. Between 2018 and 2022, 89 such events affected the US, including 4 droughts, 6 floods, 52 severe storms, 18 tropical cyclones, 5 wildfires, and 4 winter storm events. During this period, Florida had the highest total damage ($140 billion) and experienced the highest damages from a single event—Hurricane Ian ($113 billion). Over the 1980–2022 period, Texas had the highest total damages ($375 billion). While similar data are unavailable for the US-Affiliated Pacific Islands, Super Typhoon Yutu caused $500 million in property damage alone in Saipan and the northern Marianas in 2018 (NCEI 2019). Increasing costs over time are driven by changes in the assets at risk and the increase in frequency or intensity of extreme events caused by climate change.

What would 3.6°F (2°C) of global warming feel like in the United States? Compared to the historical average (1991–2020), the US is expected to experience above-average warming as the planet heats up. Figure 7.13 illustrates these projected temperature changes across the country under a 2°C global warming scenario. This warming will not be equally distributed, with some regions likely to see more dramatic temperature increases than others.

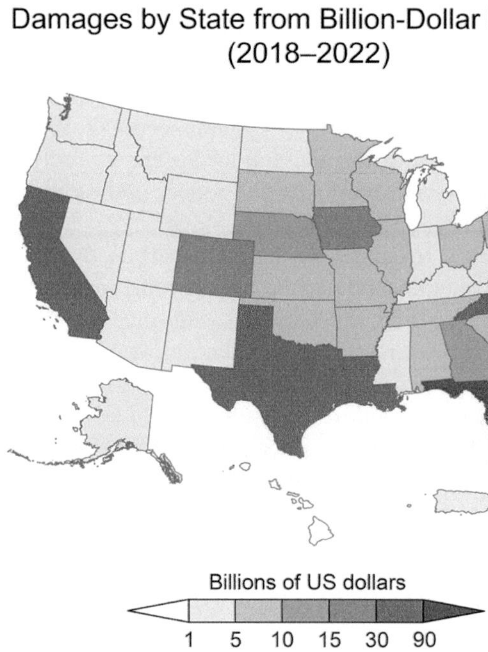

FIGURE 7.12 US State damage.

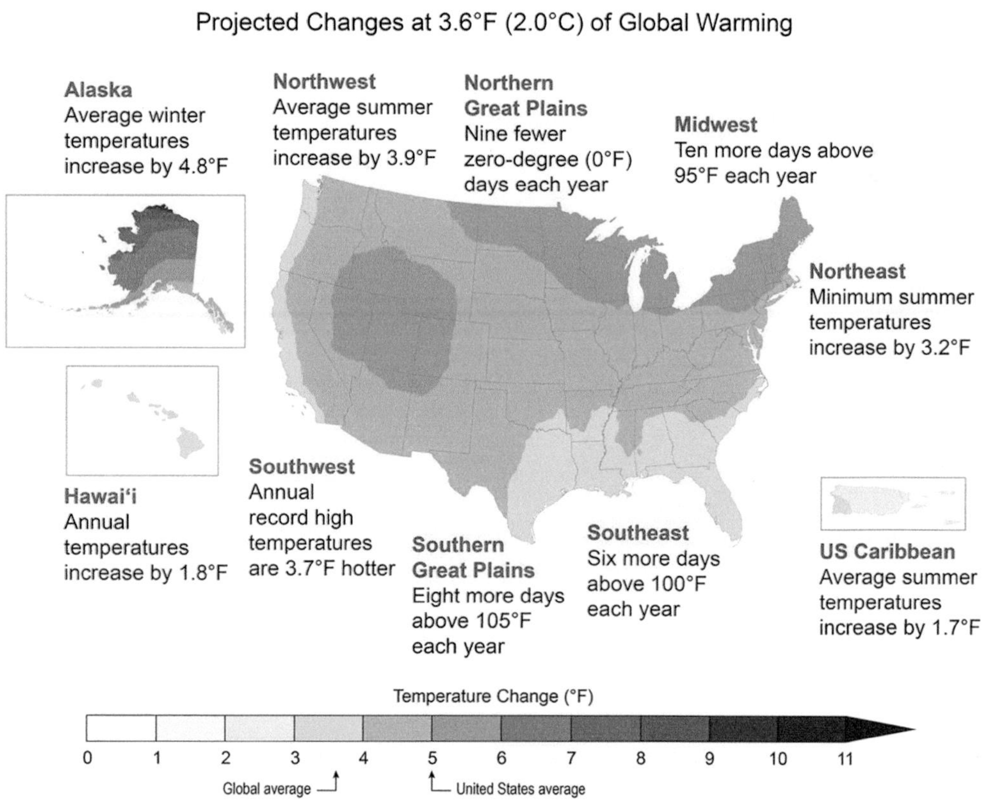

FIGURE 7.13 Projected changes at 2 degrees of global warming in US regions.

7.9 NET ZERO BY 2050

Achieving net-zero emissions in the US by 2050 requires significant cuts in carbon dioxide and other greenhouse gases, balanced by increased atmospheric CO_2 removal. This involves deep reductions in emissions, with remaining emissions offset by enhanced CO_2 removal methods. Energy efficiency, renewable energy, and electrification are vital for the energy sector. Additionally, emerging technologies like hydrogen and low-carbon fuels will reduce emissions in hard-to-decarbonize sectors and increase CO_2 removal.

Figure 7.14 outlines the pathway to achieving net-zero carbon dioxide emissions by 2050, integrating emission reduction strategies and carbon uptake enhancement. The upper red curve represents carbon dioxide emissions, which must decrease significantly through established measures such as energy efficiency improvements, electrification, decarbonized electricity generation, and optimized agricultural practices. Additional reductions can be achieved by adopting hydrogen and low-carbon fuels, carbon capture and storage, and advanced energy management systems. Simultaneously, the lower blue curve illustrates the expansion of carbon sinks, supported by nature-based solutions such as afforestation, improved land management, and avoided land conversion. Emerging opportunities include ocean carbon dioxide removal, biomass energy with carbon capture, and direct air capture technologies. The interplay of these interventions creates a declining trajectory of net emissions, achieving equilibrium between residual emissions and carbon uptake by 2050. This portfolio underscores the necessity of leveraging existing solutions while investing in innovative approaches to address the complex challenges of climate change mitigation.

In the United States, states have been grouped into distinct climate assessment regions, depending upon the geographic and climatic diversity of the nation. These regions include the Northwest, Southwest, Upper and Lower Great Plains, Midwest, Northeast, Southeast, Caribbean, Alaska, and Hawaii and Pacific Islands. Each region faces unique climate challenges, such as water scarcity, rising sea levels, extreme weather events, and shifting agricultural conditions. Coastal areas, such as the Southeast and Pacific Islands, are particularly vulnerable to sea level rise and hurricanes, while inland regions like the Great Plains and Midwest grapple with drought and extreme rainfall. Alaska experiences rapid warming and permafrost thaw, impacting ecosystems and infrastructure. This regional approach enables tailored analyses of climate risks and facilitates

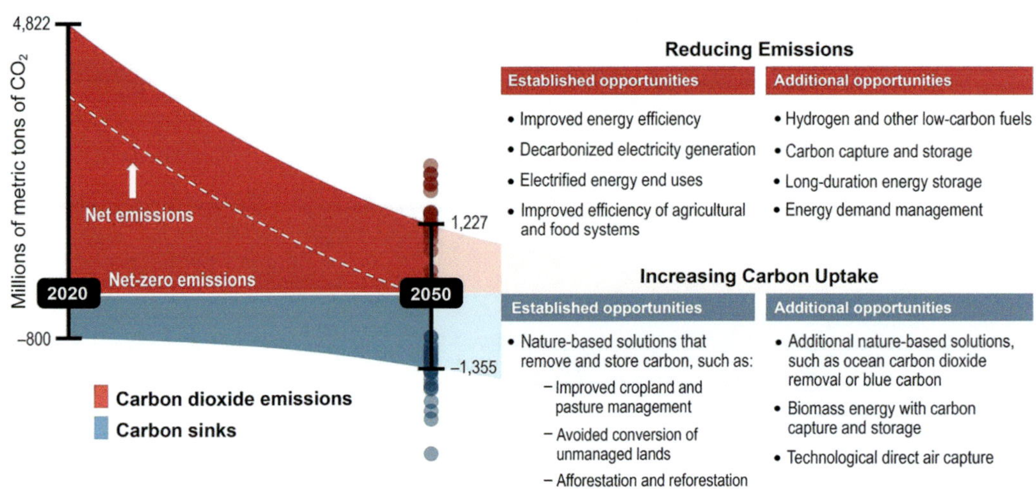

FIGURE 7.14 US net zero strategies.

the development of targeted mitigation and adaptation strategies. Across the US, various regions are implementing climate adaptation, resilience, and mitigation actions. Pittsburgh's 2022 stormwater code in the Northeast addresses increased rainfall, and the city aims for carbon neutrality by 2050. The Southeast sees flood risk reduction in Princeville, NC, and energy efficiency initiatives in Orlando, FL, including biogas generation. Puerto Rican community organizations in the US Caribbean are advancing renewable energy, disaster preparedness, and sustainable development.

The Midwest features wetland restoration in Ashtabula, OH, and carbon storage management in Michigan forests. In the Northern Great Plains, Nebraska promotes soil health practices, and wind energy has tripled regionally, with Tribes leading renewable energy efforts. The Southern Great Plains supports soil management for carbon storage and leads in wind energy generation. In the Northwest, the Confederated Tribes of the Colville Reservation focus on carbon capture in forest management.

The Southwest collaborates on water conservation amid severe drought, with Phoenix aiming for a 50% emissions reduction by 2030. Alaska integrates Indigenous Knowledge in climate adaptation, including kelp farming, to combat ocean acidification. Hawaii and US-affiliated Pacific Islands are enhancing renewable energy, with Kauai aiming for a 90% renewable portfolio by 2026. Pacific islands use blue carbon ecosystems to offset emissions and protect infrastructure.

Figure 7.15 shows that since 2018, city- and state-level adaptation plans and actions (green bars, left) increased by 32%, complemented by a 14% increase in the total number of new state-level mitigation activities (blue bars, right; 69% have updated their policies). In 2021, 271 city-level mitigation actions were in place (open circles, right), according to the Global Climate Action Tracker. Renewable energy and energy efficiency projects on Tribal lands have also expanded.

7.10 CONCLUSION

Global carbon emissions are far from evenly distributed. This chapter mapped the world's carbon hierarchy—ranking countries from highest to lowest emitters and unpacking the drivers behind their total, per capita, and lifecycle emissions. We observed that a handful of nations—China, the United States, India, the European Union, Russia, and Brazil—account for the majority of global emissions, with each country's footprint shaped by its energy mix, industrial structure, and stage of development. Per capita emissions reveal stark inequalities: wealthy nations and affluent urban centers often have much higher footprints than developing countries or rural populations, even when their total emissions are lower. Lifecycle assessments add another layer, exposing the hidden carbon costs embedded in products and international trade.

The chapter also emphasized the importance of accurate and transparent data for tracking emissions and informing fair and effective climate policies. Initiatives such as the Carbon Disclosure Project and robust national inventories are crucial for accountability and progress. Yet, the challenge remains: how to balance economic growth—especially in fast-developing nations—with the urgent need for deep decarbonization and climate justice.

As we move into the next chapter, the focus shifts from who emits and how much to how climate action is governed. Chapter 8 explores the complex web of climate regulations and policies across local, national, and international jurisdictions. It examines the roles of governments, international organizations, and non-state actors and looks at how coordination, innovation, and capacity-building are shaping the global response to climate change. The journey continues from understanding emissions to navigating the rules and alliances needed to drive real-world climate solutions.

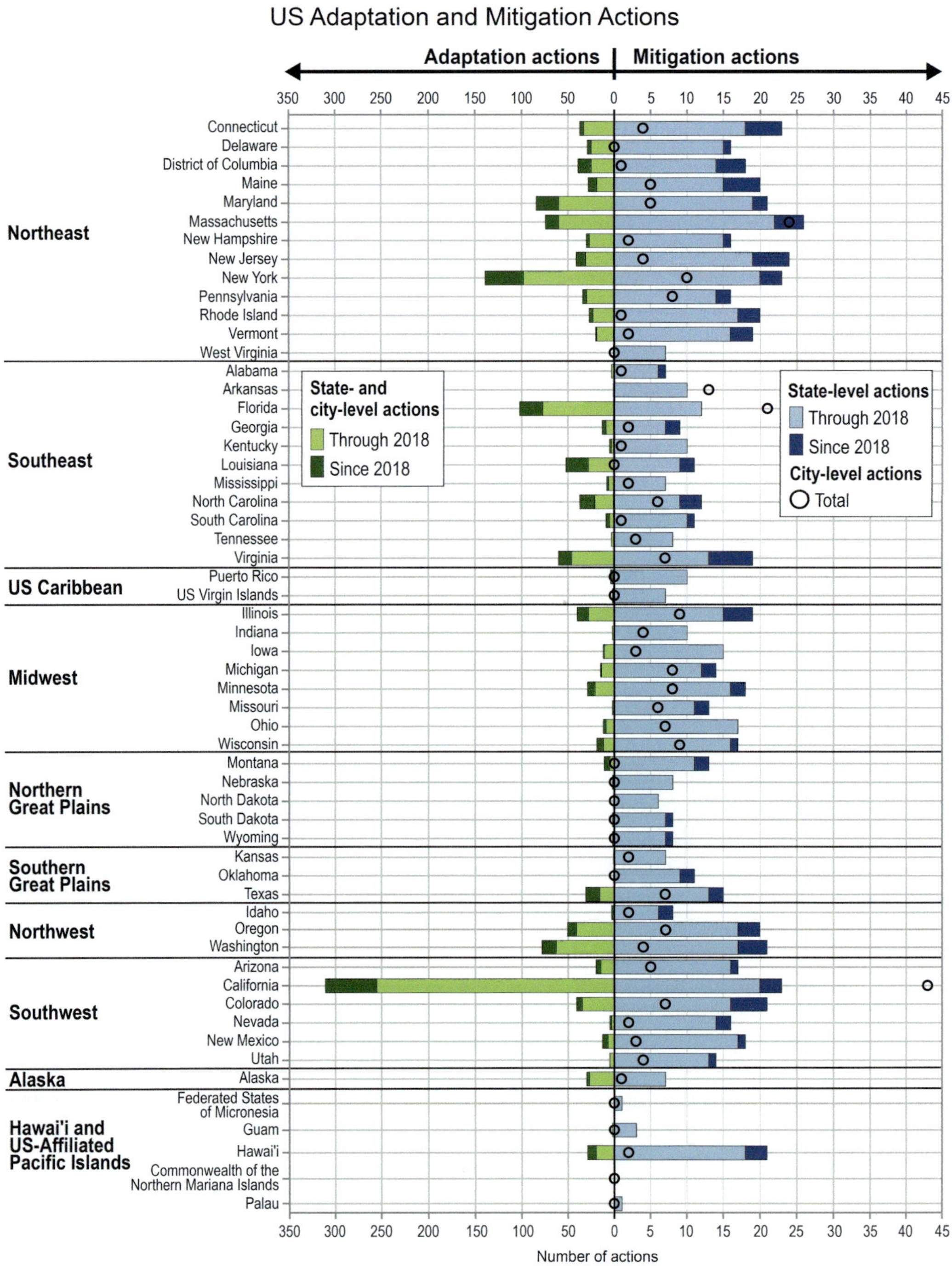

FIGURE 7.15 City- and state-level adaptation plans and actions.

REVIEW QUESTIONS

1. How carbon emissions are measured through various metrics?
2. How do data sources for emission estimations help for tracking progress and implementing effective climate policies?
3. Who are the top greenhouse gas emitters by total carbon emissions and who are the top emitters by per capita?
4. Discuss the main drivers of national per capita emissions in the United States
5. What are the challenges faced by developing nations with the low total emissions but high growth rates In terms of global warming?
6. Discuss the global grid model of carbon footprints based on the EU's population model.
7. Discuss how Net Zero by year 2015 can be achieved including the U.S. Net Zero Strategies.
8. How energy sectors contribute to the increase in greenhouse gas emissions in countries such as India and Brazil and what are the measures taken by these countries to mitigating or decreasing them?
9. Discuss the European Union's metal industries' contribution to global warming and the efforts taken by them to reduce the emissions.
10. What are the observed changes in various climate aspects illustrated in Figure 7.11?

REFERENCES

Arcand, B., & Meadowcroft, J. (2024). Net zero by 2050: The case for green industrial policy. *Environmental Politics*. https://doi.org/10.1080/09644016.2024.2375899

Arndt, H.-K., & Ewe, C. (2017). Analysis of product lifecycle data to determine the environmental impact of the Apple iPhone. In V. Wohlgemuth, F. Fuchs-Kittowski, J. Wittmann (Eds.), *Advances and New Trends in Environmental Informatics. Progress in IS* (pp. 3–13). Springer. https://doi.org/10.1007/978-3-319-44711-7_1

Cordella, M., Alfieri, F., & Sanfelix, J. (2021). Reducing the carbon footprint of ICT products through material efficiency strategies: A life cycle analysis of smartphones. *Journal of Industrial Ecology*. https://doi.org/10.1111/JIEC.13119

Ercan, M., Malmodin, J., Bergmark, P., Kimfalk, E., & Nilsson, E. (2016). *Life Cycle Assessment of a Smartphone*. Atlantis Press. https://doi.org/10.2991/ICT4S-16.2016.15

Fabrizio, K. R., & Kim, E.-H. (2016). *Voluntary Disclosure and Information Intermediaries: Evidence from the Carbon Disclosure Project. Academy of Management Proceedings,* 2016(1), Article 15005. https://doi.org/10.5465/ambpp.2016.15005abstract

Farias, L. das G. Q. de, Andrade, J. C. S., & Gomes, S. M. da S. (2017). *Estratégias de legitimidade de suchman evidenciadas pelas empresas brasileiras destinatárias do pedido do carbon disclosure project. Revista de Administração da UFSM*. https://doi.org/10.5902/1983465912677

Fezzigna, P., Borghesi, S., & Caro, D. (2019). Revising emission responsibilities through consumption-based accounting: A European and post-brexit perspective. *Sustainability*. https://doi.org/10.3390/SU11020488

Grubb, M., Jordan, N. D., Hertwich, E. G., Neuhoff, K., Das, K., Bandyopadhyay, K. R., Asselt, H. van, Sato, M., Wang, R., Pizer, B., & Oh, H. (2022). Carbon leakage, consumption, and trade. *Annual Review of Environment and Resources*. https://doi.org/10.1146/annurev-environ-120820-053625

Gurney, K. R., Kılkış, Ş., Seto, K. C., Lwasa, S., Moran, D., Riahi, K., Keller, M., Rayner, P., & Luqman, M. (2022). Greenhouse gas emissions from global cities under SSP/RCP scenarios, 1990 to 2100. *Global Environmental Change,* 73, 102478. https://doi.org/10.1016/j.gloenvcha.2022.102478

Haga, H., & Zhou, W. (2021). An empirical analysis of international carbon transfer. In W. Zhou, X. Qian, K. Nakagami (Eds.), *East Asian Low-Carbon Community* (pp. 289–305). Springer. https://doi.org/10.1007/978-981-33-4339-9_16

Hertwich, E. G., & Peters, G. P. (2009). Carbon footprint of nations: A global, trade-linked analysis. *Environmental Science & Technology*. https://doi.org/10.1021/ES803496A

Hossain, M., & Farooque, O. A. (2019). The emission trading system, risk management committee and voluntary corporate response to climate change – A CDP study. *International Journal of Accounting and Information Management*. https://doi.org/10.1108/IJAIM-04-2017-0050

Jay, A. K., Crimmins, A. R., Avery, C. W., Dahl, T. A., Dodder, R. S., Hamlington, B. D., Lustig, A., Marvel, K., Méndez-Lazaro, P. A., Osler, M. S., Terando, A., Weeks, E. S., &

Jhariya, M. K., Raj, A., Banerjee, A., Meena, R. S., Devi, A., & Kittur, B. H. (2024). Urban emission reduction and carbon management. In M.K. Jhariya, R.S. Meena, A. Raj (Eds.), *Agroforestry for Carbon and Ecosystem Management* (pp. 161–171). Elsevier. https://doi.org/10.1016/b978-0-323-95393-1.00022-1

Kopp, M. (2008). Das carbon disclosure project (CDP). *Uwf Umweltwirtschaftsforum*. https://doi.org/10.1007/S00550-008-0073-4

Lininger, C. (2015). Literature on Emissions Embodied in Trade. In: Consumption-Based Approaches in International Climate Policy. Springer Climate. Springer, Cham. https://doi.org/10.1007/978-3-319-15991-1_6

Lützkendorf, T. (2020). The role of carbon metrics in supporting built-environment professionals. *Buildings & Cities*. https://doi.org/10.5334/BC.73

Minx, J.C., Lamb, W.F., Andrew, R.M., Canadell, J.G., Crippa, M., Döbbeling, N., Forster, P.M., Guizzardi, D., Olivier, J., Pongratz, J., Reisinger, A., Rigby, M., Peters, G. P., Saunois, M., Smith, S. J., Solazzo, E., Tian, H. (2022). *A comprehensive and synthetic dataset for global, regional and national greenhouse gas emissions by sector 1970–2019* (Version v6) [Data set]. Zenodo. https://doi.org/10.5281/zenodo.6483002

Muthu, S.S. (2015). *The Carbon Footprint Handbook*. CRC Press. https://doi.org/10.1201/B18929

Pflüger, M., Gütschow, J. (2021). *UNFCCC Country-Submitted Greenhouse Gas Emissions Data Until 2021-12-03*. Zenodo. https://doi.org/10.5281/zenodo.5752337

Pflüger, M., Gütschow, J. (2022). *UNFCCC Country-Submitted Greenhouse Gas Emissions Data Until 2022-12-13*. Zenodo. https://doi.org/10.5281/zenodo.7432088

Reijnders, L. (2015). Life cycle assessment of greenhouse gas emissions. In W.Y. Chen, T. Suzuki, M. Lackner, M. (Eds.), *Handbook of Climate Change Mitigation and Adaptation* (pp. 61–91). Springer. https://doi.org/10.1007/978-3-319-14409-2_2

Rodríguez, M., Camacho, J.A., Almeida, L.D.S., Molina, J. (2021). Domestic versus foreign origin of total energy use: An analysis for Brazil. *Energy Reports*, 7, 6327–6337. https://doi.org/10.1016/j.egyr.2021.09.073

Rodríguez, M.M., Muñoz, F.F., Valentinetti, D. (2021b). Corporate impact of carbon disclosures: A non-linear empirical approach. *Journal of Financial Reporting and Accounting*. https://doi.org/10.1108/JFRA-01-2020-0002

Stanny, E. (2018). Reliability and comparability of GHG disclosures to the CDP by US electric utilities. *Social and Environmental Accountability Journal*. https://doi.org/10.1080/0969160X.2018.1456949

Yamano, N., Guilhoto, J.J.M. (2020). CO_2 emissions embodied in international trade and domestic final demand: Methodology and results using the OECD Inter-Country Input-Output Database. *Research Papers in Economics*. https://doi.org/10.1787/8F2963B8-EN

8 Navigating Regulatory Jurisdictions

This chapter delves into the intricate landscape of climate change regulations across various jurisdictions, from local to international levels. It explores the roles of different actors, including local governments, international organizations, and non-state entities, in shaping and implementing climate policies. This chapter examines the challenges and opportunities in navigating these diverse regulatory frameworks, emphasizing the importance of collaboration and innovation in achieving effective climate action. It also highlights the significance of international agreements, national regulations, and sub-national initiatives in driving the transition toward a low-carbon and climate-resilient future.

8.1 LOCAL LAWS

To combat climate change, individuals from individual clusters, cities, and villages can take various impactful actions: spread awareness and encourage others to reduce carbon emissions; pressure politicians and businesses to support eco-friendly policies; use sustainable transportation like walking, cycling, or public transit; reduce energy consumption by opting for renewable sources and efficient appliances; adopt plant-based diets; buy local and sustainable products; minimize food waste by buying only what is needed and composting; choose eco-friendly fashion by buying less and recycling clothes; participate in tree-planting initiatives to combat deforestation; and make planet-friendly financial investments. These efforts collectively contribute to reducing greenhouse gas (GHG) emissions and fostering a sustainable future.

Cities are home to almost 4.4 billion inhabitants, and by 2050, nearly 7 out of 10 people will reside in cities. Inhabitants demand energy, water, transportation, airports, infrastructure, etc. Thus, urban areas become responsible for high GHG emissions. Each city has its own characteristics on which climate change mitigation measures can be implemented. Combating climate change from the grassroots level, with a focus on proper city planning, building codes, and solid waste management, is crucial.

8.1.1 Zoning and Land-Use Regulations

Zoning and land-use regulations play a crucial role in combating GHG emissions and supporting climate change resilience. Zoning integrates residential, commercial, and recreational spaces by encouraging mixed-use development, reducing the need for long commutes, and thereby lowering vehicle emissions. Mixed-use zoning increases urban density, encourages active transportation, and fosters economic development, which collectively reduce reliance on private vehicles and cut transportation-related emissions (Moos et al., 2018). When people live closer to their workplaces and recreational areas, reliance on personal automobiles decreases, promoting walking, biking, and public transportation.

Additionally, land-use regulations can prioritize developments near public transportation hubs through transit-oriented development (TOD), reducing dependence on private vehicles and thereby lowering GHG emissions from the transportation sector. TOD, as a sustainable urban planning strategy, integrates land use and transportation, promoting walkable and environmentally sustainable communities (Jamme et al., 2019). It encourages the use of public transportation, thereby reducing motorized passenger travel demand and contributing to significant GHG reductions (Khreis et al.,

DOI: 10.1201/9781003479666-8

2018). However, TOD can also lead to unintended social consequences, such as residential displacement and gentrification, which necessitate careful planning to balance economic development with social equity (Wang et al., 2024). Additionally, TOD's impact on local environmental transformation is significant, as it influences land-use catchment areas and accessibility, which are crucial for sustainable urban development (Tong et al., 2018).

Preserving open spaces and green areas through zoning laws designates parks, green belts, and other open spaces, which serve as carbon sinks, absorb CO_2, maintain biodiversity, and mitigate urban heat island effects. This also prevents overdevelopment and reduces the carbon footprint associated with new construction. Implementing green building standards through zoning regulations mandates or incentivizes green building practices and sustainable materials, promoting energy-efficient designs and renewable energy installations, which help reduce the energy consumption of buildings and lower overall GHG emissions. These strategies collectively underscore the importance of integrating environmental, social, and economic considerations in urban planning to achieve sustainable and resilient cities (Khreis et al., 2018; Tong et al., 2018).

Controlling urban sprawl through land-use regulations limits the distance between homes, workplaces, and amenities, reducing vehicle emissions and energy consumption. By promoting higher density and infill development within existing urban areas, zoning helps reduce the GHG emissions associated with transportation and infrastructure expansion (Rani, 2024). Managing impervious surfaces through zoning can help manage stormwater runoff and reduce flood risks while ensuring the maintenance of green spaces that aid in carbon sequestration and urban cooling. This regulation also promotes permeable materials and green roofs, reducing urban heat islands and improving air quality (Gil-Ozoudeh et al., 2023).

Protecting vulnerable areas by restricting development in floodplains, coastal zones, and wildfire-prone regions reduces the risk of climate-related disasters, which often lead to higher emissions during emergency responses and rebuilding efforts. Moreover, protecting these areas maintains natural buffers that help mitigate the effects of climate change (Belokurov et al., 2010; Dudley et al., 2011). Supporting renewable energy through land-use regulations facilitates the installation of renewable energy infrastructure, such as solar farms, wind turbines, and geothermal plants. By setting aside zones specifically for renewable energy projects, communities can increase their share of clean energy, reducing reliance on fossil fuels and lowering GHG emissions.

8.1.2 Building Codes and Energy Efficiency Standards

To implement efficient building codes that mitigate climate change, a city council or municipal office can incorporate several strategies based on the United Nations Environment Program (UNEP) article on climate-resilient buildings. Firstly, to build resilience to heat waves, urban planning should include urban forests and green spaces, ensuring trees and plants are integrated into public and residential areas to mitigate the urban heat island effect. Architectural design standards should establish building orientation guidelines that optimize natural ventilation and promote using green roofs, reflective surfaces, and Trombe walls to capture solar heat. Green building standards are crucial in promoting sustainable construction practices, significantly reducing energy consumption and carbon footprints. These standards, such as the LEED rating system, encourage the adoption of energy-efficient designs and renewable energy sources, thereby minimizing environmental harm and promoting sustainability in urban landscapes. Infill development, a strategy to control urban sprawl, is essential in minimizing infrastructure expansion and the associated emissions. By focusing on the redevelopment of underutilized urban areas, infill development reduces the need for new infrastructure, thereby decreasing the carbon footprint of urban expansion.

In areas vulnerable to coastal flooding and sea-level rise, building codes should require structures to be elevated on pillars or stilts, allowing water to flow underneath and ensuring connectivity between elevated structures through passages. Exploring and approving designs for buoyant buildings can also provide emergency shelter during floods and serve community needs. To withstand cyclones and strong winds, structural integrity standards should enforce strong connections between foundations and roofs and mandate aerodynamic building shapes to reduce wind resistance. Encouraging frangible architecture, where non-essential parts of a building are designed to break away safely during strong winds, can protect the main structure.

For cold climates, building codes should set standards for thermal insulation in roofs, walls, and windows to minimize heat loss and promote energy-efficient building materials and designs. Building orientations that maximize sun exposure should be recommended, along with the use of Trombe walls and water walls to store and radiate heat efficiently. To implement these measures, city councils should draft and pass local legislation that incorporates these resilience measures into the building code. Providing education and training for architects, builders, and city planners on the new building standards and best practices for climate resilience is crucial. Offering incentives such as tax breaks or grants for buildings that exceed basic resilience requirements and providing support for retrofitting existing buildings can encourage compliance. Finally, establishing a system for regular inspections and implementing penalties for non-compliance will ensure adherence to the new building codes.

8.1.3 Waste Management Regulations

Cities and municipalities play a crucial role in mitigating climate change through effective waste management practices. They can prioritize waste prevention measures by implementing policies and initiatives that encourage reduced consumption, reuse, and recycling, thereby minimizing waste generation and associated GHG emissions. Additionally, cities can establish comprehensive recycling programs to divert recyclable materials from landfills, reducing the need for raw material extraction and manufacturing while conserving energy.

Circular waste management systems emphasize recycling, reuse, and waste prevention and have shown significant potential in mitigating GHG emissions. For instance, transitioning from traditional to circular waste management practices in China could enhance GHG mitigation as coverage increases from 35% in 2025 to 100% in 2035, although the benefits may taper off as waste quantities decline due to population shrinkage (Li, 2024). In the United States, adopting sustainable waste management practices, such as increasing recycling rates from 27% to 65% and reducing landfill use, could reduce GHG emissions by approximately 700 million metric tons annually by 2050, contributing significantly to achieving net-zero emissions targets (Halevi et al., 2023). Sustainable waste management practices, including composting and anaerobic digestion, not only reduce landfill methane emissions but also generate renewable biogas, further minimizing environmental impacts (Halevi et al., 2023; Lodh, 2024).

Some local-level examples:

Case One: Crowdsourcing apps in Indonesia: An Indonesian app, similar to Pokémon Go, encourages users to map land across the archipelago. This helps protect forests and Indigenous people by identifying areas that need conservation.

Wind-Powered Cargo Ships: A Swedish company is designing wind-driven cargo ships to reduce reliance on fossil fuels and limit climate change.

Smart Solar Pumps in Africa: High-tech solar pumps are mapping underground freshwater reservoirs across Africa, helping prevent them from running dry. These pumps are a cleaner and cheaper alternative for small-scale farmers.

8.1.4 City Initiatives

City networks such as C40, Local Governments for Sustainability (ICLEI), and the Global Covenant of Mayors for Climate and Energy are crucial in addressing climate change by fostering collaboration and innovation among urban centers. These networks facilitate the alignment of city actions with global climate targets, such as the 1.5°C target of the Paris Agreement, by promoting voluntary planning and sustainable urbanization practices. C40's Climate Action Planning program exemplifies this by encouraging cities to adopt climate-responsive measures that are not bound by normative constraints but rather driven by local commitment and awareness (Santopietro & Scorza, 2024). ICLEI supports sustainable urban development across 125 countries, emphasizing the integration of environmental, social, and economic dimensions to enhance urban resilience and reduce carbon footprints (Gürçam, 2023). The Global Covenant of Mayors, with over 10,000 participating cities, focuses on data sharing, innovation, and financing to support low-carbon urbanization pathways, which are essential for achieving national and global decarbonization goals (Platonova-Oquab & Shenvi, 2023). These networks also highlight the importance of integrated urban planning strategies that prioritize sustainable transport, renewable energy, and green spaces, as demonstrated by successful initiatives in cities like Copenhagen and Singapore (Gürçam, 2023). By leveraging the collective power of cities, these networks help bridge the gap between national climate policies and local urban development, facilitating access to finance and enabling effective implementation of climate actions (Platonova-Oquab & Shenvi, 2023). Overall, city networks are instrumental in advancing sustainable urbanization and climate resilience, contributing significantly to global sustainability goals (Santopietro & Scorza, 2024; Gürçam, 2023; Platonova-Oquab & Shenvi, 2023).

Smaller initiatives like the Climate Alliance and Carbon Neutral Cities Alliance exemplify how cities can lead climate action by integrating local challenges and supporting Indigenous populations through shared expertise and resource pooling. These networks highlight the importance of incorporating Indigenous knowledge systems (IKS) into climate strategies, as Indigenous communities are disproportionately affected by climate change yet possess unique adaptive capacities rooted in traditional practices (Datta et al., 2024; Gauer et al., 2021). Despite the recognition of IKS in international frameworks like the Paris Agreement, Indigenous participation in decision-making remains limited, underscoring the need for inclusive climate solutions that resonate with Indigenous perspectives (Datta et al., 2024; Tran & Salamanca, 2023). Furthermore, Indigenous peoples have increasingly utilized formal participatory spaces to advocate for self-determination and territorial rights, emphasizing the principle of Free, Prior, and Informed Consent in environmental governance (Schilling-Vacaflor & Gustafsson, 2022). By fostering decolonization processes and revitalizing Indigenous ways of relating to the Earth, these initiatives can contribute to more effective and equitable climate action, benefiting both Indigenous and non-Indigenous communities alike (Mcgregor & Sritharan, 2023).

8.1.4.1 C40 Cities

One of the leading global city networks, C40 Cities, is engaging mayors from nearly 100 capitals and major cities around the world to collaborate on climate pledges and action. C40 has a large set of activities and concretely supports a Climate Action Planning program for cities around the world to create and implement climate action plans in line with the 1.5-degree Celsius goal of the Paris Agreement. As of 2022, 41 cities had published plans, and 169 cities (members and non-members) were involved in the program. Guidance material on the approach is publicly available and is, for example, used in adapted form in the Danish DK2020 program, which engages almost all Danish municipalities in climate planning.

8.1.4.2 ICLEI

ICLEI – Local Governments for Sustainability is another leading global network of more than 2,500 local and regional governments committed to sustainable urban development. ICLEI is active in more than 125 countries, aiming to influence sustainability policy and drive local action for low-emission development. ICLEI is broadly focused on urban sustainability, with key areas of work being energy, transport, and climate. For example, activities include training programs for city climate planners and numerous other programs.

8.1.4.3 Compact of Mayors

The Compact of Mayors was launched in 2014 to bring together the three initiatives described above around a common set of principles with which participating cities must comply. In 2016, the Compact was formally merged with the European Union's Covenant of Mayors and became the Global Covenant of Mayors for Climate and Energy to advance the city-level transition to a low-emission and climate-resilient economy and to demonstrate the global impact of local action.

8.1.4.4 Global Covenant of Mayors for Climate and Energy (GCoM)

GCoM is the global alliance for city climate leadership, building on the commitment of more than 10,000 cities and local governments. These cities span six continents and 140 countries. GCoM has a number of initiatives, focusing on (1) data and reporting; (2) implementation and finance; and (3) joint innovation programs. GCoM functions like a cooperation framework for the other initiatives, where C40 and ICLEI continue their own programs, and the European Union (EU) Covenant of Mayors functions as a regional entity of GCoM.

8.1.4.5 Cities Race to Zero

The Cities Race to Zero is a subset of the Global Race to Zero discussed above and covers cities only. Partners include all the other city initiatives (C40, ICLEI, the Global Covenant of Mayors), which here come together on the long-term net-zero target, which was not part of their original mandates. Other institutions that are also engaged in the Cities Race to Zero include CDP, the World Resources Institute, and WWF. The collaboration aims to ensure that cities are aligned on the science-based targets required to achieve a net-zero future and provides support to cities signing on to the campaign.

8.1.4.6 Climate Alliance

The climate alliance city network represents a holistic approach to climate action, linking concrete local solutions with global responsibility. It has some 1,700 members spread across more than 25 European countries. The alliance is more of a membership and advocacy structure that feeds into the Global Covenant of Mayors, at both the global and EU levels. It has a special feature on supporting Indigenous populations, for example, in the Amazon region, and linking these with partners in the EU.

8.1.4.7 Carbon Neutral Cities Alliance

The Carbon Neutral Cities Alliance is a collaboration of around 25 cities around the world working to achieve carbon neutrality. The alliance has links with C40, and many member cities are also C40 members. The alliance is more of a peer learning and influencing network of ambitious cities, and by focusing on carbon neutrality in the next few decades, members go further than the large alliance. City planning processes: There are different planning process descriptions for each initiative, but the basic steps are the same. Over time, the reporting frameworks have evolved from individual guidance for members toward a gradual convergence taking place both on the organizational side and subsequently on the methodological side and on associated registries.

8.2 INTERNATIONAL LAWS AND INITIATIVES ADDRESSING GHG MITIGATION

8.2.1 The UN Intergovernmental Panel on Climate Change

The urgency of climate action necessitates a global response, yet the reality of climate change governance is a complex web of international and national regulations. Within this intricate landscape sits the Intergovernmental Panel on Climate Change (IPCC), a unique body tasked with providing policymakers with the scientific basis for informed decision-making. This chapter explores the IPCC's role in navigating regulatory jurisdictions, examining its strengths, limitations, and its influence on national and international climate policy. Established in 1988 by the UNEP and the World Meteorological Organization (WMO), the IPCC is an intergovernmental body that assesses the scientific, technical, and socio-economic aspects of climate change. Crucially, the IPCC does not set regulations or enforce policies. Instead, it acts as a bridge between the scientific community and policymakers.

The IPCC achieves this through a rigorous assessment process. Thousands of scientists from around the world volunteer their expertise to produce comprehensive reports that synthesize the latest climate science. These reports assess the physical science basis of climate change, its impacts and vulnerabilities, and potential mitigation and adaptation strategies.

Climate change presents a multifaceted challenge with implications across various sectors. National governments hold primary responsibility for environmental protection within their borders. However, the global nature of climate change necessitates international cooperation. Several key regulatory frameworks address climate change at the international level:

- **The United Nations Framework Convention on Climate Change (UNFCCC)**: Adopted in 1992, the UNFCCC is the cornerstone of international climate policy. It establishes a framework for international cooperation in mitigating climate change and adapting to its effects.
- **The Kyoto Protocol**: Adopted in 1997, the Kyoto Protocol set binding emission reduction targets for developed countries.
- **The Paris Agreement**: Adopted in 2015, the Paris Agreement represents a significant shift in international climate policy. It sets a long-term goal of limiting global warming to well below 2°C and pursuing efforts to limit it to 1.5°C.

The IPCC's assessments inform these international agreements and play a crucial role in shaping national climate policies. By providing objective scientific evidence, the IPCC helps policymakers understand the scientific basis of climate change and its potential consequences. This information is essential for developing effective and evidence-based climate regulations at both national and international levels. The IPCC's approach offers several strengths in navigating the complex regulatory landscape:

- **Scientific Rigor**: The IPCC reports undergo a rigorous review process involving thousands of scientists worldwide. This ensures the objectivity and credibility of the scientific findings presented.
- **International Collaboration**: The IPCC brings together scientists from various countries and disciplines. This collaborative approach fosters a global understanding of climate change and its potential solutions.
- **Policy Neutrality**: The IPCC does not advocate specific policy solutions. Instead, it presents the scientific evidence neutrally and objectively, allowing policymakers to make informed decisions based on their specific circumstances.

Despite its strengths, the IPCC approach also faces limitations:

- **Timeliness**: The IPCC assessment reports are released on a multi-year cycle, which can be challenging in a rapidly evolving field like climate science.
- **Scientific Uncertainty**: Climate science is a complex field with inherent uncertainties. While the IPCC strives to present a clear and robust understanding of climate change, some uncertainties remain.
- **Political Influence**: The IPCC operates within a complex political environment. The reports are subject to approval by all member governments, which can sometimes lead to the watering down of scientific language.

Despite these limitations, the IPCC has significantly influenced climate policy at both national and international levels. The IPCC's assessments have:

- **Provided the Scientific Basis for International Agreements**: The IPCC reports have played a crucial role in shaping international agreements like the Kyoto Protocol and the Paris Agreement. By providing clear evidence of the urgency of climate action, the IPCC has helped build consensus among nations.
- **Informed National Climate Strategies**: National governments rely on the IPCC's assessments to develop their own climate strategies. The reports provide scientific evidence to support the development of regulations that aim to reduce emissions and adapt to the impacts of climate change.
- **Raised Public Awareness**: Through its reports and media outreach, the IPCC has helped raise public awareness about climate change. This has put pressure on governments to take action and has fostered a global movement for climate action.

The global effort to mitigate climate change has evolved significantly over the years. Beginning with establishing the IPCC and the UNFCCC in the late 20th century, milestones include the adoption of the Kyoto Protocol in 1997 and the Paris Agreement in 2015, setting ambitious targets to limit global warming. Additional measures, like the Kigali Amendment to phase out HFCs and youth-led climate strikes, underscore growing awareness and activism. Despite setbacks like the COVID-19 pandemic, renewed commitments, such as the Glasgow Climate Pact in 2022 and the EU's "Fit for 55" package in 2023, demonstrate ongoing efforts to address the urgent challenge of climate change on a global scale.

Not all net-zero commitments are created equal, and many of them are lacking in crucial details, scope, and near-term accountability. This has led to the proliferation of tools to assess the credibility and quality of company climate transition plans, with a prominent example being the climate action 100+ Net-Zero Company Benchmark. This has also led to a new trend of companies putting their climate transition plans up for a shareholder vote—as popularized by the "Say on Climate" campaign. Such votes come with their own set of risks and opportunities.

Meanwhile, the global climate outlook is still as precarious as ever. While COP26 in Glasgow saw some encouraging developments, the incremental decline in absolute global emissions that is needed to keep 1.5°C within reach is still nowhere to be seen. In addition, new analyses like the IEA's long-awaited Net Zero by 2050 Scenario have demonstrated how little time the world has to make the transformational changes in economies and societies needed to limit global warming to safe levels.

8.2.2 The Conference of Parties

Conference of Parties (COP) summits address climate change and biodiversity. Established after the 1992 Rio Summit, these annual meetings bring together 197 countries to discuss climate action. Since 2015, all participating nations have acknowledged climate change and aim to limit GHG emissions while achieving net zero by the end of the century. A significant turning point came at COP21 in Paris (2015), when all parties agreed on the human-caused nature of climate change. The resulting Paris Agreement aims to limit global warming to below 2°C, ideally 1.5°C. Despite ongoing efforts, recent COPs like COP27 (2022) highlight challenges. Fossil fuel reliance remains high, and progress on mitigation has been slow. There has been some progress on financing climate change's negative impacts on vulnerable countries.

8.2.3 Paris Agreement 2015

The Paris Agreement is a historic international treaty aimed at tackling climate change. Adopted in 2015, it became legally binding in 2016 and boasts near-universal participation (196 countries).

The core objective is curbing global warming. Ideally, this means keeping the average temperature rise well below 2°C compared to pre-industrial levels, with even stronger efforts to limit it to a maximum of 1.5°C. Experts warn that exceeding this threshold could trigger much harsher climate consequences, including more frequent and intense droughts, heatwaves, and heavy rainfall.

The Paris Agreement operates on a five-year cycle to enhance global climate action, with each phase aiming for greater ambition. Countries submit their nationally determined contributions (NDCs), outlining their plans to mitigate GHG emissions and adapt to climate impacts. These NDCs are revised every five years to reflect more ambitious targets in line with the agreement's goal of limiting global warming. Additionally, countries are encouraged to develop long-term low GHG emission development strategies (LT-LEDS) to provide a broader context and direction for their climate efforts. These strategies complement NDCs by offering a vision for sustainable development beyond short-term targets.

Wealthier nations are lending a hand to developing countries by providing financial aid, better technology, and training. It helps developing countries reduce emissions and become more resilient to climate change's effects. The Paris Agreement set this framework in motion. It created a system where developed countries give money for climate action in developing countries. This money goes toward reducing emissions (mitigation) and adapting to climate change (adaptation). Technology is another area of support. The agreement established a system to share technology that helps reduce emissions and build resilience. It could be things like new solar panels or ways to grow crops more resistant to drought. Finally, the agreement recognizes that some countries need extra help building the skills and knowledge to tackle climate change. So, developed countries provide training and other capacity-building forms to developing countries.

Tracking progress on climate change is performed through a system established by the Paris Agreement called the Enhanced Transparency Framework (ETF). This system requires countries to report on their actions to reduce emissions (mitigation), adapt to the effects of climate change (adaptation), and the support they are providing or receiving for these efforts. These reports are reviewed internationally to assess how well we are doing collectively. So far, the Paris Agreement has spurred the development of new low-carbon technologies and markets. More and more countries, regions, cities, and companies are setting goals to achieve carbon neutrality. Zero-carbon solutions are becoming cost-competitive with traditional options in some sectors, like energy and transportation. It is creating new business opportunities, and by 2030, these solutions could be competitive in over 70% of the global emissions sectors.

8.2.4 COP-28 Dubai

At COP28, held in Dubai in December 2023, an agreement was reached, signaling the "beginning of the end" for the fossil fuel era, emphasizing a swift, just, and equitable transition toward renewable energy. This agreement, part of the world's first "global stocktake," aims to strengthen climate action to keep global temperature rise within 1.5°C. The stocktake recognizes the need to cut global GHG emissions by 43% by 2030 compared to 2019, urging nations to triple renewable energy capacity and double energy efficiency improvements by 2030. Additionally, it calls for a phase-down of unabated coal power and the phasing out of inefficient fossil fuel subsidies. Parties must submit ambitious emission reduction targets aligned with the 1.5°C limit in their next climate action plans by early 2025.

COP28 set four pillars aiming to achieve key outcomes in global climate action. The first pillar is fast-tracking a just, orderly, and equitable energy transition. It focused on fast-tracking a just and equitable energy transition, which is evident through initiatives like the Global Renewables and Energy Efficiency Pledge and collaborative alliances like the Powering Past Coal Alliance. The second pillar is fixing climate finance, with endorsements such as the UAE Leaders' Declaration and collaborative endeavors like the Assets to Flows II. The third pillar focuses on people, lives, and livelihoods, as demonstrated by initiatives like the Sharm el-Sheikh Adaptation Agenda and commitments to enhance resilience in various sectors. Furthermore, the fourth pillar underpins everything with full inclusivity, as seen through the appointment of a Youth Climate Champion and initiatives like the Pudong Indigenous Peoples Initiative to leverage technology and cultural sectors for inclusive climate action.

8.2.5 ASEAN Europe Efforts to Carbon Neutrality

ASEAN's preparation to address climate change and carbon emissions involves a comprehensive approach that combines adaptation and mitigation strategies.

- **Assessment and Planning**: ASEAN thoroughly assesses the region's climate change impacts, vulnerability, and risks. This information guides the development of adaptation and mitigation plans tailored to the needs of ASEAN member states (AMS).
- **Enhanced Transparency**: ASEAN prioritizes transparency in reporting and monitoring GHG emissions and climate actions. It includes strengthening scientific information bases, tracking regional progress, and adopting methodologies for transparent reporting under frameworks like the ETF.
- **Integration and Synergies**: ASEAN integrates climate actions into sectoral and development planning to promote synergies between adaptation and mitigation efforts. It involves mainstreaming climate considerations into various sectors such as energy, agriculture, and forestry.
- **Inclusive Engagement**: ASEAN promotes multi-stakeholder engagement and public-private partnerships to ensure inclusive decision-making processes and implementation of climate actions. Community involvement and cooperation among AMS are emphasized to address climate challenges effectively.
- **Technology and Capacity Building**: ASEAN focuses on technology development, diffusion, and capacity building to support climate action. It includes sharing expertise, promoting clean and resilient technologies, and enhancing human capacity for climate-related interventions.
- **Finance and Investment**: ASEAN recognizes the importance of financial support for climate adaptation and mitigation efforts. Strategies include leveraging public and private finance, integrating green recovery principles into post-pandemic economic recovery plans, and accessing regional/international funds.
- **Long-term Vision and Targets**: ASEAN aims to develop long-term visions and targets that are aligned with the Paris Agreement's goals. This involves setting ambitious goals for emissions reduction, renewable energy expansion, and sustainable land-use practices supported by regional cooperation and collaboration.

8.2.6 USA Efforts to Carbon Neutrality

The Biden-Harris Administration has announced new actions to address the climate crisis and reduce GHG emissions. These measures aim to integrate the Social Cost of Greenhouse Gases (SC-GHG) into federal agency decision-making processes, including budgeting, procurement, and environmental reviews. By considering the SC-GHG, agencies can better assess the costs of climate change impacts and make informed decisions to mitigate them. Key points of the announcement include:

- **Budgeting**: Federal agencies will incorporate SC-GHG into budget formulation and implementation. This involves measuring programmatic emissions, assessing discretionary grants, considering harm-based penalty calculation methodologies, and evaluating international assistance.

- **Procurement**: The federal government, as a significant purchaser, will integrate SC-GHG into procurement processes. This includes focusing on high-impact procurements, minimizing climate risk in acquisitions, and working with the Federal Acquisition Regulatory (FAR) Council to develop regulatory updates.
- **Environmental Reviews**: Agencies will consider SC-GHG in environmental reviews conducted under the National Environmental Policy Act (NEPA). This includes finalizing NEPA guidance on GHG emissions and using SC-GHG to compare the climate change impacts of proposed actions.

The Office of Global Change is responsible for implementing and managing U.S. international policy on climate change and representing the United States in negotiations under the UNFCCC, and in many other international fora focused on climate change.

The White House Office of Domestic Climate Policy (Climate Policy Office) implements the President's domestic climate agenda, coordinating the all-of-government approach to tackling the climate crisis, creating good-paying union jobs, and advancing environmental justice. EPA tracks and reports GHG emissions, leverages sound science, and invests in America to combat climate change. On March 20, 2024, EPA announced new, more protective final standards to reduce further harmful air pollutant emissions from light-duty and medium-duty vehicles, starting with model year 2027.

The final rule builds upon EPA's federal GHG emissions standards for passenger cars and light trucks for model years 2023 through 2026, established in 2021, and leverages advances in clean car technology to unlock benefits to Americans ranging from improving public health through reducing smog- and soot-forming pollution from vehicles, to reducing climate pollution, to saving drivers money through reduced fuel and maintenance costs. These standards will phase in over model years 2027 through 2032.

EPA issues annual reports on vehicle manufacturers' performance in meeting the GHG emissions standards and on light-duty fuel economy and GHG emission trends since 1975.

EPA's Clean Air Act protections for the oil and natural gas industry help combat climate change and reduce emissions of other harmful air pollutants, including volatile organic compounds contributing to ground-level ozone ("smog") and hazardous air pollutants such as benzene.

8.2.7 Europe's Efforts toward Carbon Neutrality

The EU boasts some of the most comprehensive environmental regulations globally. Understanding and navigating these complex legal frameworks can be daunting for businesses and organizations operating within the EU. This chapter delves into the European Environment Agency's (EEA) role in navigating these regulatory hurdles and ensuring environmental compliance. A two-tiered approach characterizes the EU environmental policy landscape. At the core lies a robust set of directives, regulations, and decisions issued by the European Commission. These instruments establish overarching environmental goals and minimum standards that all member states must implement. However, member states have some flexibility in translating these standards into national legislation, leading to a degree of variation across the EU.

Established in 1990, the EEA is the EU's central agency for environmental information and assessments. Playing a pivotal role in navigating regulatory landscapes, the EEA offers a multitude of resources and services:

- **Legislation Database**: The EEA maintains a comprehensive database of all relevant EU environmental legislation, including directives, regulations, and decisions. This database provides a one-stop shop for businesses to access and understand the regulations applicable to their activities.

- **Guidance Documents**: The EEA develops and publishes guidance documents that clarify the interpretation and implementation of EU environmental legislation. These documents offer invaluable practical advice to businesses on complying with specific regulations.
- **National Implementation Reports**: The EEA collects and analyzes reports submitted by member states on how they have transposed EU environmental legislation into national law. This information can help businesses understand the variations in national regulations across the EU.

The European Green Deal sets the overall roadmap for achieving EU climate neutrality by 2050 by tackling the threat of climate change while growing economically and protecting people's well-being. With the European Climate Law, the EU made climate neutrality by 2050 a legally binding goal, set an interim target of a net 55% emission reduction by 2030, and is working on setting the 2040 target. The Fit for 55 proposal aims to bring EU legislation aligned with the 2030 goal. The impacts of the 2022 gas and energy security crisis highlighted the importance of transitioning faster toward a clean and secure EU energy system. Under the wider umbrella of the European Green Deal, Europe's 2030 policy ambitions include:

- Reducing net GHG emissions to at least 55% below 1990 by 2030.
- Improving the share of renewable energy use.
- Increasing energy efficiency.
- Setting a more ambitious and cost-effective path to achieving climate neutrality by 2050.
- Stimulating green job creation and continuing the EU's track record of cutting GHG emissions while growing its economy.
- Providing more opportunities to recharge or fuel vehicles with alternative fuels and providing alternative power supply for ships and planes.
- Encouraging international partners to increase their ambition to limit the rise in global temperature to 1.5°C and avoid the most severe consequences of climate change.
- Introducing a waste hierarchy for proper waste management includes prevention, reuse, recycling, recovery, and disposal.

The European Union Emissions Trading System (EU ETS) is a cornerstone of the EU's strategy to combat climate change. This cap-and-trade scheme places a limit on GHG emissions from specific industries and sectors, forcing them to reduce their environmental footprint. Understanding the intricacies of the EU ETS is crucial for businesses operating within the EU or with a supply chain that touches member states. This chapter will equip you with the knowledge to navigate this complex regulatory landscape. The EU ETS operates on a simple principle—it caps total emissions and allows companies to trade emission allowances within that limit. These allowances, essentially permits to emit one ton of CO_2-eq, are distributed through a combination of free allocation and auctions. Companies must surrender enough allowances each year to cover their verified emissions.

The EU ETS casts a wide net, encompassing a variety of installations and activities. The main categories include:

- **Power Stations and Combustion Plants**: These include facilities that generate electricity or heat by burning fossil fuels.
- **Energy-intensive Industries**: Sectors like steel, cement, glass, lime, and paper production are all subject to the EU ETS.
- **Aviation**: The system has been in place since 2012 for airlines operating flights within the European Economic Area (EEA).

Compliance with the EU ETS requires a multi-step process:

- **Monitoring and Reporting**: Companies must meticulously monitor and report their emissions to a competent authority. This process necessitates a robust monitoring, reporting, and verification (MRV) plan.
- **Allowance Acquisition**: Businesses can obtain allowances through a free allocation from the government (though this is gradually being phased out) or by purchasing them on the EU ETS market.
- **Surrender of Allowances**: By the year's deadline, companies must surrender enough allowances to cover their verified emissions. Failing to do so incurs hefty fines.

The EU ETS allowance market plays a crucial role in driving emission reductions. The total number of available allowances is gradually reduced over time, creating scarcity and increasing prices. This encourages companies to invest in cleaner technologies and processes to reduce emissions and reduce reliance on purchased allowances.

8.2.8 International Renewable Energy Agency

The global transition toward renewable energy presents a complex landscape for project developers and investors. Beyond the technical considerations of integrating renewables into the grid and ensuring efficient project operation, navigating the regulatory frameworks of different countries is paramount for success. This chapter delves into the intricacies of navigating regulatory jurisdictions for renewable energy projects, specifically focusing on the insights and resources offered by the International Renewable Energy Agency (IRENA).

The renewable energy sector is characterized by a dynamic and evolving regulatory environment. Unlike traditional fossil fuel-based generation, renewable energy projects often require new regulatory frameworks to address grid integration, power purchase agreements (PPAs), and environmental impact assessments. This complexity is further amplified by the international dimension, where individual countries establish regulatory frameworks tailored to their specific needs and contexts.

For project developers and investors venturing into new markets, navigating this regulatory labyrinth can be daunting. Inconsistent regulatory frameworks, bureaucratic hurdles, and a lack of transparency can create significant challenges, impacting project feasibility, bankability, and overall investment attractiveness. IRENA, an intergovernmental organization dedicated to promoting the widespread adoption and sustainable use of all forms of renewable energy, is crucial in assisting stakeholders in navigating regulatory complexities. IRENA's work focuses on:

- **Knowledge Sharing**: IRENA provides a comprehensive knowledge base on renewable energy regulations across different countries. This includes resources such as the "Rethinking Regulation" platform, which offers detailed information on regulatory frameworks, best practices, and case studies from around the world.
- **Capacity Building**: IRENA supports governments in developing and implementing effective regulatory frameworks for renewables. This includes capacity-building workshops, technical assistance programs, and the development of model regulatory guidelines.
- **Policy Advocacy**: IRENA advocates for harmonization and streamlining of renewable energy regulations across different countries. This promotes regulatory predictability and transparency, ultimately fostering a more attractive investment environment for the sector.

When considering a renewable energy project in a new jurisdiction, project developers and investors should carefully assess the following regulatory aspects:

- **Project Permitting**: Understanding the permitting process for renewable energy projects—including timelines, required documentation, and potential environmental impact assessments—is crucial.
- **Grid Integration Regulations**: The regulations governing grid access, interconnection requirements, and potential wheeling charges for renewable energy projects need to be thoroughly evaluated.
- **Power Purchase Agreements (PPAs)**: The structure and terms of PPAs, including feed-in tariffs, auctions, and power pool arrangements, significantly impact project economics.
- **Fiscal Incentives**: Available fiscal incentives such as tax breaks, accelerated depreciation allowances, and feed-in tariffs can improve project viability and attract investment.
- **Local Content Requirements**: Some countries may have local content requirements mandating the use of domestically produced materials or services in renewable energy projects.

IRENA offers a multitude of resources to assist stakeholders in navigating regulatory complexities. Here are some key resources to explore:

- **Rethinking Regulation Platform**: This online platform provides comprehensive information on renewable energy regulations across different countries, including legal frameworks, best practices, and case studies (https://www.irena.org/Energy-Transition/Policy).
- **Regulatory Frameworks for Renewable Energy Auctions**: This report guides the design and implementation of effective auction schemes for renewable energy projects (https://www.irena.org/-/media/Files/IRENA/Agency/Publication/2015/Jun/IRENA_Renewable_Energy_Auctions_A_Guide_to_Design_2015.pdf).
- **Renewable Energy Auctions**: A Guide to Design: This guide offers a step-by-step approach to designing and implementing successful renewable energy auctions (https://www.irena.org/publications/2015/Jun/Renewable-Energy-Auctions-A-Guide-to-Design).

8.2.9 Global Environment Facility

The global environment facility (GEF) plays a critical role in financing projects that address global environmental challenges like climate change, biodiversity loss, and land degradation. However, navigating the complex web of regulatory frameworks across participating countries can be a significant hurdle for project developers and implementers. This chapter delves into the GEF's regulatory landscape, equipping stakeholders with a roadmap for navigating these complexities.

The GEF is a unique partnership among 183 member governments, international institutions, civil society organizations, and private sectors. It operates under a multilateral framework established by international agreements like the Convention on Biological Diversity (CBD) and the UNFCCC. These conventions set broad environmental goals and principles, but the specific regulatory frameworks for achieving them are often defined at the national level.

Several key areas require attention when navigating the regulatory landscape of the GEF:

- **National Environmental Laws**: Each participating country has its environmental laws and regulations that govern activities related to project implementation. These laws may address environmental impact assessments (EIAs), protected areas, and waste management. Understanding these national laws is crucial for ensuring project compliance and avoiding delays.
- **Social Safeguards**: The GEF has social and environmental safeguards that aim to minimize project risks and ensure positive social impacts. These safeguards address involuntary resettlement, Indigenous peoples' rights, and gender equality. Project developers must demonstrate compliance with both national and GEF safeguard policies.

- **Access and Benefit Sharing (ABS)**: The Nagoya Protocol [Nagoya Protocol on Access to Genetic Resources and the Fair and Equitable Sharing of Benefits Arising from their Utilization to the CBD] governs access to genetic resources and the fair and equitable sharing of benefits arising from their utilization. Projects involving the use of genetic resources need to comply with the Nagoya Protocol and relevant national regulations.
- **Procurement**: The GEF has its procurement guidelines that project implementers must follow. These guidelines address issues like transparency, competition, and value for money. Understanding these guidelines is essential for ensuring efficient and accountable project execution.

Several strategies can help project developers and implementers navigate the complex regulatory landscape:

- **Early Engagement**: Stakeholders should engage with relevant national and local authorities early in the project development process. This allows for early identification of potential regulatory hurdles and facilitates discussions on achieving compliance.
- **Capacity Building**: It is crucial to build the capacity of project implementers on relevant environmental laws and GEF safeguards. Training programs and workshops can equip project teams with the knowledge and skills to navigate the regulatory framework effectively.
- **Seeking Guidance**: The GEF Secretariat and its implementing agencies (World Bank, UNDP, UNEP) offer guidance on navigating regulatory requirements. They can provide resources and expertise to support project developers in understanding and complying with relevant regulations.
- **Partnerships**: Partnering with local NGOs and legal experts can be invaluable in navigating national regulatory complexities. These partners can provide insights into specific regulations and facilitate dialogue with relevant authorities.

8.2.10 World Bank Climate Investment Funds

The World Bank's Climate Investment Funds (CIF) are pivotal in channeling global climate finance toward developing countries. These funds support projects that aim to reduce GHG emissions, strengthen climate resilience, and promote sustainable development.

The CIF is an innovative financing mechanism housed within the World Bank. It comprises several thematic funds, each with a specific focus area, such as clean technology, forestry, and climate resilience. While the CIF provides overarching investment criteria and environmental and social safeguards, recipient countries are ultimately responsible for translating these principles into action. This creates a multi-layered regulatory environment that requires careful consideration.

Several key areas require particular attention when navigating the regulatory landscape of the CIF:

- **National Policies and Strategies**: Developing countries receiving CIF support typically have their own national climate change policies, strategies, and action plans (NAPs). These documents outline national priorities and mitigation and adaptation goals. Aligning CIF projects with these national frameworks is crucial for ensuring project coherence and effectiveness.
- **Environmental and Social Safeguards**: The CIF adheres to a robust set of environmental and social safeguards that aim to minimize project risks and maximize positive social impacts. These safeguards address issues like EIAs, involuntary resettlement, and labor standards. Project proponents must demonstrate compliance with both CIF safeguards and relevant national regulations.

- **Procurement**: The World Bank's procurement guidelines govern the selection of goods, works, and services for CIF-funded projects. These guidelines emphasize transparency, competition, and value for money. Understanding these guidelines is essential for efficient and accountable project implementation.
- **Access to Finance**: Recipient countries often have established procedures for accessing and managing international climate finance. These procedures may involve national designated authorities (NDAs) or specific channels for project approval. Understanding these access mechanisms is crucial for project proponents to navigate the funding landscape effectively.

8.2.11 Green Climate Fund

The Green Climate Fund (GCF) plays a pivotal role in channeling climate finance toward developing countries. However, navigating the GCF's regulatory landscape can be complex, with various jurisdictional considerations at play. This chapter explores the key regulatory frameworks governing the GCF, providing guidance for project developers and recipient countries to ensure successful project implementation.

The GCF was established by the COP to the UNFCCC at the 2010 Cancun Climate Change Conference. Its mandate is to support developing countries in their efforts to undertake transformative climate action through a combination of grants, loans, guarantees, and equity.

The GCF operates under the authority of the UNFCCC and is governed by a Board comprised of an equal number of representatives from developed and developing countries. The Board sets strategic policies, approves funding proposals, and oversees the Fund's operations.

Several key regulatory frameworks influence how the GCF operates and how project developers and recipient countries must engage with the Fund. These include:

- **UNFCCC Decisions**: The COP guidance and decisions form the core foundation for the GCF's operations. These decisions outline the Fund's objectives, eligibility criteria, and environmental and social safeguards.
- **Governing Instrument for the Green Climate Fund**: This document, adopted by the Board in 2012, establishes the GCF's legal framework, governance structure, and operational modalities.
- **Environmental and Social Safeguards (ESS)**: The GCF adheres to a robust set of ESS standards that aim to minimize environmental and social risks associated with funded projects. These standards encompass environmental assessment, social impact assessment, stakeholder engagement, and grievance redress mechanisms.
- **National Designated Authorities (NDAs)**: Each developing country government designates an NDA as its focal point for the GCF. The NDA serves as the official channel for communication between the country and the GCF, including endorsement of funding proposals.
- **Accredited Entities (AEs)**: The GCF works through AEs, which are national, regional, or international institutions with experience in climate change project development and implementation. AEs partner with project developers to prepare and submit funding proposals to the GCF.

8.2.12 Climate Action Network

The fight against climate change is a complex one, demanding action on multiple fronts. While international agreements and global movements play a crucial role, effective climate action often hinges on navigating the intricate web of regulations at national and sub-national levels. This chapter equips the climate action network (CAN) with a framework to navigate these regulatory jurisdictions, maximizing its impact in advocating for robust climate policies. Climate change regulations encompass a vast array of legislative frameworks, policies, and standards. Here's a breakdown of the key levels:

- **International Agreements**: These set global goals for emission reduction and adaptation, such as the Paris Agreement. While not directly binding on individual countries, they create a strong international framework for national action.
- **National Regulations**: These translate international agreements into domestic policy. National governments enact laws mandating emission reductions, promoting renewable energy, and establishing carbon pricing mechanisms.
- **Sub-national Regulations**: Many countries empower local governments to set stricter environmental regulations or implement innovative climate solutions. This includes city-level building codes promoting energy efficiency or regional emissions trading schemes.

8.2.13 International Standards Organization

ISO 14064 is an international standard for GHG accounting and verification, providing organizations with a framework to measure, quantify, and report their GHG emissions. It is divided into three parts. The first part, "Specification with Guidance at the Organization Level for Quantification and Reporting of Greenhouse Gas Emissions and Removals," focuses on helping organizations quantify and report their GHG emissions and removals. It involves setting organizational boundaries for GHG accounting, identifying and categorizing GHG sources and sinks, selecting appropriate methodologies for quantifying GHG emissions and removals, reporting these emissions and removals, and ensuring transparency, consistency, and accuracy in the GHG inventory.

The second part, "Specification with Guidance at the Project Level for Quantification, Monitoring, and Reporting of Greenhouse Gas Emission Reductions or Removal Enhancements," addresses GHG projects by providing guidelines for quantifying, monitoring, and reporting GHG emission reductions or removal enhancements. It includes defining project boundaries, identifying baseline scenarios and determining baseline emissions, selecting and applying methodologies for quantifying GHG emission reductions or removals, monitoring and reporting project performance, and ensuring the credibility and accuracy of reported data.

The third part, "Specification with Guidance for the Validation and Verification of Greenhouse Gas Assertions," outlines the principles and requirements for validating and verifying GHG assertions. It covers establishing the validation and verification process, defining the roles and responsibilities of verifiers, setting criteria for validation and verification, planning and conducting validation and verification activities, reporting findings, and ensuring the independence, impartiality, and competence of the verification process.

8.2.14 Greenhouse Gas Protocol

The GHG Protocol is a comprehensive global standardized framework used to measure and manage GHG emissions from private and public sector operations, value chains, and mitigation actions. Widely recognized and used by governments, businesses, and other organizations, it helps account for and report on GHG emissions. The protocol was developed through a partnership between the World Resources Institute (WRI) and the World Business Council for Sustainable Development (WBCSD), resulting from a broad stakeholder engagement process including governments, NGOs, businesses, and other organizations.

To ensure that your inventory represents a faithful, true, and fair account of your company's emissions, adhere to the GHG accounting and reporting principles:

- Relevance
- Completeness
- Consistency
- Transparency
- Accuracy.

The GHG Protocol consists of several standards and guidelines. The Corporate Standard provides a framework for companies to measure and manage their GHG emissions, categorizing them into three "scopes": Scope 1 (direct emissions from owned or controlled sources), Scope 2 (indirect emissions from the generation of purchased electricity, steam, heating, and cooling consumed by the reporting company), and Scope 3 (all other indirect emissions that occur in a company's value chain). The Scope 3 Standard offers guidance for organizations to measure and report their full value chain emissions, including upstream and downstream activities. The Corporate Value Chain (Scope 3) and Product Life Cycle Standards help companies assess their value chain and product-level emissions to identify reduction opportunities. Additionally, the Mitigation Goal Standard provides a standardized approach for designing, tracking, and reporting progress toward GHG reduction goals, and the Project Protocol offers a methodology for quantifying and reporting GHG reductions from specific mitigation projects.

The GHG Protocol ensures consistency and transparency in GHG accounting and reporting, offering flexibility that can be applied to a variety of sectors and organizations. Its credibility builds trust among stakeholders by providing rigorous and well-established methodologies. Implementing the GHG Protocol helps organizations with environmental management by identifying emissions sources, setting reduction targets, and tracking progress. It aids in risk management by recognizing and mitigating climate-related risks, supports compliance with regulatory and voluntary reporting requirements, and enhances credibility and demonstrates leadership in sustainability and climate action.

The GHG Protocol provides a range of tools, training, and resources to help organizations implement its standards, including software solutions that integrate its guidelines to facilitate data collection, analysis, and reporting. Fundamental to the global effort to address climate change, the GHG Protocol offers a consistent and transparent approach to measuring and managing GHG emissions across diverse sectors and activities. While the GHG Protocol identifies, explains, and provides options for GHG inventory best practices, ISO 14064 establishes minimum standards for compliance with these best practices.

8.3 NON-STATE ACTORS

Non-state actors, encompassing a diverse array of entities such as NGOs, civil society organizations, local governments, and business entities, play a vital role in advancing climate strategies alongside governmental efforts. General non-state actors, including academic institutions, investors, and sub-national bodies like municipalities and provinces, contribute through advocacy, implementation, and the mobilization of resources. Civil society organizations focus on public engagement, equitable solutions, and fostering accountability in climate action. Local governments translate national policies into actionable initiatives, employing urban planning, waste management, and zoning regulations to reduce emissions and enhance resilience. Business actors, including corporations and private enterprises, drive innovation and sustainability through investments in renewable energy, low-carbon technologies, and sustainable supply chain practices. Together, these actors provide essential expertise, resources, and local-level action, amplifying the global response to climate change and addressing challenges beyond the capacity of state-led initiatives alone.

8.3.1 Finance Sector Initiatives

Many parts of the finance sector have come together in alliances, reflecting different types of institutions such as banks, asset owners, investors, and insurance. Some of the main initiatives are presented below. While several other smaller national, regional, and global initiatives are engaging in the finance sector, the ones listed are dominant and often include smaller initiatives as members.

8.3.1.1 Glasgow Financial Alliance for Net Zero

Building on the already described Global Race to Zero initiative led by the Climate Champions, a new sector-wide coalition, the Glasgow Financial Alliance for Net Zero (GFANZ), was launched at COP 26 to broaden, deepen, and raise the net-zero ambitions across the financial system. The alliance is mainly focusing on raising commitments and engaging members, with a requirement to develop credible roadmaps for a net-zero transition. It brings together the alliances established by different financial actors, as discussed below.

8.3.1.2 Net-Zero Banking Alliance

The Net-Zero Banking Alliance (NZBA) is industry-led and is convened by the United Nations Environment Programme's Finance Initiative. It brings together banks from a broad set of countries that have signed a formal commitment statement to align their lending and investment portfolios with net-zero emissions by 2050. Combining near-term action with accountability, this ambitious commitment sees banks setting an intermediate target for 2030 or sooner, using robust, science-based guidelines. The NZBA was launched in April 2021 and, after one year, had more than 100 banks as members. It works to reinforce, accelerate, and support the implementation of members' decarbonization strategies and has published a set of Guidelines for Climate Target Setting for Banks. In addition to its links to the GFANZ, the NZBA constitutes the climate part of the wider UNEP initiative on Principles for Responsible Banking, which focuses on engaging banks in a commitment to the wider Sustainable Development Goals. As of 2022, the initiative had around 129 signatories from 41 countries representing almost half of global banking assets.

8.3.1.3 Net-Zero Asset Owner Alliance

The Net-Zero Asset Owner Alliance (NZAOA), launched in September 2019, is organized in a similar manner as the NZBA, with the UNEP Finance Initiative as the convener and secretariat. Members sign up to a common commitment statement, including transitioning investment portfolios to net-zero GHG emissions by 2050, and regularly report on progress, including establishing intermediate targets every five years in line with Article 4.9 of the Paris Agreement. As of 2023, the NZAOA had 86 members with more than $10 trillion in assets. The NZAOA has had a dedicated MRV function since its establishment in 2019 and requires annual target, progress, and emissions reporting for its members, selected from across more than 100 metrics or key performance indicators. All key performance indicators of the alliance have now been integrated across the Principles for Responsible Investment reporting framework, and more than 5,000 investors now have the option to report against the same metrics and key performance indicators. The banking and insurance initiatives presented below will move toward a similar approach modeled on the AOA work.

8.3.1.4 Net Zero Asset Managers Initiative

The Net Zero Asset Managers Initiative (NZAMI) is also a partner of the GFANZ. It functions as an umbrella structure for six investor network partners that cover different regional groups of investors:

1. The Investor Group on Climate, with members in Australia and New Zealand
2. A wider Asia Investor Group on Climate Change covering investors active in China, Hong Kong, India, Indonesia, Japan, Malaysia, the Republic of Korea, Singapore, Taiwan, Thailand, the Philippines, Viet Nam and the Pacific

3. The Institutional Investors Group on Climate Change, which is a European membership body for investor collaboration on climate change
4. The Principles for Responsible Investment initiative, with its more than 5,000 signatories
5. CERES as a knowledge and engagement partner, bringing the CERES Investor Network
6. CDP as the disclosure partner.

In total, the NZAMI has 273 signatories with a total asset management portfolio of over $60 trillion. The same members also come together in the Climate Action100+ initiative, which has a similar mandate but a longer history and wider reach, with around 700 companies engaged.

8.3.1.5 Net-Zero Insurance Alliance

The Net-Zero Insurance Alliance (NZIA) is a group of over 20 leading insurers representing more than 11% of the world insurance premium volume globally. The alliance currently represents the insurance business in the GFANZ, and members have committed to transitioning their insurance and reinsurance underwriting portfolios to net-zero GHG emissions by 2050. The NZIA is collaborating with the Partnership for Carbon Accounting Financials to develop the first global standard to measure and disclose emissions attributable to insurance underwriting portfolios—or "insurance-associated emissions." This pioneering global standard is scheduled to be launched later in 2023. NZIA members are required to publish their respective first interim science-based targets within six months after the publication of an NZIA Target-Setting Protocol, which will build on the NZIA's work with the Partnership for Carbon Accounting Financials to develop a global standard to measure insured emissions.

8.3.2 Industry Sector Initiatives

The picture for the industry sector is very similar, with several global convening partnerships and many sub-sector-specific partnerships and engagement initiatives. The Global Race to Zero campaign mobilizes more than 5,000 private companies through these partnerships and initiatives. The main ones are briefly introduced below, again without trying to be exhaustive.

8.3.2.1 Global Cement and Concrete Association

The Global Cement and Concrete Association announced a net-zero pledge and roadmap before COP 26 in 2021, with 40 major companies worldwide coming together to pledge a 25% reduction in GHG emissions by 2030 and net-zero emissions by 2050.

8.3.2.2 Responsible Steel

Responsible Steel is a not-for-profit organization that provides a sustainability certification scheme for the steel industry worldwide. Together with the Climate Group, Responsible Steel has established SteelZero as a convener of the steel industry, with an aim to accelerate a transition to net-zero emissions in 2050. The initiative also engages steel users in, for example, the auto and shipping industry to create a demand requirement for net-zero steel in the future. International Wineries for Climate Action (IWCA) is a collaboration among environmentally committed wineries taking a science-based approach to reducing carbon emissions across the wine industry. The initiative develops methodologies to measure GHG emissions specifically targeted to the wine industry, building on the GHG Protocol (discussed later). IWCA membership requires that members regularly carry out GHG emissions inventories covering scopes 1–3 of their operations and have those inventories verified to the ISO 14064-1 standard.

8.3.2.3 Pledge to Net Zero

Pledge to Net Zero similarly engages the wider environment industry in a pledge and reporting initiative, providing guidance material that builds on the GHG Protocol and is adapted to specific needs in the sub-sector.

8.3.2.4 The Oil & Gas Methane Partnership 2.0

The Oil & Gas Methane Partnership 2.0 (OGMP 2.0) is a multi-stakeholder initiative launched in 2020 by UNEP and the Climate and Clean Air Coalition. The OGMP 2.0 provides a comprehensive, measurement-based reporting framework for the oil and gas industry to help improve the accuracy and transparency of reporting on methane emissions. Over 100 companies with assets on five continents, representing a significant share of the world's oil and gas production, have joined the partnership. OGMP 2.0 members also include natural gas transmission and distribution pipeline operators, gas storage capacity, and liquefied natural gas terminals. The OGMP 2.0 Reporting Framework provides a standard for methane reporting, requiring companies to report methane emissions from all sources at both operated and non-operated ventures across the oil and gas value chain. With the Global Methane Pledge launched at COP 26 in Glasgow in 2021, more than 150 countries have agreed to reduce methane emissions by 30% by 2030 collectively, and, in this way, provide a strong focus on private sector methane reductions in key sectors like oil and gas, waste handling, and agriculture.

8.3.3 Food, Agriculture and Forestry Sector Initiatives

8.3.3.1 Food Emissions 50

Food Emissions 50 is convening an investor-led initiative to accelerate progress toward a net-zero future in the food sector. Engaging 50 of North America's highest-emitting public food companies, investors seek to move companies to improve their GHG emission disclosures, set ambitious emission reduction targets, and implement ambitious climate transition plans in line with the Paris Agreement.

8.3.3.2 Alliance for Responsible Forest Management supports

Alliance for Responsible Forest Management supports green initiatives in forestry along with other institutions, such as WWF and Conservation International. The alliance has a vision to accelerate responsible forest management across the commercial tropical world and contribute to reducing GHG emissions from entering the atmosphere from the forest sector. This is done by developing best-practice training manuals at the regional level and providing localized training in Africa, the Asia-Pacific region, and Latin America.

8.4 CONCLUSION

Climate action is shaped by a complex web of regulations and policies that span local, national, and international jurisdictions. This chapter has unpacked how cities, countries, and global organizations each play a distinct role in setting and enforcing climate rules—from zoning laws and building codes at the city level to international agreements like the Paris Agreement and the European Green Deal. We have seen how local governments drive innovation through urban planning, waste management, and city networks while national policies and international frameworks set the targets and provide the resources for large-scale change. Non-state actors—businesses, NGOs, and the financial sector—are becoming increasingly influential, advocating for higher standards, greater transparency, and increased investment in climate solutions.

The path to effective climate action is not straightforward. Overlapping regulations, varying enforcement, and the need for coordination across sectors and borders make implementation challenging. However, collaboration, innovation, and capacity-building are emerging as key factors in aligning efforts and accelerating progress. As climate risks intensify, the ability to navigate this regulatory landscape—adapting to new standards, leveraging financial mechanisms, and building partnerships—will determine how quickly and equitably we move toward a sustainable future.

The next chapter shifts from policy frameworks to the frontiers of technology and innovation. Chapter 9 examines the latest advancements in carbon reduction, including carbon capture and green hydrogen, as well as AI-driven solutions and waste valorization. By connecting regulatory action with technological progress, the discussion moves from "what rules shape climate action" to "what tools can drive it further," equipping readers with insight into the innovations that will define the next era of climate solutions.

REVIEW QUESTIONS

1. What are zoning and land-use regulations, and how do they impact carbon emissions?
2. How do building codes influence energy efficiency in urban development?
3. What is the role of waste management regulations in reducing emissions?
4. How do local city initiatives contribute to climate change mitigation?
5. How does the UNFCCC support international climate policy?
6. What are the challenges of enforcing the Paris Agreement on a global scale?
7. How do non-state actors, like financial institutions, influence climate policy?
8. What is the significance of the GHG Protocol in emissions reporting?
9. How does the EU's carbon neutrality goal affect global climate action?
10. What is the role of the International Renewable Energy Agency (IRENA) in promoting clean energy?

REFERENCES

Belokurov, A., Dudley, N., Krueger, L., Lopoukhine, N., MacKinnon, K., Sandwith, T., Sekhran, N., Stolton, S. (2010). Natural *Solutions: Protected Areas Helping People Cope with Climate Change (Korean Version).* IUCN.

Datta, R., Chapola, J., Owen, K., Hurlbert, M., Foggin, A. (2024). Indigenous land-based practices for climate crisis adaptions. *Explore-the Journal of Science and Healing*. https://doi.org/10.1016/j.explore.2024.103042

Dudley, N., Stolton, S., Belokurov, A., Krueger, L., Lopoukhine, N., MacKinnon, K., Sandwith, T., Sekhran, N. (2011). Natural solutions: Protected areas helping people cope with climate change.*Oryx*. https://doi.org/10.1017/S0030605311001608

Gauer, V.H., Schaepe, D.M., Welch, J.R. (2021). Supporting indigenous adaptation in a changing climate: Insights from the Stó:lō Research and Resource Management Centre (British Columbia) and the Fort Apache Heritage Foundation (Arizona). *Elementa: Science of the Anthropocene*. https://doi.org/10.1525/ELEMENTA.2020.00164

Gil-Ozoudeh, I., Iwuanyanwu, O., Okwandu, A.C., Ike, C.S. (2023). Sustainable urban design: The role of green buildings in shaping resilient cities. *International Journal of Applied Research in Social Sciences*. https://doi.org/10.51594/ijarss.v5i10.1481

Gürçam, S. (2023). Paving the way for climate resilience through sustainable urbanization: A comparative study. *Lectio Socialis*. https://doi.org/10.47478/lectio.1396483

Halevi, S.J., Giordano, C.R., Brunt, M.E.V. (2023). Moving up the waste hierarchy: Modeled emissions reductions from applying sustainable waste management practices in the United States. Proceedings of the 2022 EEC/WTERT Conference. Ed. Castaldi, MJ, & Ali, SS. ASME, 2023

Jamme, H.T., Rodriguez, J., Bahl, D., Banerjee, T. (2019). A twenty-five-year biography of the TOD concept: From design to policy, planning, and implementation. *Journal of Planning Education and Research*. https://doi.org/10.1177/0739456X19882073

Khreis, H., Sudmant, A., Gouldson, A., Nieuwenhuijsen, M.J. (2018). Transport policy measures for climate change as drivers for health in cities. *Journal of Transport and Health*. https://doi.org/10.1016/J.JTH.2018.05.035

Li, R. (2024). Greenhouse gas mitigation potential via a large-scale compulsory circular waste management system. *Journal of Environmental Management*. https://doi.org/10.1016/j.jenvman.2024.122425

Lodh, D. (2024). Sustainable waste management practices and their profound impact on environmental conservation. In *Futuristic Trends in Agriculture Engineering & Food Sciences* (Vol. 3, Book 21, pp. 7–18). IIP Series. https://www.doi.org/10.58532/V3BCAG21P1CH2

Mcgregor, D., Sritharan, M. (2023). Decolonising the dialogue on climate change. In S. Lamalle, P. Stoett (Eds.), *Indigenous Knowledges, Legal Orders and Ethics* (pp. 66–86). Cambridge University Press. https://doi.org/10.1017/9781108769327.005

Moos, M., Vinodrai, T., Revington, N., Seasons, M. (2018). Planning for mixed use: Affordable for whom? *Journal of the American Planning Association*. https://doi.org/10.1080/01944363.2017.1406315

Platonova-Oquab, A., Shenvi, A. (2023). The *Power of Cities: Harnessing Low-carbon Urbanization for Climate Action*. World Bank. https://doi.org/10.1596/40635

Rani, P. (2024). Sustainable urban infrastructure: Innovations in green building technologies and urban planning. *Journal of Sustainable Solutions*. https://doi.org/10.36676/j.sust.sol.v1.i1.04

Santopietro, L., Scorza, F. (2024). Voluntary planning and city networks: A systematic bibliometric review addressing current issues for sustainable and climate-responsive planning. *Sustainability*. https://doi.org/10.3390/su16198655

Schilling-Vacaflor, A., Gustafsson, M.-T. (2022). Indigenous-led spaces in environmental governance. In K. Ruckstuhl, I.A. Velásquez Nimatuj, J.-A. McNeish, N. Postero (Eds.), *The Routledge Handbook of Indigenous Development*. Routledge. https://doi.org/10.4324/9781003153085-22

Tong, X., Tong, X., Wang, Y., Chan, E.H.W., Zhou, Q. (2018). Correlation between transit-oriented development (TOD), land use catchment areas, and local environmental transformation. *Sustainability*. https://doi.org/10.3390/SU10124622

Tran, M., Salamanca, A. (2023). Advancing climate policy: Harnessing Indigenous knowledge at the science-policy interface. In *SEI Working Paper*, Stockholm.

Wang, X., Xie, Y., Xia, L., He, J., Lin, B. (2024). Investigating the effect of transit-oriented development (TOD) on social equity—Examining the displacement of Footscray, Melbourne. *Buildings*. https://doi.org/10.3390/buildings14030824

9 Innovations in Carbon Reduction

This chapter explores innovative technologies and approaches in carbon reduction, highlighting the potential of emerging solutions to mitigate climate change. It delves into cutting-edge technologies like carbon capture, utilization, and storage (CCUS), renewable energy advancements, and energy efficiency innovations. This chapter also discusses the role of carbon offsetting, artificial intelligence (AI), electrification, green hydrogen, and big data in reducing carbon emissions. It further explores waste valorization, bio-based fuels and polymers, and emission monitoring as crucial components of a comprehensive carbon reduction strategy. Additionally, this chapter examines the significance of accurate and transparent carbon data, including primary, secondary, and dynamic data, in driving effective climate action. It also addresses the limitations and drawbacks of relying solely on carbon removal technologies, emphasizing the importance of emissions reduction across all sectors.

9.1 CATEGORIES OF CARBON REDUCTION TECHNOLOGIES

The urgent need to mitigate carbon emissions has catalyzed the development of diverse technologies, each tailored to address specific facets of the global climate challenge. These innovations span a wide spectrum of approaches, from capturing and storing carbon emissions to transitioning to renewable energy sources, leveraging biological and chemical processes, and utilizing data-driven solutions for optimization. Categorizing these technologies based on their technological approach, stage of development, and scalability provides critical insights into their potential to reduce carbon emissions and their readiness for deployment. This structured perspective underscores the multifaceted strategies required to combat climate change effectively.

9.1.1 Technological Approach

9.1.1.1 Capture and Storage Technologies

1. Carbon Capture, Utilization, and Storage (CCUS)
2. Direct Air Capture (DAC)
3. Cryogenic Carbon Capture (CCC).

These technologies focus on removing CO_2 directly from emission sources or the atmosphere and storing it or converting it into useful products.

9.1.1.2 Energy-Based Technologies

1. Renewable Energy Advancements
2. Electrification
3. Green Hydrogen
4. Energy-Efficient Integrations

These approaches aim to reduce carbon emissions by transitioning to cleaner energy sources and improving energy efficiency.

DOI: 10.1201/9781003479666-9

9.1.1.3 Biological and Chemical Technologies

- Bio-based Fuels and Polymers
- Waste Valorization
- Electrochemical Ocean Carbon Dioxide Removal.

These technologies leverage biological processes or chemical reactions to reduce carbon emissions or create sustainable alternatives to carbon-intensive products.

9.1.1.4 Data and Artificial Intelligence-Driven Technologies

- Artificial Intelligence
- Big Data Analytics
- Digital Twin Technology
- Emission Monitoring Systems.

These technologies use advanced data processing and analysis to optimize energy use, predict emission patterns, and improve overall efficiency in carbon reduction efforts.

9.1.2 Stage of Development

9.1.2.1 Established Technologies

- Renewable Energy (solar, wind)
- Carbon Offsetting
- Energy-Efficient Integrations.

These technologies are already widely implemented and have proven their effectiveness in reducing carbon emissions.

9.1.2.2 Emerging Technologies

- Green Hydrogen
- Advanced CCUS Systems
- Direct Air Capture.

These technologies show significant promise but are still in the process of scaling up and proving their long-term viability.

9.1.2.3 Experimental Technologies

- Electrochemical Ocean Carbon Dioxide Removal
- Advanced Biopolymers
- Cryogenic Carbon Capture.

These technologies are in earlier stages of development and require further research and testing before widespread implementation.

9.1.3 Complexity and Scale

9.1.3.1 Large-Scale Industrial Solutions

- CCUS for Power Plants
- Green Hydrogen Production
- Direct Air Capture Facilities.

These technologies often require significant infrastructure and are typically implemented at an industrial scale.

9.1.3.2 Modular and Scalable Solutions

- AI-driven Energy Optimization
- Waste Valorization Systems
- Emission Monitoring Technologies.

These solutions can be implemented at various scales, from individual buildings to entire cities or industrial complexes.

9.1.3.3 Localized and Specific Applications

- Bio-based Fuels and Polymers
- Energy-Efficient Integrations for Buildings
- Carbon Offsetting Projects.

These technologies can be applied to specific sectors or localized areas to address particular carbon emission challenges.

9.2 CARBON CAPTURE, UTILIZATION, AND STORAGE

CCUS technologies, encompassing carbon capture, utilization, and storage, are pivotal in mitigating CO_2 emissions from industrial processes and power plants, thereby addressing climate change. These technologies capture CO_2 emissions and either store them underground or convert them into valuable products such as building materials and synthetic fuels (Wu et al., 2024; Nath et al., 2024). Recent advancements in membrane separation and solvent-based techniques have significantly enhanced the efficiency of these systems, while innovations like modular photobioreactors using algae have shown promise in converting emissions into biomass (Shammi & Akter, 2024). Despite these technological strides, challenges such as scalability and cost remain significant barriers to widespread adoption. The cement industry, for instance, could achieve near climate-neutrality by integrating CCUS with biomass and decarbonized electricity, potentially mitigating up to 10_6% of its operational emissions. However, the lack of large-scale application in industries like cement production highlights the need for further development and investment (Dávila et al., 2023). Additionally, integrating CCUS with new energy sources could enhance energy conversion and storage, offering a promising pathway for future energy systems systems (Hu, 2023).

Addressing these challenges requires sustained investment, robust regulatory frameworks, and public support to fully unlock the potential of CCUS technologies in achieving global carbon neutrality goals.

Startups utilize a range of technologies, some using membranes and others using solvents, to capture CO_2. The captured carbon dioxide is then concentrated either chemically or in biological processes, which further serves as a feedstock or is simply stored underground. By preventing the CO_2 from industrial operations from reaching the atmosphere, CCUS stabilizes CO_2 levels. This provides carbon-neutral feedstock alternatives to industries like oil and gas, aviation, construction, and chemical. CO_2 storage solutions, on the other hand, facilitate the use of depleted oil or gas fields or saline formations.

9.2.1 Algiecel Advances Algal-Based Carbon Capture

The use of microalgae for carbon capture is advantageous due to its ability to sequester CO_2 through photosynthesis, thereby mitigating greenhouse gas emissions and contributing to climate change mitigation efforts. The integration of microalgae cultivation with industrial CO_2 emissions, such as those from power plants, enhances the feasibility of large-scale applications, as it provides a continuous CO_2 supply necessary for high growth rates. Furthermore, microalgae can be genetically improved to enhance CO_2 sequestration efficiency and biomass productivity, which is crucial for sustainable development and the production of value-added products like biofuels and food-grade compounds. The economic viability of such systems is supported by technoeconomic analyses, which suggest that the production of biomass at competitive prices is achievable, although further research is needed to reduce costs and improve energy efficiency. Additionally, microalgae photobioreactors can simultaneously treat wastewater, capturing CO_2 and recovering valuable resources, thus offering a multifaceted solution to environmental challenges. This aligns with the potential of microalgae to replace petroleum-derived fuels sustainably, without impacting the food supply chain.

UK-based startup Algiecel develops modular photobioreactors that convert CO_2 emissions and LED light into microalgae biomass and oxygen. The startup's compact mobile microalgae photobioreactors and downstream units fit into standard shipping containers. This enables the transformation of industrial CO_2 emissions into biomass rich in food-grade omega-3, proteins, vitamins, and carotenoids. The carbon capture solution in this way allows industrial plants to reduce their carbon emissions and enable a new revenue stream.

9.2.2 Aqualung Achieves Absorbant-Free Carbon Capture

Aqualung's innovative carbon capture technology leverages membrane separation, a method that has gained attention for its compactness, energy efficiency, and environmental benefits compared to traditional methods. This approach aligns with recent advancements in membrane technology, which focus on enhancing performance through novel materials like thermally rearranged polymers and ionic liquid membranes, aiming to improve sustainability and cost-effectiveness. Aqualung's system is particularly notable for its elimination of absorbents, which reduces the energy typically required for regeneration and avoids the use of toxic additives, thereby lowering operational risks and maintenance costs. This is consistent with the broader trend in carbon capture technologies that seek to minimize chemical use and enhance energy efficiency, as seen in other innovative approaches like cryogenic technology that also avoid chemical reagents.

Furthermore, Aqualung's modular design allows for deployment in environments with low CO_2 concentrations, a feature that is crucial for expanding the applicability of carbon capture technologies across various industrial settings. This adaptability is essential for integrating carbon capture into existing systems and achieving carbon neutrality, as emphasized in discussions on carbon-negative technologies. Overall, Aqualung's technology represents a significant step forward in the development of sustainable and efficient carbon capture solutions, aligning with current research trends and addressing key challenges in the field.

9.3 CARBON OFFSETTING

Carbon offsetting allows net-positive emitters to neutralize their greenhouse gas (GHG) emissions by trading negative carbon credits. The purchase of offset tokens facilitates the funding of initiatives that reduce carbon emissions, such as reforestation projects. Carbon-negative projects include technologies that capture GHG at industrial sites, improve soil management, and leverage energy-efficient integrations. This allows individuals, organizations, and governments to compensate for their polluting activities and mitigate them through the use of carbon credits. Further, cloud- and blockchain-based bookkeeping improves the transparency of carbon credit records and makes them fully traceable. By offsetting emissions, individuals and organizations support projects that reduce or remove emissions from the atmosphere, mitigating the impact of climate change.

9.3.1 Abatable Simplifies Carbon Intelligence and Procurement

UK-based startup Abatable offers a carbon intelligence and procurement platform. It aids the purchase of high-quality offsets for hard-to-abate emissions and maximizes the social impact of climate contributions. The platform provides insights into carbon projects and project developers by evaluating the fundamental factors contributing to the quality of carbon credits. Abatable continuously monitors every project to mitigate risks stemming from infrastructure, systems, and policies, ensuring the committed quality of carbon credits for organizations.

9.3.2 Tanso Offers Automated Carbon Accounting

German startup Tanso automates carbon accounting and offers analytical solutions for effective emission reduction. The startup's software simplifies data collection and emissions calculation through process automation, Excel/CSV uploads, application programming interface (API), and a single-source-of-truth approach. Tanso's single-source-of-truth approach and automated processes provide organizations with their carbon footprint and set clear reduction goals. This allows the planning of decarbonization initiatives, like carbon offsetting and benchmarking, against science-based target standards.

9.4 ARTIFICIAL INTELLIGENCE

AI allows organizations to collect and derive insights from massive amounts of data and identify patterns that humans fail or take time to recognize. Deeper insights into carbon emissions provide organizations with higher agility to adopt processes that cut down GHG emissions. This accelerates sustainable transformation and reduces expenses in a period of lower cash flow. AI-powered solutions also analyze the environmental impact of emissions and provide forecasts on storms, wildfires, and droughts, which enables better mitigation plans. Finally, AI aids the optimization of technologies related to carbon capture and renewable energy that improves efficiency and reduces emissions.

9.4.1 AI Shading Ensures Low Building Carbon Footprint

AI Shading is a Canadian smart and green technology startup developing economical solutions that allow commercial buildings to achieve emission reduction targets. The startup's solution also accomplishes this by regulating indoor temperatures through smart indoor shades, sensors, mobile apps, and energy management platforms. AI Shading equips building managers with energy-saving reports to better analyze energy usage and track emission reduction goals.

9.4.2 GAIA Provides Automated Emission Tracking

Dutch startup GAIA leverages AI to simplify and automate GHG emissions reduction for businesses. The startup's solution measures and tracks emissions by analyzing the data from the input of choice—manual, linked, or integrated. The AI algorithm utilizes the raw data to extrapolate measurements and approximate any missing data. This way, GAIA's solution pinpoints main emission sources as well as customizes optimization roadmaps and emission reduction priorities.

9.5 ELECTRIFICATION

Advancements in renewable energy technologies, particularly solar and wind, are pivotal in reducing dependence on fossil fuels and facilitating the electrification of various industries, including transportation and manufacturing. The integration of renewable energy into electric vehicle (EV) charging networks is crucial for enhancing sustainability and reducing the carbon footprint of electric mobility. This integration faces challenges such as the variability of renewable power and grid capacity constraints, which can be addressed through smart grids and vehicle-to-grid (V_2G) systems (Abdullahi & Adnan, 2024). The adoption of EVs, supported by optimal power management strategies, contributes significantly to environmental sustainability by reducing greenhouse gas emissions and fossil fuel reliance. These strategies ensure efficient charging and discharging of EV batteries, integrating seamlessly with existing power grid infrastructure, and enhancing grid stability and reliability (Dasireddy et al., 2023). The transition to EVs, powered by renewable energy, is essential for decarbonizing the transport sector, as demonstrated by studies in various regions with different carbon dioxide emission factors (Ali et al., 2023).

Smart grid technologies and bidirectional power interfaces further improve the efficiency and reliability of EVs, making them more accessible and economically viable for consumers (Gayathri, 2021). The integration of EVs with high-speed charging networks and smart grids exemplifies the scalability of electrification, driven by innovations in battery storage and grid optimization. Smart grids facilitate efficient and sustainable electricity distribution, enhancing the interaction between EVs and renewable energy sources, thereby reducing reliance on fossil fuels and promoting environmental sustainability (Zhang, 2024; Mukherjee et al., 2024). The synergy between EVs and renewable energy, such as solar and wind, is crucial for energy security and emission reduction, although it faces challenges due to the intermittent nature of these energy sources (Mukherjee et al., 2024; Manohar et al., 2024). Innovative solutions, such as the hybrid model combining solar and wind power for EVs, address these challenges by ensuring continuous energy production and reducing dependence on conventional grids (Manohar et al., 2024). Furthermore, smart grid technologies, including smart metering and intelligent charging station scheduling, optimize EV charging and energy management, minimizing energy waste and peak grid loads (Zhang, 2024; Zhao & Liang, 2023).

Advanced approaches like the integrated GA-GRU-RL method enhance charging schedule optimization and energy management, significantly reducing grid energy costs and promoting green energy adoption (Zhao & Liang, 2023). The deployment of EVs also contributes to greenhouse gas emission reduction, aligning with sustainable development goals and supporting the transition to a cleaner energy landscape (Kaur, 2024). Overall, the electrification of commercial and industrial sectors through these innovations not only cuts emissions but also enhances energy efficiency, paving the way for a sustainable future in transportation (Zhang, 2024; Kaur, 2024).

9.5.1 Zevx Accelerates Fleet Electrification

Zevx's development of battery electric power systems and data intelligence for e-mobility applications aligns with several innovative strategies and technologies. The integration of data-driven approaches, as seen in the battery-lifetime-aware scheduling system for electric bus fleets, can significantly reduce battery degradation while maintaining service quality, which is crucial for the longevity and efficiency of commercial fleet operations. Furthermore, the intelligent networking integrated system for electric commercial vehicles highlights the importance of cloud-based platforms for fleet management, enabling seamless data interaction and remote monitoring, which enhances operational efficiency and safety. Zevx's focus on power management is supported by systems that optimize power flow and manage charging activities, ensuring efficient energy use and cost reduction, which are essential for sustainable fleet operations. Additionally, the use of predictive analytics and renewable energy sources in optimizing EV fleet operations can lead to substantial reductions in greenhouse gas emissions and operational costs, aligning with Zevx's goal of decarbonizing fleet operations. These technologies collectively provide a comprehensive framework for improving the operational efficiency and sustainability of electric commercial fleets, supporting Zevx's mission to accelerate the electrification of commercial fleet assets.

9.5.2 Leaf Energy Allows Seamless Energy Transition

Leaf Energy, a US-based startup, is at the forefront of providing electrification solutions, including EVs and charging stations, with a focus on high-speed charging through proprietary terminal designs compatible with existing grid networks. This aligns with the current trends and challenges in EV charging infrastructure, where the placement and scheduling of charging stations are critical for enhancing EV adoption. The startup's technology likely incorporates advanced systems similar to those described by Johansen and Naesje, which involve multipurpose charging systems with solid-state transformers that efficiently manage energy transfer from the grid to the vehicle. Furthermore, Leaf Energy's platform for real-time environmental tracking is indicative of a broader trend toward integrating smart technologies in EV infrastructure, as highlighted by Lee et al., who discuss a multipurpose real-time tracking system that monitors battery status and charger availability, facilitating services like reservation and emergency rescue. Additionally, the startup's approach may benefit from smart charging infrastructure frameworks that utilize bidirectional interfaces and wireless communication technologies to optimize energy flow and reduce grid stress, as proposed by Ul-Haq et al. By leveraging these advanced technologies and frameworks, Leaf Energy not only addresses the immediate needs of EV users but also contributes to the sustainable expansion of renewable energy integration, supporting businesses in tracking and achieving their energy transition goals.

9.6 GREEN HYDROGEN

Green hydrogen, produced through renewable energy-powered electrolysis, is increasingly recognized as a pivotal component in achieving deep decarbonization across various sectors, including transportation and heavy industry. This zero-emission alternative is primarily generated by splitting water into hydrogen and oxygen using electricity derived from renewable sources like solar and wind, which significantly reduces carbon emissions compared to traditional hydrogen production methods (Aissani et al., 2024; Yin et al., 2024). The integration of advanced technologies such as solid oxide electrolyzers (SOECs) enhances production efficiency by utilizing waste heat from existing industrial processes, thereby lowering overall energy consumption and costs (Sahak et al., 2024). Despite its potential, the widespread adoption of green hydrogen faces challenges, including high production costs, the need for substantial infrastructure for storage and distribution, and stringent safety regulations (Bhardwaj, 2024). Innovative storage solutions, such as metal hydrides and carbon nanostructures, are being explored to address these challenges, alongside the development of advanced materials for more efficient hydrogen storage (Bhardwaj, 2024). Furthermore, the socio-economic benefits of green hydrogen, such as job creation and energy independence, underscore its importance in fostering a sustainable global economy. However, to realize its full potential, significant investments in infrastructure and supportive regulatory frameworks are essential to drive large-scale adoption and integration into existing energy systems (Chowdhury et al., 2024). As technological advancements continue, green hydrogen is poised to play a crucial role in the transition to a low-carbon future, necessitating ongoing research and policy support to overcome current barriers (Yin et al., 2024; Chowdhury et al., 2024).

9.6.1 Azolla Generates Fuel-Cell Spec Green Hydrogen

Azolla's Biodrome technology represents a significant advancement in the production of green hydrogen, particularly for the power and transportation industries. This technology utilizes a blend of bio-methanol and de-ionized water to generate hydrogen suitable for fuel cells through catalytic reformation, which is a process that aligns with the broader trend of using renewable sources for hydrogen production. The use of bio-methanol as a hydrogen carrier is particularly advantageous because it simplifies the transportation, storage, and handling compared to traditional high-pressure hydrogen or ammonia. This approach not only facilitates on-site low-carbon hydrogen production but also integrates well with existing renewable energy systems, thereby contributing to the reduction of greenhouse gas emissions. The catalytic processes involved in Azolla's technology are supported by recent advancements in carbon-based catalysts, which are known for their efficiency and eco-friendliness, making them suitable for sustainable hydrogen production. Furthermore, the economic feasibility of such technologies is increasingly being demonstrated, particularly when using biomass-derived materials like bio-methanol, which can be integrated into existing industrial processes such as methanol synthesis. Overall, Azolla's approach exemplifies the potential for innovative hydrogen production methods to support a sustainable energy future by leveraging renewable resources and advanced catalytic technologies.

9.6.2 Faraday Fuels Offers Solid Oxide Electrolyzers

Faraday Fuels' development of scalable, high-throughput solid oxide technology for green hydrogen production aligns with the broader advancements in Solid Oxide Electrolysis Cells (SOECs) as a promising method for generating green hydrogen. SOECs are recognized for their high electrical efficiency and the ability to operate without noble metal catalysts, making them a cost-effective and sustainable option for hydrogen production. The use of $La_{1-x}Sr_xGa_{1-y}Mg_yO_3$-δ (LSGM) as an electrolyte in SOECs, due to its superior ionic conductivity at intermediate temperatures, enhances the durability and efficiency of hydrogen production, achieving a faradaic efficiency of up to 80%. This technology is particularly advantageous for industries like steel, chemical, and aviation, which require deep decarbonization, as hydrogen serves as a clean energy carrier with the potential to significantly reduce greenhouse gas emissions. The high-temperature operation of SOECs, ranging from 700°C to 1000°C, allows for faster reaction kinetics and the potential integration of renewable energy sources, further reducing the electric energy demand and enhancing overall energy efficiency. The U.S. legislative support, through acts like the Bipartisan Infrastructure Law and the Inflation Reduction Act, provides financial incentives for low-carbon hydrogen production, which could benefit startups like Faraday Fuels in scaling their technology. Overall, the integration of advanced materials and innovative designs in SOECs, such as those developed by Faraday Fuels, is crucial for overcoming current challenges in cost and durability, paving the way for a sustainable hydrogen economy.

9.7 ENERGY-EFFICIENT INTEGRATIONS

Industrial excess energy is a largely untapped resource, for which there is potential for reuse as well as benefits to industries and society. The utilization of excess energy in the form of heat, power, and fuel provides a way to reduce the use of primary energy. This contributes to the mitigation of CO_2 emissions. The most common recovery solutions reuse discarded heat in secondary processes on site or use it to preheat incoming water and combustion air. Other power recovery solutions include the reclamation of pressure energy from processes operating at elevated pressures, using even small pressure differences to produce electricity through pressure recovery turbines. In addition to solutions that recover wasted energy, innovations in materials and processes avoid energy wastage and cut down on the associated emissions.

9.7.1 Sapphire Technologies Simplifies Excess Pressure Energy Recovery

US-based startup Sapphire Technologies creates energy recovery systems that harness the power of expanding gas to produce reliable and clean electricity. The startup's FreeSpin In-line Turboexpander Generator specifically converts energy wasted in pressure reduction processes into electric power without interrupting operations. Sapphire Technologies' solution caters to hydrogen and natural gas industrial applications. It maximizes efficiencies, improves productivity, reduces carbon emissions, offsets electrical costs, and achieves financial returns.

9.7.2 HT Materials Science Facilitates Efficient Heat Transfer

Irish startup HT Material Science offers smart heat transfer fluids for commercial and industrial heating and cooling systems. The startup's patented nanofluid additive, Maxwell, is a drop-in solution that enhances the heat transfer of a base refrigerant, improving the efficiency of HVAC systems. This reduces energy consumption and facilitates a proportional reduction in carbon emissions. The higher heat transfer rate enables the adoption of smaller, less costly, and more efficient equipment that further adds to operational sustainability.

9.8 BIG DATA

AI and big data are significantly transforming carbon reduction strategies by offering detailed insights into emission patterns and system inefficiencies, thereby optimizing energy systems and enhancing environmental management. AI-driven tools are pivotal in the energy sector, where they enhance strategic management through optimization, predictive analytics, and automated decision systems, leading to improved operational efficiency and environmental outcomes (Zülfikaroğlu, 2024). In the oil and gas industry, AI-powered automation processes enhance carbon capture efficiency by enabling real-time monitoring and predictive maintenance, which boosts CO_2 sequestration rates and minimizes operational interruptions (Anyebe et al., 2024). Furthermore, AI and big data analytics are crucial in climate resilience, processing vast environmental data to improve climate forecasting and develop early warning systems, thus supporting disaster risk reduction and sustainable resource management (Pasupuleti, 2024). In the healthcare sector, AI technologies, including machine learning and deep learning, are applied to reduce carbon footprints by optimizing supply chains and predictive management, thereby making healthcare services more eco-friendly (Upadhyay et al., 2024). Additionally, AI's impact on enterprise energy management is evident in its ability to reduce carbon emission intensity, particularly in private enterprises, by promoting green development standards and technological innovation (Yang et al., 2024). Collectively, these technologies facilitate informed decision-making and targeted mitigation efforts across various sectors, driving significant advancements in carbon reduction strategies.

9.8.1 Grid Matrix Aids Transport Emission Reduction

Grid Matrix's approach to leveraging existing traffic infrastructure aligns with the broader trend of integrating advanced technologies into traffic management systems to reduce congestion, emissions, and accidents. The use of big data and analytics, as highlighted in the startup's platform, is a common theme in contemporary traffic management solutions. For instance, the Smart Traffic Management (STM) system utilizes IoT and AI to dynamically adjust traffic flows, thereby optimizing efficiency and enhancing emergency response, which can significantly alleviate urban congestion and improve safety. Similarly, the integration of Graph Representation Learning (GRL) in traffic forecasting and route optimization provides real-time insights, enabling proactive decision-making and efficient resource allocation, which are crucial for reducing congestion and emissions. The Smart Traffic for Congestion Avoidance (ST-CA) framework further emphasizes the importance of smart systems that utilize expert systems for decision-making, enhancing flexibility and efficiency in traffic management. Additionally, the Intelligent Traffic Management System (ITMS) leverages cloud computing, IoT, and data analytics to predict optimal routes and reduce pollution, aligning with Grid Matrix's goals of creating a safer and greener urban environment. These systems collectively demonstrate the potential of technology-driven solutions to transform urban mobility, supporting the transition of cities into more efficient and sustainable places to live and work.

9.8.2 CarbonSense simplifies Carbon Footprint Management

CarbonSense's software solution exemplifies the convergence of green and digital technologies, a trend that significantly contributes to carbon emission reduction by enhancing energy efficiency and promoting renewable energy adoption. This convergence, as highlighted in the literature, plays a crucial role in improving carbon productivity and facilitating low-carbon development, particularly in regions with advanced digital infrastructure and moderate environmental regulations. The software's ability to collect and analyze data from various sensors and equipment aligns with the findings that energy digitalization can enhance carbon productivity through improved energy technology innovation and utilization efficiency. Moreover, the integration of big data and AI in CarbonSense's system supports the transition to carbon neutrality by enabling comprehensive energy consumption analysis and early detection of equipment inefficiencies, which are essential for optimizing energy management and reducing greenhouse gas emissions. The insights provided by such systems are crucial for making informed decisions that minimize energy consumption and accelerate the shift toward renewable energy sources, thereby supporting the broader goals of carbon neutrality and sustainable development. This approach not only aids individual enterprises in managing their carbon assets more effectively but also contributes to regional and national efforts to achieve strategic low-carbon transformation targets, such as China's "3060" decarbonization goal.

9.9 WASTE VALORIZATION

Waste valorization is a transformative approach that converts waste materials into valuable products, thereby reducing landfill contributions and associated methane emissions. Techniques such as anaerobic digestion and pyrolysis play crucial roles in this process. Anaerobic digestion is effective in producing biogas from organic waste, contributing to renewable energy generation and reducing reliance on fossil fuels (Kumar et al., 2023). Pyrolysis, particularly catalytic pyrolysis, is instrumental in converting plastic waste into low-carbon fuels and other value-added products. For instance, polyethylene waste can be transformed into microbial lipids through a tandem process involving catalytic pyrolysis and biological conversion, showcasing a sustainable strategy for plastic waste management (Qian et al., 2023). Furthermore, pyrolysis and gasification systems are highlighted for their potential to convert hazardous plastic waste into hydrocarbons and fuels, offering a faster and more realistic solution to the growing plastic waste problem (Ahmad et al., 2023). These valorization techniques support the principles of a circular economy by minimizing resource extraction and promoting sustainable production methods. They enable the development of circular products by using waste as raw materials, thus aligning with sustainable production and consumption goals (Rossi et al., 2024). The integration of these processes into industrial-scale operations is expected to enhance the production of desirable components, such as carbon nanomaterials, further contributing to environmental sustainability (Ahmad et al., 2023). Overall, waste valorization not only addresses environmental challenges but also opens new opportunities for sustainable economic growth (Balu et al., 2022).

9.9.1 Alimentary Advances Domestic Waste Valorization

Alimentary's integrated waste treatment plant (IWTP) in New Zealand leverages a patented process that combines biomimicry and anaerobic co-digestion to efficiently manage organic waste, reduce greenhouse gas emissions, and generate revenue through biogas production. The use of anaerobic digestion, as highlighted in various studies, is a key component of this process. For instance, the design of a novel intestinal reactor improves the uniformity and solid content in high-solid organic waste anaerobic digestion, which can enhance urban waste treatment capacity by addressing issues like solid-liquid stratification and energy consumption in traditional reactors.

Additionally, the concept of decentralized energy recovery through domestic anaerobic digesters, such as the Boitatá, demonstrates the viability of treating household organic waste and generating biogas to meet a portion of energy demands. The startup of demo-scale anaerobic digestion plants further illustrates the importance of process stability and optimization, with strategies like sequencing batch operations enhancing biogas yield and system resilience. Moreover, the integration of biogas plants in agricultural settings shows potential for increased efficiency and methane content, contributing to a closed biological cycle and reduced energy consumption. Finally, the application of anaerobic co-digestion on islands underscores the environmental and economic benefits of using local organic waste as a renewable energy source, aligning with Alimentary's goals of reducing waste disposal costs and emissions while generating high-value biogas.

9.9.2 Pyro-Degrade Facilitates Plastic Waste to Fuel Conversion

Kenyan startup Pyro-degrade's development of pyrolysis technology to convert plastic waste into fuel aligns with the broader research and advancements in waste-to-energy technologies. Pyrolysis, a thermochemical process that decomposes waste in the absence of oxygen, is increasingly recognized for its potential to address plastic pollution and produce eco-friendly energy alternatives. The startup's Pyro-diesel, a sulfur-free and low-carbon diesel substitute, exemplifies the benefits of pyrolysis-derived fuels, which typically have lower sulfur content and can improve engine performance while reducing emissions like carbon monoxide and particulate matter. The process employed by Pyro-degrade is self-sufficient, utilizing the fuel it produces to operate off the grid, which is a significant advantage in reducing reliance on traditional energy sources. This approach not only mitigates the environmental impact of plastic waste but also provides a sustainable energy solution for agricultural and domestic applications, as highlighted by the potential of pyrolysis to produce high-efficiency diesel substitutes with lower emissions. Furthermore, the use of catalysts, such as indigenous clay, in catalytic pyrolysis can enhance fuel yield and efficiency, making it a viable alternative to imported catalysts and supporting local resource utilization. Overall, Pyro-degrade's technology represents a promising advancement in sustainable waste management and energy production, contributing to the reduction of fossil fuel dependence and environmental pollution.

9.10 BIO-BASED FUELS AND POLYMERS

Bio-based solutions, such as biofuels and biodegradable polymers, present promising renewable alternatives to fossil fuels and petrochemical plastics, with innovations in algae-based biodiesel and polymer fermentation technologies offering scalable, low-carbon substitutes. Biodegradable polymers, derived from renewable sources like plants and microbes, reduce reliance on fossil fuels and minimize plastic pollution through composting or natural breakdown, providing enhanced end-of-life options (Samuel et al., 2024). The demand for bioplastics, which decompose into carbon dioxide, water, and organic compounds, is increasing due to their eco-friendly nature and potential to alleviate the energy crisis by reducing dependence on nonrenewable energy sources (Nisar et al., 2024). Agricultural by-products and organic waste, such as pineapple processing waste and rice straw, serve as feedstocks for producing biomaterials and bioproducts, contributing to a sustainable bioeconomy that lowers greenhouse gas emissions and promotes a green environment (Ummalyma, 2024). Algal biopolymers, which utilize polysaccharide molecules for pollutant binding and removal, further enhance sustainability by reducing plastic waste accumulation and supporting a circular, zero-carbon economy (Mani et al., 2023). These bio-based solutions not only offer environmental benefits but also hold significant potential for economic sustainability, emphasizing the need for continued research and development in this field to optimize processes and expand applications (Ummalyma, 2024; Mani et al., 2023).

9.10.1 SeaH4 Produces Carbon-Neutral Fuels

SeaH4's innovative approach to producing sustainable biodiesel and liquefied natural gas (LNG) from algae aligns with South Africa's broader efforts to transition toward a sustainable energy future. The startup's circular system, which minimizes operational costs and maximizes yield, contributes to the CO_2-neutral operation of combustion engines, thereby supporting the reduction of greenhouse gas emissions. This initiative is particularly relevant given South Africa's heavy reliance on fossil fuels and the associated environmental challenges. As highlighted by Udeagha and Ngepah, technological innovation is essential for reducing CO_2 emissions and achieving energy security. Furthermore, the integration of renewable energy sources, as explored by Mathebula et al., demonstrates the potential to optimize energy systems for cost and emissions reduction—critical factors for industrial applications such as those targeted by SeaH4. The potential of renewable energy in South Africa is further underscored by studies on hybrid tidal-wind systems and floating photovoltaic installations, which offer sustainable alternatives to traditional energy sources. Collectively, these studies emphasize the importance of innovative clean technologies like those developed by SeaH4 in supporting South Africa's transition to a green economy and addressing climate change challenges. Additionally, the technoeconomic dataset provided by Dixon et al. can assist in modeling and planning for such sustainable energy solutions, ensuring that initiatives like SeaH4's are effectively integrated into the national energy strategy.

9.10.2 Dionymer Offers Petrochemical Plastic Alternatives

Dionymer's innovative approach to upcycling industrial organic waste into biodegradable polyhydroxyalkanoates (PHA) aligns with broader trends in bioplastic development, as highlighted in recent research. Utilizing organic waste as a sustainable feedstock for bioplastics is gaining momentum, with PHA standing out due to its mechanical properties comparable to synthetic plastics and its inherent biodegradability, which helps address the global challenge of plastic waste accumulation. Dionymer's fermentation technology mirrors methodologies found in studies employing Bacillus mycoides for PHA production from food waste, demonstrating strong potential for high productivity and mechanical stability in bioplastic applications.

Additionally, the incorporation of agricultural waste, such as tropical fruit peels, into bio-based materials further highlights the potential of organic residues to serve as sustainable alternatives to conventional materials. The development of biodegradable composite films from brewers' spent grain demonstrates the versatility of organic waste in reducing environmental impacts and fostering a circular economy. These innovations not only provide low-carbon substitutes for petrochemical plastics but also create valuable recovery pathways for bio-waste across diverse industries, including agriculture and pharmaceuticals, thereby enhancing the sustainability of both material production and waste management practices.

9.11 EMISSION MONITORING

Emission monitoring and management reduce emissions by enabling organizations to accurately measure and track their emissions and identify opportunities for reduction. Monitoring systems detect emissions from various sources and provide real-time data on emissions levels. Management systems then analyze and interpret this data, allowing organizations to identify improvement areas and implement emission reduction strategies. By regularly monitoring and managing emissions, organizations ensure their emissions are within legal limits, reduce their environmental impact, and improve their sustainability. Moreover, organizations now leverage a comprehensive and systematic approach to emission reduction, monitoring, and management. This helps them to achieve emission reduction goals and ensure carbon-neutral operations.

9.11.1 GAIT Delivers Greenhouse Gas Measurement

Singaporean startup GAIT offers climate tech solutions that facilitate the deployment of GHG measurement systems. The startup's technology combines machine learning, flux sensors, and spatial data to deliver real-time measurements of greenhouse gas emissions. Additionally, it accesses historical flux data to provide more accurate models that measure the effectiveness of climate action projects. The solution also provides recommendations on actions required to further reduce emissions and achieve the set air quality targets.

9.11.2 Blue Sky Advances Fugitive Methane Emission Detection

Blue Sky is a US-based startup that develops an infrared optical sensor technology for the continuous detection and measurement of fugitive methane emissions. NIRview is the startup's emission monitoring solution that employs a spectroscopy-based approach to automatically scan a user-defined grid and provide a detailed 2D concentration map. The solution, in this way, offers oilfields a cost-effective detection technology that pinpoints the location of leaks, facilitating quick repairs and substantially reducing emissions.

9.12 CRYOGENIC CARBON CAPTURE

CCC is a highly innovative CO_2 capture technology that has the potential to reduce carbon emissions from fossil-fueled power plants by 95%–99%, at half the cost and energy of current state-of-the-art (Amine based) carbon capture processes. CCC also removes other pollutants, such as SO–, NO–, and mercury.

The Cryogenic CO_2 Capture (CCC) process dries and cools flue gas from existing systems, modestly compresses it, cools it to a temperature slightly above the point where CO_2 forms a solid, expands the gas to further cooling, precipitates an amount of CO_2 as a solid that depends on the final temperature, pressurizes the CO_2, and lastly reheats the CO_2 and the remaining flue gas by cooling the incoming gases. The final result is the CO_2 in a liquid phase and a gaseous nitrogen stream. CO_2 capture efficiency depends primarily on the pressure and temperature at the end of the expansion process. In addition to the cost and energy efficiency advantages, the CCC process enjoys several ancillary benefits, including leveraging of existing capital investments, pollutant control, water savings, and potential energy storage.

9.13 DIGITAL TWIN TECHNOLOGY

Digital twin technology is significantly transforming various sectors by creating virtual replicas of physical systems, which integrate real-time data to simulate performance and sustainability impacts. In manufacturing, digital twins enhance operational efficiency by enabling real-time dynamic optimization and adaptive control, as demonstrated in flexible manufacturing systems (FMS), where productivity increased by 14.53%, energy consumption decreased by 13.9%, and quality improved by 15.8% through intelligent machine coordination. Additionally, digital twins facilitate better decision-making in discrete manufacturing by synchronizing material flow data, thus improving production planning and on-time delivery. In the context of the circular economy, digital twins provide life cycle data that support sustainable value creation and enable the monitoring and implementation of circular economy strategies, such as optimal end-of-life decision-making for vehicles. Furthermore, in the solid-wood-panel processing industry, digital twins have led to enhanced production efficiency and better material use, illustrating their potential to transform traditional industries into smart manufacturing environments. The integration of AI with digital twins in 3D printing processes further underscores their role in achieving greater efficiency and sustainability in Industry 4.0/5.0 technologies.

9.13.1 IKEA's Digital Twin in East Asian Stores

IKEA's digitalization efforts in East Asia, which involved digitalizing 37 retail stores in nine months, exemplify the transformative potential of digital technologies in enhancing energy efficiency and reducing greenhouse gas emissions. By integrating 7,000 data points from various building and energy management systems into a single digital twin, IKEA has leveraged the Industrial Internet of Things (IIoT) and digital twin architecture to optimize energy use and minimize carbon emissions. This approach aligns with broader trends in digitalization, where technologies such as AI, IoT, big data analytics, and cloud computing are employed synergistically to achieve carbon emission reduction goals. The digital twin serves as a bridge between the physical and digital worlds, enabling real-time monitoring and predictive analytics that inform energy-efficient decision-making and operational improvements. Furthermore, digitalization in retail, as seen in IKEA's case, involves interpreting, interrelating, and integrating digital processes to enhance business operations and sustainability. This digital transformation not only supports energy efficiency but also contributes to sustainable development goals by fostering innovation in energy management and promoting low-carbon systems. The integration of digital technologies into IKEA's operations reflects a growing trend among companies to adopt circular economy principles, where digitalization and IoT play crucial roles in achieving sustainability and long-term environmental benefits.

9.14 DIRECT AIR CAPTURE

DAC technology represents a promising approach to mitigating climate change by directly removing CO_2 from the atmosphere, as highlighted in several studies. Innovations such as electrochemical steps and the use of amino acid-functionalized ionic liquid-based deep eutectic solvents (DESs) have been explored to enhance efficiency and reduce costs. Electrochemical DAC processes, for instance, have shown potential in achieving lower energy demands compared to traditional methods (Rosen et al., 2024). Similarly, DESs have demonstrated high CO_2 absorption efficiency, offering a promising alternative for capturing CO_2 from the air (Bera et al., 2024). Additionally, carbonate electrolysis has been developed to improve energy efficiency, with advancements in gas-diffusion layers and cell geometry contributing to reduced operating voltages and increased CO_2 purity (Tong et al., 2024). Despite these advancements, the scalability of DAC technology remains limited by its economic and energy constraints, necessitating further research and integration with renewable energy sources to enhance its viability and reduce its carbon footprint (Ozkan, 2024; Li & Yao, 2024).

9.14.1 South Texas DAC

The South Texas DAC hub exemplifies the potential benefits of DAC systems in carbon removal by demonstrating the scalability and integration of advanced CO_2-scrubbing technologies. This hub, led by 1PointFive, a subsidiary of Occidental Petroleum, is part of a broader initiative funded by the Biden administration, which allocated \$1.2 billion from the 2021 federal infrastructure bill to support the development of DAC facilities. The South Texas hub, along with another in Lake Charles, Louisiana, is designed to capture 1 million metric tons of carbon dioxide annually from the atmosphere, which will then be sequestered underground or beneath the ocean floor for permanent storage. This large-scale implementation aims to validate the effectiveness of DAC technology, which remains largely unproven at such a scale. By utilizing technology from Carbon Engineering, the project seeks to enhance the efficiency and cost-effectiveness of carbon capture processes. The success of the South Texas DAC hub could pave the way for more widespread adoption of DAC systems, contributing significantly to global carbon reduction efforts and helping to mitigate climate change impacts by removing substantial amounts of CO_2 from the atmosphere (Bettenhausen, 2023).

Electrochemical methods for extracting CO_2 from seawater and stabilizing it as bicarbonate ions or solid carbonates present a promising approach to enhancing ocean alkalinity and reducing atmospheric CO_2 concentrations. These methods, such as the SEA MATE process, utilize electrochemistry to convert seawater components into acid and base, increasing seawater pH and alkalinity, which shifts dissolved inorganic carbon speciation toward carbonate, thereby enhancing CO_2 uptake from the atmosphere (Ringham et al., 2024). The development of symmetric electrochemical systems, which utilize reversible reactions at both the anode and cathode, has shown potential for energy-efficient seawater alkalization, achieving significant reductions in energy consumption compared to traditional methods (Guan et al., 2024). Additionally, coupling electrochemical processes with biological methods, such as enhancing marine phytoplankton growth through acid addition, can further increase CO_2 sequestration, demonstrating the potential for integrated approaches to maximize carbon removal (Hibbeln et al., 2024). However, challenges remain in optimizing these technologies for efficiency and environmental sustainability. For instance, the Bipolar Membrane Electrodialysis (BPMED) process requires further research to improve long-term operation efficiency and mitigate scaling effects (Aliaskari et al., 2024). Moreover, the electrochemical ocean iron fertilization approach, which combines alkalinity enhancement with iron release to stimulate biological carbon pumps, offers a scalable solution but requires careful regulation of iron concentrations and system parameters (Taqieddin et al., 2024). Overall, while these electrochemical methods hold significant promise, continued research and development are essential to optimize their performance and ensure their environmental sustainability.

9.14.2 BPMED Seawater Extraction

Electrochemical oceanic carbon capture using BPMED is a process designed to extract CO_2 from seawater. The BPMED cell is composed of two end electrodes, charged membranes, and flow paths between the membranes. The core of this technology is the bipolar membrane (BPM), which consists of a cation-exchange membrane (CEM) and an anion-exchange membrane (AEM) laminated together with a water dissociation catalyst. The BPMED method captures CO_2 by converting bicarbonate ions (HCO_3–) present in seawater into gaseous CO_2 or carbonate ions (CO_3^{2}–) using a pH-swing facilitated by the BPM. This process involves the dissociation of water into acid and base at the BPM, which creates the necessary pH conditions to transform bicarbonate into carbonate or CO_2 gas.

9.15 IMPORTANCE OF CO_2 DATA FOR CARBON REDUCTION

Accurate and transparent carbon data are crucial for effective emission reduction strategies, as they enable precise identification of emission sources and support comprehensive carbon assessments. Primary CO_2 data, derived from direct measurements, offer precision in pinpointing emission sources, while secondary data fill gaps where direct measurements are unavailable, thus supporting a holistic understanding of carbon emissions. The International Energy Agency (IEA) emphasizes the importance of data in mapping progress in clean energy transitions and achieving global greenhouse gas reduction goals (Birol, 2023). Innovations such as the Global Gridded Daily CO_2 Emissions Datasets (GRACED) provide near-real-time, fine-grained data on CO_2 emissions from various sectors, including power, industry, and transportation, at a high spatial and temporal resolution (Ciais et al., 2023; Dou et al., 2021). These datasets, which integrate multiple data sources like satellite NO_2 data, offer timely insights into emission patterns and are crucial for designing localized and adaptive climate policies, especially in the post-COVID era (Ciais et al., 2023). The reliability of these datasets is enhanced through uncertainty analysis, ensuring that the high resolution does not compromise data accuracy (Ciais et al., 2023). Furthermore, the continuous updates of these dynamic data platforms reflect technological advancements, providing real-time insights that are essential for decision-making and policy adaptation (Dou et al., 2021). Validated by third-party organizations, these innovations in carbon accounting systems enhance reliability and encourage the widespread adoption of data-driven solutions, ultimately facilitating more effective climate change mitigation strategies.

In carbon reduction through the evolution and application of different types of CO_2 data, primary CO_2 data is highlighted as the most accurate and reliable form of emissions data, being directly measured or calculated by the emissions sources themselves. This push for greater transparency and accuracy in emissions reporting allows companies to identify specific sources of emissions and target them more effectively for reduction, particularly in direct (Scope 1) and energy-related indirect (Scope 2) emissions. Secondary CO_2 data complements primary data by filling gaps where primary data is unavailable, allowing for broader and more comprehensive carbon footprint assessments. By utilizing multiple sources such as scientific research, national statistics, and existing databases, companies can estimate their emissions with reasonable accuracy even in the absence of direct measurements. Although secondary data can sometimes be outdated, it provides a baseline that companies can use to compare with primary data, identifying areas for potential improvement and innovation.

Dynamic CO_2 data represents a significant innovation by continuously updating information based on new inputs, ensuring its relevance and currency. Unlike static databases, dynamic data reflects the latest changes in technology, energy mixes, and emissions calculation methodologies, allowing companies to work with the most current data possible. Detailed, process-oriented calculations in dynamic data more accurately reflect the specific emissions of different materials and processes. Platforms like sustamize's Product Footprint Engine (PFE) and sustamizer® utilize dynamic data to provide real-time insights, driving ongoing carbon reduction efforts.

The article also underscores how dynamic databases, validated by third-party organizations like GUTcert, contribute to innovations in carbon reduction. Regular updates to the dynamic database ensure that companies can continuously improve their emissions data and reduction strategies. By incorporating the latest technological developments and innovations in materials and processes, companies can more accurately track their progress and optimize their strategies. Third-party validation ensures that the data and methodologies used are robust and trustworthy, encouraging more companies to adopt these innovative approaches.

9.16 CHALLENGES AND LIMITATIONS OF CARBON DIOXIDE REMOVAL

In theory, carbon dioxide removal (CDR) technologies are the answer to many climate and emissions problems. But to make a dent in global warming, there's a long way to go.

Carbon removal technologies, while essential for achieving net-zero emissions, present significant limitations that necessitate a balanced approach prioritizing emissions reductions across all sectors. High costs, energy intensity, and scalability challenges are prominent issues associated with CDR technologies, which are not yet proven at scale and could potentially deter conventional mitigation efforts by creating a false sense of security regarding future emissions reductions (Anderson et al., 2023; Brad & Schneider, 2023). Over-reliance on CDR can perpetuate fossil fuel dependence, delay energy transitions, and impede the achievement of climate goals, as seen in scenarios where excessive CDR deployment leads to reduced efficiency improvements and electrification rates (Ampah et al., 2023). Ethical concerns also arise, particularly regarding land and water resource allocation, as land-based CDR options compete with food production and biodiversity protection (Voget-Kleschin et al., 2024). Moreover, the potential environmental impacts of large-scale CDR deployment necessitate careful consideration of alternative pathways, such as lifestyle changes and demand-side management, which could reduce the need for CDR while being less ethically problematic than initially perceived (Voget-Kleschin et al., 2024). The integration of renewable energy and electrification, supported by policy and capital investments, offers a viable path for deep decarbonization, highlighting the interconnectedness of sustainable development goals and the need for a systems dynamics approach to climate policy (Morte et al., 2023). Therefore, while

CDR technologies are crucial for offsetting residual emissions, they should complement rather than replace systemic decarbonization efforts, ensuring a comprehensive strategy that addresses both technological and ethical dimensions of climate action.

9.17 CONCLUSION

Innovation is rapidly reshaping the landscape of carbon reduction. This chapter explored a wide spectrum of emerging technologies—from carbon capture, utilization, and storage (CCUS) to direct air capture, green hydrogen, and digital twin systems—that are pushing the boundaries of what's possible in emissions mitigation. We have seen how artificial intelligence, big data, and advanced monitoring tools are making emissions tracking smarter and more actionable, while waste valorization and bio-based materials are transforming environmental challenges into new economic opportunities. Electrification, renewable energy, and energy efficiency remain at the core of decarbonization; however, the chapter also emphasizes the crucial role of accurate and transparent carbon data in guiding decisions and measuring real progress.

Yet, the chapter also underscored the limits of relying solely on technological fixes. Many carbon removal solutions face challenges in terms of cost, scalability, and potential environmental side effects. The most effective path forward combines these innovations with aggressive emissions reductions across all sectors, robust policy support, and sustainable practices. A multifaceted approach—blending technology, data, and behavioral change—is essential for achieving a low-carbon, climate-resilient future.

Looking ahead, the next chapter shifts focus from technological frontiers to the urgent need for rethinking energy management. Chapter 10 will examine how moving away from conventional energy sources and embracing renewables, smart grids, and decentralized systems can accelerate the global transition to sustainable energy. The journey continues from "what's possible" with innovation to "how" we can transform energy systems at scale for a cleaner, more equitable world.

REVIEW QUESTIONS

1. What are the most promising technologies for carbon capture and storage?
2. How can AI contribute to emission reductions?
3. What role does green hydrogen play in a low-carbon economy?
4. How does big data improve emissions tracking and reporting?
5. What are some innovations in energy-efficient building systems?
6. How does waste valorization contribute to circular economy goals?
7. What are the pros and cons of direct air capture technology?
8. How does cryogenic carbon capture work, and where is it applied?
9. What role does digital twin technology play in carbon reduction?
10. How does bio-based fuel production differ from traditional fossil fuels?

REFERENCES

Abdullahi, U.O., Adnan, A. (2024). Integration of renewable energy into electric vehicle (EV) charging networks. *World Journal of Advanced Engineering Technology and Sciences*. https://doi.org/10.30574/wjaets.2024.13.2.0554

Ahmad, M.A., Oh, W.-D., Low, S.C. (2023). Valorization of hazardous plastic wastes into value-added resources by catalytic pyrolysis-gasification: A review of techno-economic analysis. *Renewable & Sustainable Energy Reviews*. https://doi.org/10.1016/j.rser.2023.113346

Aissani, N., Guetarni, I.H.M., Chafaa, K., Rouidi, I., Merzougui, B., Lounis, Z. (2024). Renewable energy for green hydrogen production: Experimentation and predictive tool. *Revue Des Énergies Renouvelables*. https://doi.org/10.54966/jreen.v1i3.1292

Ali, Q., Silvestre, M.L.D., Lombardi, P., Sanseverino, E.R., Zizzo, G. (2023). The role of renewable energy generation in electric vehicles transition and decarbonization of transportation sector. In *2023 IEEE International Conference on Environment and Electrical Engineering and 2023 IEEE Industrial and Commercial Power Systems Europe (EEEIC/I&CPS Europe)*, Madrid, Spain, 2023, pp. 1–6. https://doi.org/10.1109/eeeic/icpseurope57605.2023.10194766

Aliaskari, M., Wezstein, J., Saravia, F., Horn, H. (2024). A systematic analysis of operating parameters for CO_2 capture from seawater by Bipolar Membrane Electrodialysis (BPMED). *Separation and Purification Technology*. https://doi.org/10.1016/j.seppur.2024.126679

Ampah, J.D., Ampah, J.D., Jin, C., Yao, M., Afrane, S., Adun, H., Fuhrman, J., Ho, D.T., McJeon, H. (2023). Deployment expectations of multi-gigatonne scale carbon removal could have adverse impacts on Asia's energy-water-land nexus. *Nature Communications* 15, 6342 (2024). https://doi.org/10.1038/s41467-024-50594-5

Anderson, K., Buck, H.J., Fuhr, L., Geden, O., Peters, G., Tamme, E. (2023). Controversies of carbon dioxide removal. *Nature Reviews Earth & Environment*. https://doi.org/10.1038/s43017-023-00493-y

Anyebe, A.P., Yeboah, O.K.K., Bakinson, O.I., Adeyinka, T.Y., Okafor, F.C. (2024). Optimizing carbon capture efficiency through AI-driven process automation for enhancing predictive maintenance and CO_2 sequestration in oil and gas facilities. *American Journal of Environment and Climate*. https://doi.org/10.54536/ajec.v3i3.3766

Balu, R., Dutta, N.K., Choudhury, N.R. (2022). Plastic waste upcycling: A sustainable solution for waste management, product development, and circular economy. *Polymers*. https://doi.org/10.3390/polym14224788

Bera, N., Sardar, P., Hazra, R., Samanta, A.N., Sarkar, N. (2024). Direct air capture of CO_2 by amino acid-functionalized ionic liquid-based deep eutectic solvents. *ACS Sustainable Chemistry & Engineering*. https://doi.org/10.1021/acssuschemeng.4c05090

Bettenhausen, C. (2023). Biden picks two DAC hubs for federal funding. *C&EN Global Enterprise*. https://doi.org/10.1021/cen-10127-buscon1

Bhardwaj, K. (2024). Technological advancements in adopting green hydrogen in the power and energy sector. *2017 IEEE Region 10 Symposium (TENSYMP)*, New Delhi, India. https://doi.org/10.1109/tensymp61132.2024.10752303

Birol, F. (2023). Why data is vital to build a better energy future for the world. *Responsabilité et Environnement*. https://doi.org/10.3917/re1.110.0009

Brad, A., Schneider, E. (2023). Carbon dioxide removal and mitigation deterrence in EU climate policy: Towards a research approach. *Environmental Science & Policy*. https://doi.org/10.1016/j.envsci.2023.103591

Chowdhury, M.F.I., Bari, M.F.S., Islam, M., Tanim, W.S., Meraz, R.M. (2024). Unlocking the potential of green hydrogen for a sustainable energy future: A review of production methods and challenges. *Future Energy*. https://doi.org/10.55670/fpll.fuen.3.4.2

Ciais, P., Liu, Z., Chevallier, F., Janssens-Maenhout, G., Dou, X., Ke, P., Tang, M. (2023). *Near-real-time global gridded daily CO_2 emissions. EGU General Assembly 2023*, Vienna, Austria, 24–28 Apr 2023, EGU23-3303. https://doi.org/10.5194/egusphere-egu23-3303

Dasireddy, H., Manoj, V., Tharun, G.S., Harika, C. (2023). A review on integration of electric vehicles and optimal power management strategy for sustainable development. *Indian Scientific Journal of Research in Engineering and Management*. https://doi.org/10.55041/ijsrem27288

Dávila, J.G., Sacchi, R., Pizzol, M. (2023). Preconditions for achieving carbon neutrality in cement production through CCUS. *Journal of Cleaner Production*. https://doi.org/10.1016/j.jclepro.2023.138935

Dou, X., Wang, Y., Ciais, P., Chevallier, F., Davis, S.J., Crippa, M., Janssens-Maenhout, G., Guizzardi, D., Solazzo, E., Yan, F., Huo, D., Bo, Z., Deng, Z., Zhu, B., Wang, H., Zhang, Q.G., Gentine, P., Liu, Z. (2021). Global gridded daily CO_2 emissions. *arXiv: Atmospheric and Oceanic Physics*.

Gayathri, M.N. (2021). A smart bidirectional power interface between smart grid and electric vehicle. In B. Vinoth Kumar, P. Sivakumar, M. Rajan Singaravel, K. Vijayakumar (Eds.), *Intelligent Paradigms for Smart Grid and Renewable Energy Systems. Algorithms for Intelligent Systems* (pp. 103–137). Springer. https://doi.org/10.1007/978-981-15-9968-2_5

Guan, X., Zhang, G., Li, J., Kim, S.C., Feng, G., Li, Y., Cui, T., Brest, A., Cui, Y. (2024). Seawater alkalization via an energy-efficient electrochemical process for CO_2 capture. *Proceedings of the National Academy of Sciences of the United States of America*. https://doi.org/10.1073/pnas.2410841121

Hibbeln, C.F., Marsh, P., Myers, C.R., Valdez, P.J., Edmundson, S.J., Subban, C.V. (2024). Maximizing marine carbon removal by coupling electrochemical and biological methods. *Environmental Science & Technology Letters*. https://doi.org/10.1021/acs.estlett.4c00107

Hu, Y. (2023). Opportunities and potential of CCUS technology advancement under the new energy and carbon neutrality background. *Highlights in Science Engineering and Technology*. https://doi.org/10.54097/hset.v59i.10067

Kaur, A. (2024). A bibliometric analysis on 'Role of electric vehicles in the smart grids.' In G. Panda, T.N. Ramasamy, S. Ben Elghali, S. Affijulla (Eds.), *Digital Communication and Soft Computing Approaches Towards Sustainable Energy Developments. ISSETA 2023. Innovations in Sustainable Technologies and Computing* (pp. 217–229). Springer. https://doi.org/10.1007/978-981-99-8886-0_20

Kumar, V., Lakkaboyana, S.K., Sharma, N. (2023). *Waste Valorization for Value-added Products*. Bentham Science Publishers. https://doi.org/10.2174/97898151230741230101

Li, G., Yao, J. (2024). Direct air capture (DAC) for achieving net-zero CO_2 emissions: Advances, applications, and challenges. *Eng*. https://doi.org/10.3390/eng5030069

Mani, M., John, S.P., Ekambaram, G., Kuppusamy, E. (2023). Eco-friendly biopolymers and biosorbents from algae to combat pollution. In J. Samuel, A. Kumar, J. Singh (Eds.), *Relationship between Microbes and the Environment for Sustainable Ecosystem Services* (pp. 207–219). Elsevier. https://doi.org/10.1016/b978-0-323-89936-9.00009-6

Manohar, P., Sai Pavani Kalyani, B., Narmada, B., Pavan Kumar Reddy, C., Prakash, C., Venu, P. (2024). Electric vehicle unified with wind and solar power. *Indian Scientific Journal of Research in Engineering and Management*. https://doi.org/10.55041/ijsrem29661

Morte, I.B.B., Araújo, O. de Q.F., Morgado, C. do R.V., de Medeiros, J.L. (2023). Electrification and decarbonization: A critical review of interconnected sectors, policies, and sustainable development goals. *Energy Storage and Saving*. https://doi.org/10.1016/j.enss.2023.08.004

Mukherjee, S., Dhar, A., Adhikari, S., Mukherjee, R., Dutta, S., Saha, P.K. (2024). A review on electric vehicles and renewable energy synergies in smart grid. *Journal of Mines, Metals and Fuels*. https://doi.org/10.18311/jmmf/2023/43063

Nath, F., Mahmood, Md.N., Yousuf, N. (2024). Recent advances in CCUS: A critical review on technologies, regulatory aspects and economics. *Geoenergy Science and Engineering*. https://doi.org/10.1016/j.geoen.2024.212726

Nisar, B., Pahalvi, H.N., Gulzar, A., Rashid, S., Majeed, L.R., Kamili, A.N. (2024). Bioplastics: solution to a green environment and sustainability. In A.L. Srivastav, A.S. Grewal, T.D. Pham (Eds.), *Role of Green Chemistry in Ecosystem Restoration to Achieve Environmental Sustainability* (pp. 261–269). Elsevier. https://doi.org/10.1016/b978-0-443-15291-7.00021-3

Ozkan, M. (2024). Atmospheric alchemy: The energy and cost dynamics of direct air carbon capture. *MRS Energy & Sustainability*. https://doi.org/10.1557/s43581-024-00091-5

Pasupuleti, M.K. (2024). AI and big data for climate resilience: Predictive analytics in environmental management. *International Journal of Academic and Industrial Research Innovations*. https://doi.org/10.62311/nesx/46601

Qian, X., Xu, A., Zhou, J., Dong, W. (2023). Valorization of PE plastic waste into lipid cells through tandem catalytic pyrolysis and biological conversion. *Journal of Environmental Chemical Engineering*. https://doi.org/10.1016/j.jece.2023.111016

Ringham, M., Hirtle, N., Shaw, C., Lu, X., Herndon, J.M., Carter, B.R., Eisaman, M.D. (2024). A comprehensive assessment of electrochemical ocean alkalinity enhancement in seawater: kinetics, efficiency, and precipitation thresholds. *EGUsphere*. https://doi.org/10.5194/egusphere-2024-108

Rosen, N., Welter, A., Schwankl, M., Plumeré, N., Staudt, J., Burger, J. (2024). Assessment of the potential of electrochemical steps in direct air capture through techno-economic analysis. *Energy & Fuels*. https://doi.org/10.1021/acs.energyfuels.4c02202

Rossi, D., Lermen, F.H., Echeveste, M.E.S. (2024). Circular product development strategies for sustainable production and consumption based on waste valorization. *Management of Environmental Quality: An International Journal*. https://doi.org/10.1108/meq-04-2024-0179

Sahak, M.Z.M., Shaffee, S.N.A., Thant, M.M.M., Aminuddin, F.A., Rawat, D., Calzadilla, P.A.F., Krishnan, R.K., Saiffudin, N. (2024). *Advancement in Green Hydrogen Production: Integrating Solid Oxide Electrolysis Cells (SOECs) into Existing Offshore Facilities for Sustainable Energy* [Conference paper]. APOGCE 2024, Perth, Australia. https://doi.org/10.2118/221154-MS

Samuel, H.S., Ekpan, F.-D.M., Ori, M.O. (2024). Biodegradable, recyclable, and renewable polymers as alternatives to traditional petroleum-based plastics. *Asian Journal of Environmental Research*. https://doi.org/10.69930/ajer.v1i3.86

Shammi, M., Akter, B. (2024). CO_2 sequestration for conventional utilization and industrial application. In M. N. V. Prasad, L. E. Erickson, F. C. Nunes, & B. S. Ramadan (Eds.), *Decarbonization strategies and drivers to achieve carbon neutrality for sustainability* (pp. 299–319). Elsevier. https://doi.org/10.1016/B978-0-443-13607-8.00011-0

Taqieddin, A., Sarrouf, S., Ehsan, M.F., Buesseler, K., Alshawabkeh, A.N. (2024). Electrochemical ocean iron fertilization and alkalinity enhancement approach toward CO_2 sequestration. *Npj Ocean Sustainability*. https://doi.org/10.1038/s44183-024-00064-8

Tong, E., Kastelic, M., McCatty, S., Lattimer, J. (2024). Carbonate electrolysis for energy efficient direct air capture of CO_2. *Meeting Abstracts*. https://doi.org/10.1149/ma2024-01372174mtgabs

Ummalyma, S.B. (2024). Bioeconomy for sustainable biomaterials and bioproducts development. In V.K. Garg, N. Kataria (Eds.), *Bioeconomy for Sustainability* (pp. 107–124). Springer. https://doi.org/10.1007/978-981-97-1837-5_5

Upadhyay, A., Yadav, A., Aggarwal, G., Maan, V. (2024). Comparative analysis of sustainable AI techniques used to reduce the carbon footprints in the healthcare sector. *Indian Scientific Journal of Research in Engineering and Management*. https://doi.org/10.55041/ijsrem38583

Voget-Kleschin, L., Baatz, C., Heyward, C., van Vuuren, D.P., Mengis, N. (2024). Reassessing the need for carbon dioxide removal: moral implications of alternative climate target pathways. *Global Sustainability*. https://doi.org/10.1017/sus.2023.21

Wu, Q., Gao, Y., Qi, Y., Chen, H., Li, D. (2024). Research on CCUS emission reduction methodology. *Advances in Computer and Engineering Technology Research*. https://doi.org/10.61935/acetr.3.1.2024.p386

Yang, W., Yang, G., Yang, G. (2024). The impact of AI on enterprise energy management: From the perspective of carbon emissions. *Science and Technology for Energy Transition*. https://doi.org/10.2516/stet/2024096

Yin, K., Wang, Y., Zhang, X., Chen, Y., Zhong, J. (2024). *Toward Green Hydrogen Generation*. In M. Sankir & N. D. Sankir (Eds.), *Green hydrogen production and storage technologies* (Chapter 3). Wiley. https://doi.org/10.1002/9781394234110.ch3

Zhang, D. (2024). Recent advances in smart grid architecture and grid-electric vehicle-charging station interaction. *Journal of Physics*. https://doi.org/10.1088/1742-6596/2786/1/012017

Zhao, X., Liang, G. (2023). Optimizing electric vehicle charging schedules and energy management in smart grids using an integrated GA-GRU-RL approach. *Frontiers in Energy Research*. https://doi.org/10.3389/fenrg.2023.1268513

Zülfikaroğlu, S. (2024). AI-driven strategic management and decision making for energy sector. *Next Frontier for Life Sciences and AI*. https://doi.org/10.62802/q7rkdb54

10 Rethinking Energy Management

This chapter explores the need to rethink energy management due to the limitations of conventional resources and the environmental consequences of their use. It discusses the challenges posed by fossil fuels and nuclear power, including climate change, resource depletion, and pollution. This chapter then delves into alternative solutions, emphasizing the importance of renewable energy sources, energy efficiency, smart grids, and demand-side management in achieving a more sustainable energy future. It also highlights the role of technology in improving energy management and the efforts of international organizations like United Nations Economic Commission for Europe (UNECE) in promoting sustainable energy policies and practices.

The world is facing a critical juncture in its relationship with energy. Our reliance on conventional resources has fueled economic growth for centuries, but it has also come at a significant cost to the environment. Climate change, resource depletion, and pollution are stark reminders that the old ways of managing energy are simply unsustainable. This chapter will explore the current state of energy management, examining both the limitations of conventional resources and the promise of alternative solutions. Ultimately, we will chart a path toward a more sustainable future, one where energy use is efficient, clean, and accessible to all.

10.1 CONVENTIONAL RESOURCES

Conventional energy resources refer to those sources of energy that have been traditionally used for powering various human activities. These resources are typically derived from natural sources and have been utilized for centuries, often forming the backbone of industrialization and modernization. For generations, our energy needs have been met primarily by exploiting fossil fuels—coal, oil, and natural gas. These resources are undeniably effective at generating power, but their combustion releases greenhouse gases that trap heat in the atmosphere, contributing to global warming. Additionally, fossil fuels are finite resources, and their continued extraction raises concerns about peak oil and resource depletion. Nuclear power has also played a significant role in energy production, offering a high-density energy source with a lower carbon footprint than fossil fuels. However, nuclear energy comes with its own set of challenges. Concerns about nuclear accidents, radioactive waste disposal, and weapons proliferation continue to cast a shadow over this technology. The consequences of our dependence on conventional resources are becoming increasingly apparent. Climate change is disrupting weather patterns, causing extreme weather events, and leading to rising sea levels. Air pollution from fossil fuel combustion is a major health hazard, contributing to respiratory illnesses and heart disease. These environmental and health impacts represent a significant economic burden, with the costs of climate change mitigation and adaptation expected to rise in the coming decades.

10.1.1 Fossil Fuels

Coal is one of the oldest and most abundant fossil fuels. The International Energy Agency's (IEA) latest market update highlights a significant shift in global coal consumption, driven by strong growth in Asian economies that offsets declines in Europe and North America. In 2022, global coal consumption reached a record 8.3 billion tons, a 3.3% increase, and is expected to remain near this peak in 2023 due to robust demand in Asia. China, India, and Southeast Asia will account for three-quarters of global coal consumption in 2023 (Figure 10.1). Conversely, coal demand in

DOI: 10.1201/9781003479666-10

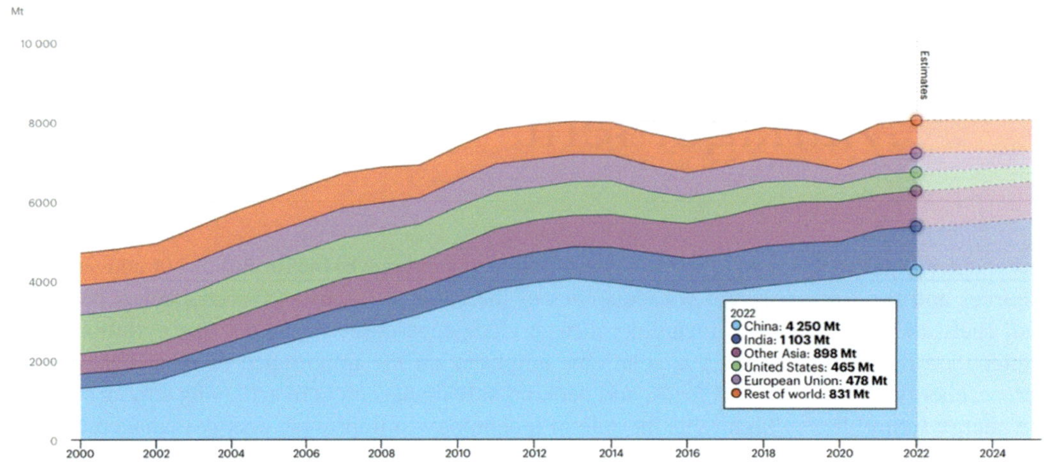

FIGURE 10.1 Global coal consumption, 2000–2025.

the European Union and the United States is declining due to the expansion of renewable energy, increased hydropower, and lower natural gas prices. Despite these regional differences, global coal demand grew by about 1.5% in the first half of 2023, driven by a 1% increase in power generation and a 2% rise in industrial use.

Fossil fuels, particularly coal, oil, and natural gas, have been integral to industrialization and economic growth, yet they pose significant environmental challenges due to their contribution to greenhouse gas (GHG) emissions and climate change. Coal combustion is a major source of CO_2 emissions, which are closely linked to economic activities and have historically driven climate change by releasing toxic levels of carbon dioxide into the atmosphere (Trương, 2023; Bess, 2022). Despite a decline in coal use in regions like Europe and the U.S., demand remains strong in emerging economies such as China and India, driven by their rapid industrialization and energy needs (Rao et al., 2023). The environmental impact of fossil fuels is profound, contributing to global warming, severe weather events, and ecological disruptions, which threaten the viability of human civilization (Bess, 2022).

The transition to sustainable energy models is urgent, as highlighted by the need for a green economy that prioritizes social equity, resource efficiency, and environmental sustainability (Moudene & Ridouane, 2024). Additionally, climate change and increasing international climate legislation pose risks to the economic growth of resource-dependent countries, necessitating improved natural resource management to ensure sustainable development (Salooja & Butler, 2023). Policymakers are urged to consider the asymmetric spillovers of climate risk and carbon emissions on energy markets, which affect energy prices and demand, and to differentiate their strategies across short- and long-term horizons (Rao et al., 2023). Addressing these challenges is crucial for fostering social well-being and managing economic costs while mitigating the adverse effects of climate change (Moudene & Ridouane, 2024; Salooja & Butler, 2023).

Oil, extensively used in transportation and manufacturing, is a significant source of GHG and criteria pollutants, contributing to environmental and health issues. The combustion of fossil fuels, including oil, releases GHGs such as CO_2, CH_4, and N_2O, alongside criteria pollutants like NO_x, CO, SO_2, and particulate matter (PM_{10} and $PM_{2.5}$) (Nielsen et al., 2022; Sun et al., 2019). Transportation, heavily reliant on fossil fuels, is a major contributor to these emissions, accounting for a substantial portion of global pollution, with road transportation being particularly impactful (D'Agosto, 2019; Zhang, 2023). In the U.S., refineries processing crude oil emit both GHGs and criteria pollutants, with emissions varying by product and process, highlighting the complexity of managing these emissions (Sun et al., 2019). Published data, as per Figure 11.4, indicates the USA has produced more oil in the years 2020–2022 than major oil-producing countries—which means more GHG generation.

The environmental impact of these pollutants is profound, affecting air quality on multiple scales, from local to global, and contributing to phenomena such as acid rain and climate change (Hidy, 2002). Methane, a potent GHG, is of particular concern due to its high warming potential, necessitating advanced monitoring systems to manage emissions effectively (Nielsen et al., 2022). The pressing need to mitigate these emissions has led to the exploration of alternative fuels, such as biofuels, which offer a potential reduction in pollution and a shift toward more sustainable energy sources (Zhang, 2023). Overall, the integration of technological advancements and policy measures is crucial in addressing the environmental challenges posed by oil-derived pollutants.

Natural gas is often promoted as a cleaner alternative to coal and oil due to its lower CO_2 emissions during combustion. However, its environmental impact is significantly influenced by CH_4 emissions with a much higher global warming potential than CO_2. Methane leakage occurs at various stages of the natural gas lifecycle, from extraction to transportation and utilization. Studies indicate that methane leakage rates can vary widely, with some systems exhibiting leakage rates as high as 66.2% globally, which can negate the climate benefits of switching from coal to natural gas if not properly managed (Gordon et al., 2023). In the context of residential heating in rural Beijing, the "coal-to-gas" transition has shown a reduction in GHG emissions by 44.8%, despite a measured methane leakage rate of 0.22% from gas boilers (Zhang et al., 2022). The lifecycle assessment of natural gas pipelines further highlights that methane emissions during transportation contribute significantly to the overall carbon footprint, with emissions ranging from 11.41 to 30.27 t $CH_4/km/10^8$ m^3 annually (Xu et al., 2022). Additionally, the production-stage methane emissions in the U.S. vary by region, with some areas exhibiting emissions intensities that add significantly to the overall GHG impact of natural gas consumption (Burns & Grubert, 2021). Despite these challenges, natural gas remains a relatively low-carbon energy source, but its environmental risks, particularly methane leakage, must be effectively managed through improved monitoring and regulatory measures to ensure its role in a sustainable energy transition (Fu et al., 2021). Figures 10.2, 10.3, and 10.4 show GHG emissions produced by fuel types and global per capita.

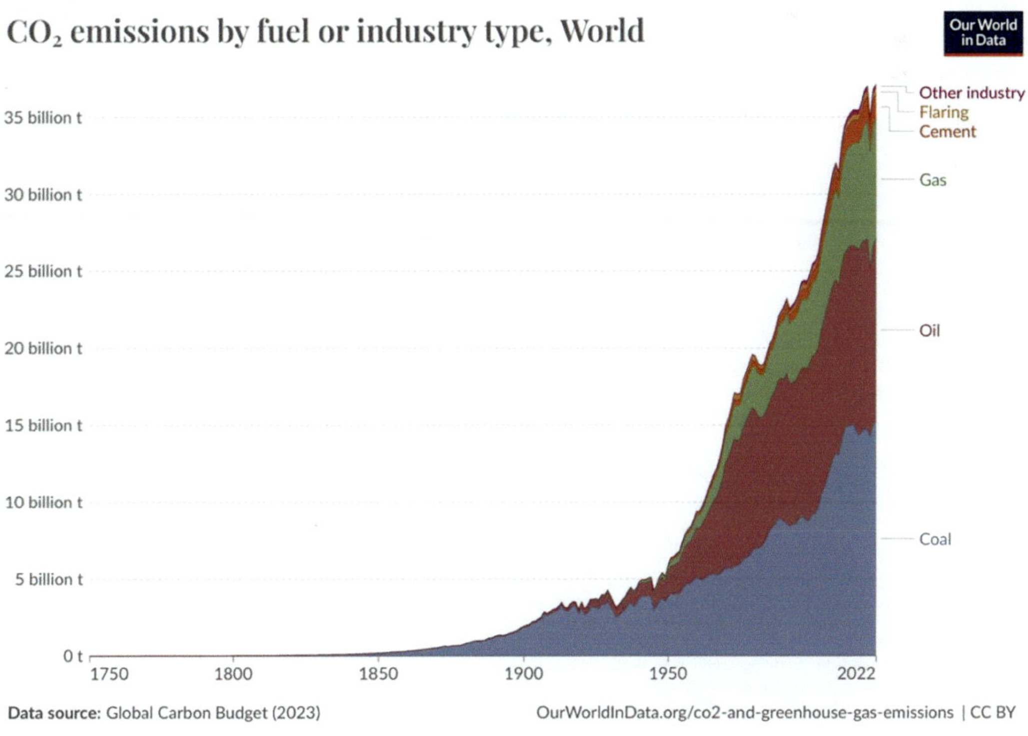

FIGURE 10.2 Emission by fuel.

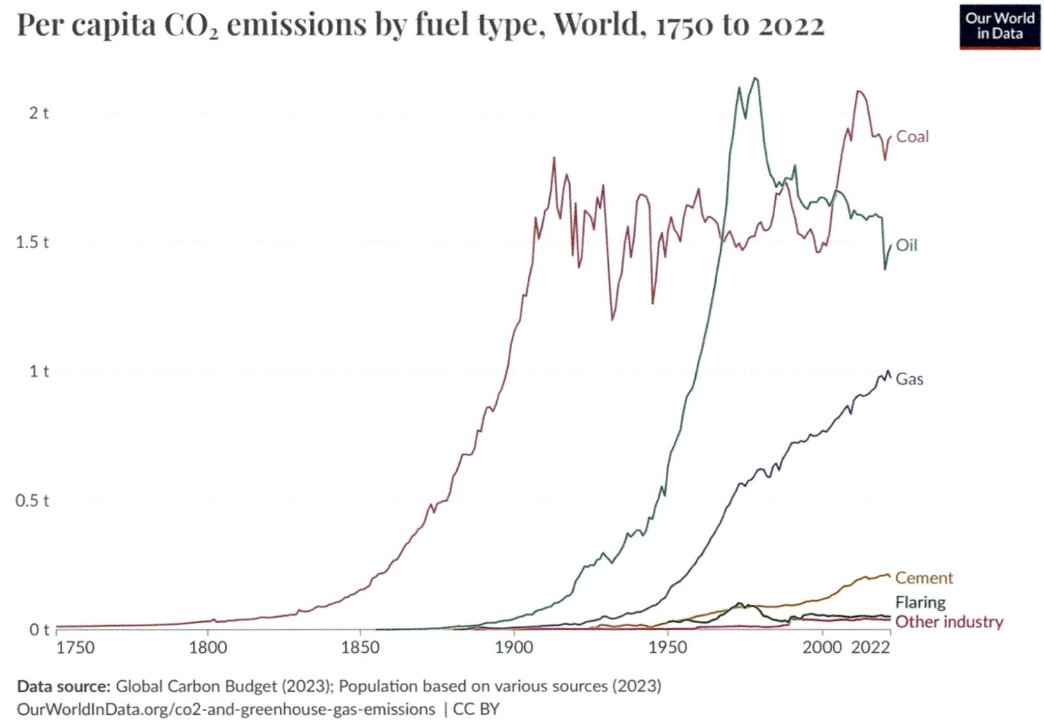

FIGURE 10.3 Per capita emissions.

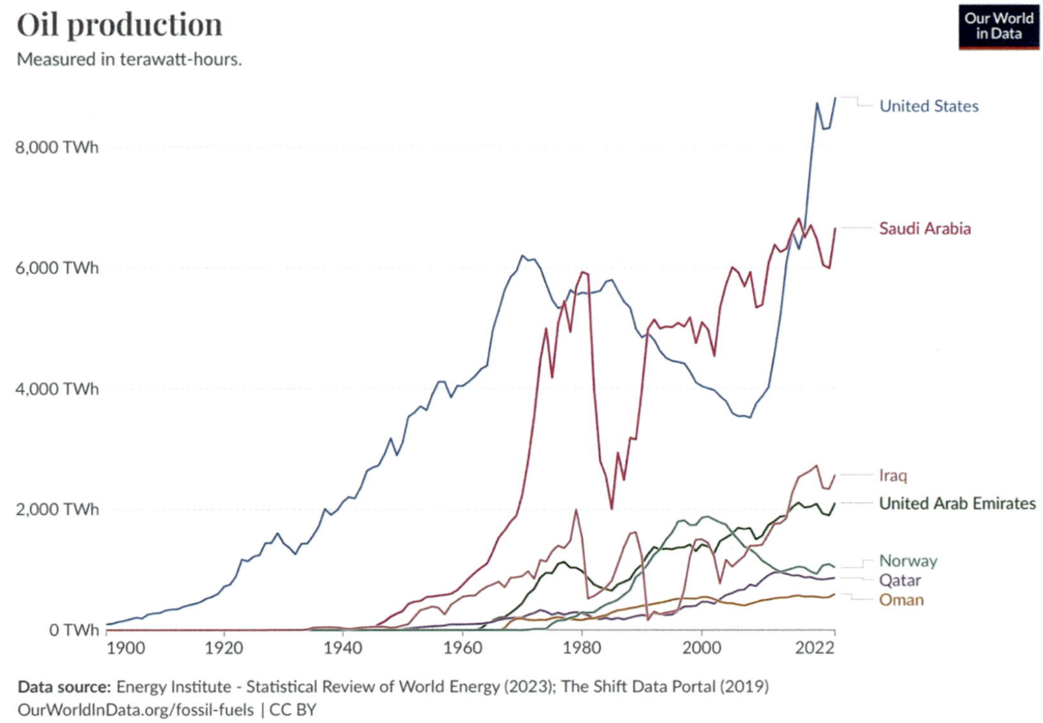

FIGURE 10.4 Oil production capacity.

10.1.2 Nuclear Energy

Nuclear energy is produced through nuclear fission, in which the nucleus of an atom is split into smaller parts, releasing a large amount of energy. This energy generates steam, which drives turbines to produce electricity. Uranium and plutonium are the primary fuels used in nuclear power plants. While nuclear energy does not produce GHG emissions during electricity generation, it does generate radioactive waste, which requires careful management and disposal.

Nuclear energy is a critical component in transitioning to a low-carbon economy, offering a high-density energy source with minimal carbon emissions compared to fossil fuels. It plays a significant role in reducing reliance on traditional energy sources, which are projected to become inadequate by 2050–2100 due to depletion and environmental concerns (Sadekin et al., 2020). Nuclear energy faces challenges such as radioactive waste management, safety concerns, and high development costs despite its benefits. Advances in reactor technology have improved safety and efficiency, yet managing radioactive waste remains a significant hurdle, necessitating sustainable solutions for long-term disposal. Public apprehension is fueled by the historical context of nuclear accidents and the potential for nuclear proliferation, which complicates its social and political acceptance (Kraft, 2024).

In countries like France, nuclear energy has been successfully integrated into the energy mix, demonstrating its potential to significantly reduce GHG emissions (Bréchet, 2023). However, the high upfront costs and the need for substantial government regulation and support hinder broader adoption (Ganea, 2022). The future of nuclear energy may also hinge on developing alternative technologies, such as thorium-based reactors and nuclear fusion, which could address some of these challenges (Ganea, 2022). Overall, while nuclear energy offers substantial benefits in combating climate change, its role remains contentious due to the complex interplay of its advantages and inherent risks (Bréchet, 2023; Sadekin et al., 2020).

10.1.3 Hydroelectric Power

Hydropower, also known as hydroelectric power, harnesses the energy of flowing water to generate electricity. It typically involves building dams to create reservoirs and then releasing water through turbines to produce electricity. Hydropower is a renewable resource, as it relies on the water cycle for replenishment. Large-scale hydropower projects can have significant environmental and social impacts, including habitat disruption, altered river flow patterns, and displacement of communities.

Hydroelectricity, while a significant contributor to global renewable energy, faces challenges related to GHG emissions, particularly from reservoirs. These emissions, primarily CO_2, CH_4, and N_2O, have been a subject of debate due to their potential impact on global warming (Ai et al., 2022; Kumar et al., 2022). The emissions are influenced by various factors, including the reservoir's age, morphology, and geographical location, with tropical regions often exhibiting higher emissions than temperate zones (Amanatidou et al., 2023; Gemechu & Kumar, 2022). Methodological challenges in accurately quantifying these emissions arise from the limited availability of measurement tools, significant spatio-temporal variability, and inconsistent modeling approaches (Kumar et al., 2022, 2023).

Life cycle assessments (LCA) of hydropower projects often overlook reservoir emissions, leading to a wide range of reported emissions values, from 1.5 to 3747.8 g CO_2-eq per kWh, with reservoir emissions potentially accounting for over 90% of total life cycle emissions in tropical regions (Gemechu & Kumar, 2022). To address these challenges, future research priorities include improving remote sensing applications, standardizing measurement methods, and enhancing international cooperation for better data collection and analysis (Ai et al., 2022). Additionally, developing comprehensive models and mitigation strategies is crucial for managing and reducing GHG emissions from hydroelectric reservoirs, thereby aligning hydropower more closely with global climate change mitigation goals (Kumar et al., 2023).

10.1.4 Biomass

Biomass refers to organic materials from plants and animals, such as wood, crop residues, and animal dung. Biomass, a versatile and renewable energy source, plays a crucial role in the global energy landscape by providing approximately 10% of the world's primary energy consumption. It encompasses a wide range of organic materials, including lignocellulosic materials, organic waste, and algae, which can be directly combusted for heat or converted into biofuels such as ethanol and biodiesel, essential for transportation and electricity generation (Srivastava & Pandey, 2024; Shahbeik & Peng, 2024). The conversion of biomass into energy involves various technologies, including thermochemical processes like pyrolysis, gasification, and combustion, as well as biochemical methods such as anaerobic digestion and transesterification (Shahbeik & Peng, 2024; Patra et. al., 2022). These technologies enable the transformation of biomass into diverse energy products, including solid char, bio-oil, bioethanol, and biomethane, contributing to a sustainable energy system. Biomass is considered carbon-neutral and is abundantly available as a coproduct from agricultural, forestry, and agro-industrial sectors, as well as from dedicated crops and organic municipal wastes (Cotana et al., 2022). The integration of biomass-based systems into the energy mix supports the transition toward a bio-based economy, addressing environmental and socio-economic challenges by reducing reliance on fossil fuels and mitigating associated environmental impacts (Shahbeik & Peng, 2024; Cotana et al., 2022). With ongoing technological advancements and supportive policies, biomass has the potential to become a cornerstone of renewable energy strategies, significantly contributing to global energy transitions and carbon emission reductions (Srivastava & Pandey, 2024).

10.2 ALTERNATIVE SOLUTIONS

Rethinking energy management involves exploring alternative approaches to conventional energy resources, such as fossil fuels, and considering more sustainable and efficient methods of energy production, distribution, and consumption. Conventional resources like coal, oil, and natural gas have been the primary sources of energy for centuries, but their finite nature and environmental impact have prompted a shift toward renewable energy sources and innovative energy management strategies.

Advancements in technology have provided businesses and corporations with the tools and innovations that will significantly improve their energy efficiency. From smart devices to data analytics, businesses now have more variety in how to monitor and optimize their energy efficiency more effectively. More sophisticated programs take advantage of technology. For example, utility tracking software predicts future energy usage and plans energy budgets, which helps a company's strategic decision makers ensure its energy management strategy correlates with its objectives and financial planning.

Energy management software helps control a corporation's budget and reduce the risk associated with energy price increases that can impact a business's ability to operate. Tracking utility costs and energy efficiency allows corporations to budget more efficiently and gain better insight into overall operational costs. According to Energy Star, decreasing energy use by 10% can lead to a 1.5% increase in net operating income. Enterprise management software uses Internet of Things (IoT), advanced connectivity, and big data, allowing a corporation to take advantage of energy data analytics for better facility management, and helps with energy consumption and energy management challenges. Intelligent asset management uses technology such as AI, IoT, and analytics to help inspect and monitor a building's efficiency, calculate potential impacts to the grid, anticipate failure, and better plan maintenance procedures.

Companies that use this technology can increase their productivity and make their facilities more energy-efficient, reducing emissions, mitigating climate risk, and extending asset lifecycles. The companies gain operational insights into clean energy sources, efficient waste management, and decarbonization. Smart devices have contributed a lot to the technology used by companies to be more energy-efficient. Here are some key devices that play a big role in improving energy efficiency: smart thermostats, energy monitoring systems, LED lighting, power management solutions, and renewable energy sources.

Smart thermostats provide accurate data on energy usage, enabling businesses to identify areas of high consumption and recommend strategies on how to reduce it. These devices allow businesses to regulate heating and cooling systems more efficiently. They learn temperature preferences, automatically adjust settings based on occupancy, and can be remotely controlled. The main idea here is the ability to proactively manage energy consumption based on real-time insights. Energy monitoring systems provide businesses with a comprehensive platform on how to monitor and control their energy usage. These systems analyze data from smart devices and provide insights on identifying waste energy and implementing ways to reduce it. Key advantages of the system include identifying energy-intensive equipment or processes for targeted improvements, setting benchmarks and goals for energy reduction, and monitoring energy usage trends to detect abnormalities or inefficiencies.

For example, features such as automated reporting and energy optimization suggestions help businesses to easily track and manage their energy consumption effectively. LED lighting (smart lighting) reduces energy consumption and lowers maintenance costs compared to traditional lighting. It lasts longer, emits less heat, and consumes up to 75% less energy. This allows businesses to automate lighting schedules and adjust brightness levels. It reduces energy usage without sacrificing the quality of illumination. Power management solutions, such as smart power strips and automatic shutdown systems, can help businesses eliminate energy waste caused by idle electronics. These devices cut off power supply to devices not in use, reducing standby power consumption. Power strips can detect when devices are in standby mode and automatically cut off power. Renewable energy sources, like solar panels, wind turbines, or geothermal systems, can help offset electricity costs and reduce reliance on fossil fuels.

10.2.1 Renewable Energy Sources

Renewable energy sources, such as solar and wind power, are superior to conventional fossil fuels for several compelling reasons. Despite the initial environmental impact associated with their manufacturing, renewable energy technologies exhibit a significantly lower carbon footprint over their operational lifespan. Photovoltaic panels and wind turbines, once installed, do not emit GHGs or other pollutants during energy generation, in stark contrast to coal and natural gas plants, which continuously release substantial quantities of carbon dioxide and methane, both potent GHGs. Innovations like grid-forming technologies facilitate the integration of these technologies into power systems, which enhance grid stability and resilience by allowing decentralized resources to operate autonomously, thus supporting the smooth integration of renewable energy sources.

Despite the initial environmental impacts associated with manufacturing renewable energy technologies, advancements in recycling processes and strategic planning in the manufacturing sector are reducing their overall environmental footprint (Usman et al., 2024). The transition to renewable energy is crucial for achieving global climate objectives, as evidenced by the significant negative impact of renewable energy on carbon emissions in major fossil fuel-consuming countries (Rahman et al., 2024). In the United States and Canada, the shift toward renewable energy, particularly solar photovoltaics, is projected to drastically reduce the levelized costs of electricity and final energy, demonstrating the economic viability of a Power-to-X economy and the potential for achieving net-zero emissions by 2050 (Lopez et al., 2024). These developments underscore the importance of continued innovation and policy support to overcome existing challenges and fully realize the benefits of renewable energy on a large scale.

Effective recycling processes for reducing the environmental footprint of renewable energy technologies involve a combination of advanced waste management techniques, innovative recycling methods for energy storage devices, and the integration of smart technologies. Traditional waste management processes, such as anaerobic digestion, gasification, and pyrolysis, are crucial for converting waste'p0 into renewable energy, thereby minimizing environmental pollution (Haque et al., 2024). For energy storage devices, recycling processes like heat treatment, chemical treatments, and metallurgy methods are essential for recovering valuable materials such as metals and carbon-based components from batteries and supercapacitors, which helps mitigate environmental hazards associated with toxic materials (Thakur et al., 2023). In the context of photovoltaic (PV) panels, cradle-to-cradle recycling is vital for managing material flows and addressing challenges related to the recycling of silicon, aluminum, and copper, which are economically significant (Peters, 2024). Additionally, innovative solutions for the end-of-life management of PV panels are necessary to reduce pollutant emissions and improve the overall sustainability of solar energy technologies (Calì & Acri, 2023). These processes and technologies collectively contribute to a more sustainable and environmentally friendly approach to managing the lifecycle of renewable energy technologies. Life cycle CO_2 equivalent from various renewable energy footprints is illustrated in Figure 10.5.

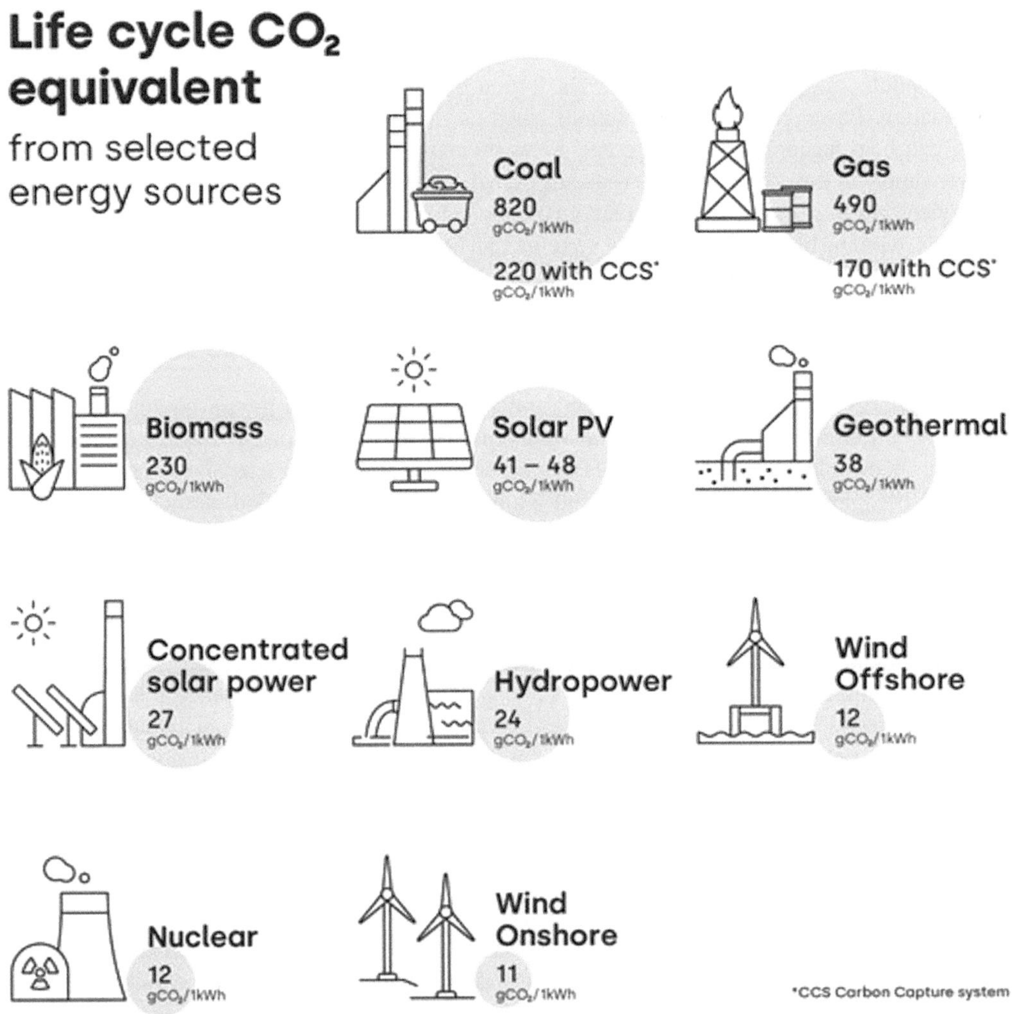

FIGURE 10.5 The footprint of renewable energy.

10.2.2 Energy Efficiency

Improving energy efficiency across various sectors, including buildings, transportation, and industry, can significantly reduce energy consumption and costs while simultaneously lowering carbon emissions. Energy efficiency sources offer significant advantages over conventional energy sources, primarily through their ability to reduce GHG emissions and enhance environmental sustainability. Unlike conventional energy sources that rely heavily on fossil fuels, which contribute directly to global warming and pollution, energy-efficient technologies minimize energy consumption and emissions (Anarene, 2024). According to the IEA, energy efficiency plays a crucial role in mitigating climate change by enabling the same level of energy services with lower energy use. Improving energy efficiency across buildings, transportation, and industrial sectors is pivotal in reducing energy demand and GHG emissions. In the building sector, energy-efficient technologies such as advanced insulation materials, high-efficiency HVAC systems, and LED lighting are crucial in minimizing energy consumption while maintaining or enhancing service quality. For instance, HVAC systems in commercial buildings account for approximately 40% of total energy consumption, highlighting the importance of efficient systems in reducing energy use and emissions. Retrofitting existing buildings with these technologies, as demonstrated by projects like the Empire State Building's retrofitting, can lead to significant energy savings, with a projected 38% reduction in energy use (Anarene, 2024).

In the industrial sector, energy efficiency is essential for enhancing energy security and reducing operational costs, with policies and initiatives targeting manufacturing industries to achieve cost-effective reductions in energy usage and GHG emissions (Khaturia et al., 2024). Additionally, energy-efficient technologies such as variable frequency drives (VFD) and smart lighting systems offer substantial potential in cutting energy consumption across various sectors (Khattra et al., 2024). Policies encouraging the adoption of these technologies, alongside carbon trading mechanisms, further support emission reductions and environmental protection (Khattra et al., 2024). Overall, energy efficiency is a cost-effective strategy to combat climate change, reduce energy costs, and improve competitiveness, making it a vital component in achieving net-zero emissions (Kiranmai et al., 2024).

10.2.3 Smart Grids

Smart grids represent a significant advancement over conventional electrical grids due to their integration of two-way communication and distributed intelligent devices. By facilitating bidirectional flows of both electricity and information, smart grids offer a more efficient and responsive delivery network. One of the key advantages lies in the enhanced flexibility they afford, enabling greater integration of highly variable renewable energy sources like solar and wind power without necessarily requiring additional energy storage infrastructure. Smart grids, through their integration of digital and communication technologies, enable two-way communication between energy producers and consumers, optimizing electricity delivery and integrating variable renewable sources. These systems enhance grid reliability by employing artificial intelligence (AI) technologies for optimization, forecasting, and energy management, which improve energy efficiency and sustainability.

Moreover, smart grids leverage IoT technologies to enhance energy management in urban environments, allowing for real-time adjustments based on data from smart meters and other sensing devices (Udayakumar et al., 2023; SwornaKokila et al., 2023). Characteristics of traditional systems compared to the smart grid are presented in Figure 10.6.

Machine learning techniques further aid in predicting peak demand and adjusting energy usage proactively, achieving significant reductions in peak demand and energy savings (SwornaKokila et al., 2023). Despite challenges such as user awareness, automated solutions like energy management controllers (EMCs) effectively address these issues, ensuring efficient energy management and improved user comfort (Ahmad et al., 2023).

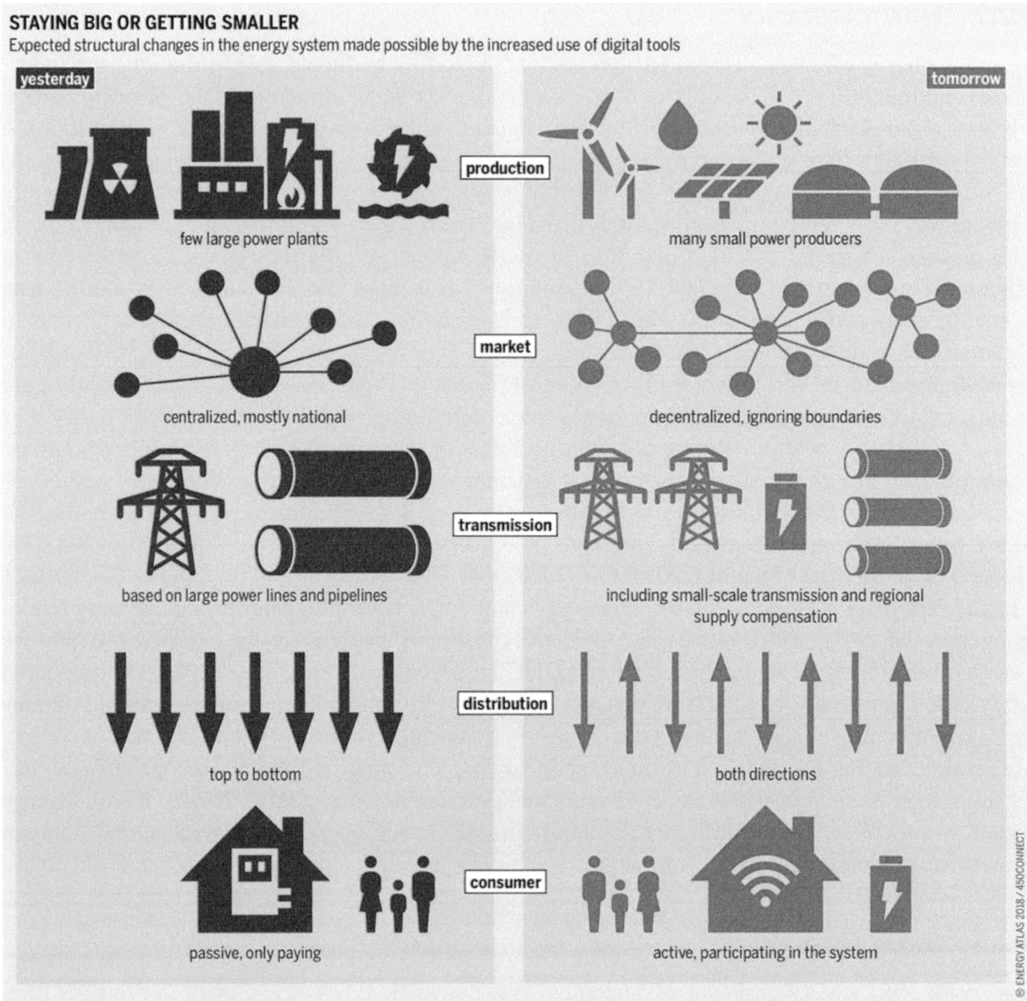

FIGURE 10.6 Characteristics of a traditional system (left) versus the smart grid (right).

10.2.4 Demand-Side Management

Encouraging consumers to adjust their energy consumption patterns through demand-side management programs can help balance energy supply and demand, reduce peak load, and lower overall energy consumption. Demand-side management (DSM) strategies, such as demand response (DR) programs, incentivize consumers to adjust their energy use during peak hours, thereby balancing supply and demand and reducing overall energy consumption (Shah et al., 2024; Udayakumar et al., 2023). These programs utilize advanced algorithms, such as the Hybrid Genetic Wind-Driven Optimization (HGWDO) and the Hybrid Genetic Bacteria Foraging Optimization Algorithm (HGBFOA), to schedule smart appliances efficiently, minimizing peak-to-average demand ratios and optimizing utility costs (Shah et al., 2024; Ahmad et al., 2023).

The HGWDO and the HGBFOA are advanced algorithms integrating principles from multiple optimization techniques to enhance performance in complex problem-solving scenarios. The HGWDO combines the Genetic Algorithm (GA) with the Wind Driven Optimization (WDO) algorithm, leveraging both strengths to improve DSM in smart grids. This hybrid approach is particularly effective in scheduling residential loads to optimize electricity costs and user comfort, outperforming other heuristic algorithms in various performance metrics. On the other hand, the HGBFOA merges the GA with the Bacterial Foraging Optimization Algorithm (BFOA), inspired by the social foraging behavior of *Escherichia coli* bacteria. This hybridization aims to enhance the global optimization capabilities of BFOA by incorporating genetic operations such as mutation and crossover, which are known to improve convergence and solution accuracy in complex optimization problems. Both HGWDO and HGBFOA exemplify the trend of hybridizing different optimization algorithms to address the limitations of individual methods, such as premature convergence and sensitivity to parameters, thereby broadening their applicability across various domains.

10.2.5 Decentralized Energy Management

DSM offers a multitude of benefits that make it a superior approach for optimizing energy consumption compared to traditional supply-side methods. Firstly, DSM provides significant financial savings by allowing businesses to strategically shift their energy usage to times when electricity prices are lower. This not only reduces operational costs but also enhances competitiveness in the market. Cement plants adopt DSM and, as shown in Figure 10.7, reduce operating costs substantially. Additionally, DSM contributes to environmental sustainability by reducing carbon emissions associated with electricity consumption. By aligning energy usage with periods of lower carbon intensity in the grid, DSM helps mitigate the environmental impact of industrial operations.

Furthermore, DSM fosters flexibility and responsiveness in energy consumption patterns, enabling businesses to adapt to dynamic market conditions and regulatory changes. This adaptability is crucial in an era where renewable energy sources, such as wind and solar, introduce variability into the grid. DSM allows industries to optimize their energy usage in tandem with renewable energy generation, maximizing the utilization of clean energy and minimizing reliance on fossil fuels.

Moreover, DSM empowers industries to enhance their operational efficiency by identifying and eliminating unnecessary energy consumption. By leveraging data-driven insights and optimization algorithms, businesses can streamline their energy usage, reduce waste, and improve overall productivity. This not only results in cost savings but also enhances the sustainability and resilience of industrial operations.

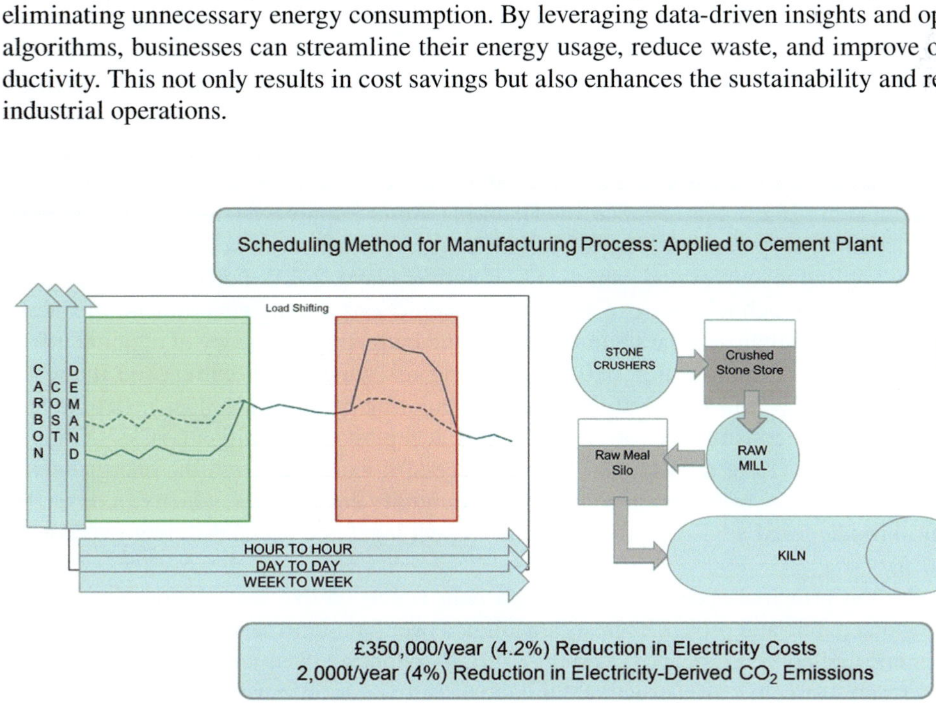

FIGURE 10.7 Expenses and emission reductions in cement plant.

10.3 THE PATH TO SUSTAINABILITY

10.3.1 Challenges and Opportunities

The transition to sustainable energy is a multifaceted challenge that involves overcoming high upfront costs, infrastructure limitations, and resistance from entrenched interests. Yet, it also presents significant opportunities through declining renewable energy costs, technological advancements, and increasing public and political support for climate action. The advancements in solar and wind technologies have notably reduced costs and improved efficiency, making them pivotal in transforming the energy landscape (Khan, 2024; Lopez et al., 2024). However, the transition is not without its socio-economic challenges, such as job displacement in fossil fuel-dependent industries and the unequal distribution of resources, which necessitates a careful balancing act to ensure a just transition (Imam & Madkaikar, 2024; Ishola, 2024).

Sustainable finance mechanisms, including green bonds and investment funds, play a crucial role in mobilizing capital for green energy projects, although regulatory uncertainties and the need for standardized ESG criteria remain barriers (Ikevuje et al., 2024). Collaborative efforts among governments, businesses, and civil society are essential to address these challenges and accelerate progress. For instance, policy frameworks and international cooperation are vital for aligning global and local goals, as seen in the renewable energy strategies of countries like Germany and India (Ishola, 2024; Nassar, 2024). Moreover, emerging technologies such as blue hydrogen and floating photovoltaic systems offer promising avenues for reducing CO_2 emissions and fostering sustainable development (Coccia, 2024). The integration of renewable energy into existing infrastructures, supported by innovations like energy storage systems and smart grids, is crucial for optimizing the transition (Adewnmi et al., 2023). Overall, a comprehensive approach that includes technological innovation, sustainable finance, and policy support is necessary to overcome the barriers to sustainable energy transition and capitalize on the opportunities it presents for a cleaner, more resilient future (Jasińska et al., 2023).

Governments and businesses can effectively collaborate to overcome infrastructure limitations and resistance from entrenched interests through strategically implementing Public-Private Partnerships (PPPs). These partnerships leverage the strengths of both sectors, combining public oversight and private sector efficiency to address funding and resource constraints in infrastructure development (Alfianto & Gayo, 2021; Leruth, 2012). The success of PPPs hinges on several collaborative governance elements, including shared principles, a robust governance structure, and a flexible collaborative process that accommodates the diverse contexts of multiple stakeholders (Nusriadi et al., 2023). For instance, the Drinking Water Supply System in Umbulan, East Java, exemplifies how government cooperation with business entities can accelerate infrastructure projects by pooling resources and expertise (Alfianto & Gayo, 2021). Furthermore, integrating modern information technologies can streamline interactions among government, business, and local communities, enhancing the socio-economic and budgetary efficiency of regional infrastructure projects (Gabidullina et al., 2020). Addressing resistance from entrenched interests, such as local communities and First Nations, requires building multi-stakeholder networks and engaging in meaningful consultations to secure social acceptance and mitigate risks of litigation or political conflict (Hale & Belanger, 2015). Successful examples from the energy sector demonstrate the importance of early and continuous community engagement, which can be facilitated by multi-jurisdictional advisory processes that respect legal and social differences (Hale & Belanger, 2015).

Additionally, the principal-agent framework can help conceptualize the relationships within PPPs, ensuring that public and private interests are aligned over long-term service contracts, despite shifting economic and political contexts (Leruth, 2012). By adopting these strategies, governments and businesses can overcome infrastructure limitations and foster an environment conducive to sustainable development and innovation.

10.3.2 Solutions to Attain Sustainable Energy in the UNECE Region

The "Pathways Project" was initiated by the UNECE Committee on Sustainable Energy (the Committee). This project aims to support UNECE countries in developing and implementing national sustainable energy policies. It is structured into three phases:

- **Phase I**: "Futurise" involves discussions with experts and stakeholders to identify critical factors shaping the energy future. Thirteen proposals were grouped into three categories: global politics and economic situation, technology, and energy policy and market development.
- **Phase II**: "Modeling" includes constructing an integrated energy and climate analytical architecture based on complex modeling with leading institutions. This phase involves creating a Technology Map to inform countries about energy technologies and costs, and developing an early warning system to track sustainable energy indicators.
- **Phase III**: "Informed Dialogue" focuses on stakeholder consultations at regional and subregional levels to assess modeling results and prepare policy recommendations. Results are prepared for both the aggregated UNECE level and seven subregions.

The project encompasses three distinctive scenarios: the Reference (REF) scenario, based on the Shared Socio-economic Pathway (based on SSP2); the Nationally Determined Contributions (NDC) scenario; and the P2C scenario, which aims to achieve the 2°C target of the Paris Agreement by 2100. The three scenarios, along with variations in technology costs, were clustered within the two axes of the scenario space, as presented in Figure 10.8.

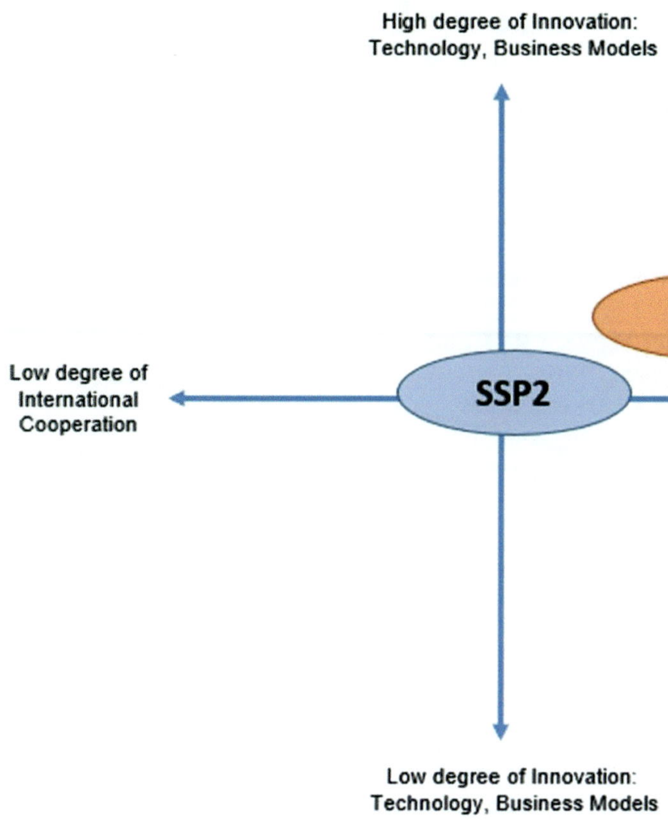

FIGURE 10.8 From storylines to policy pathways.

Figure 10.9 illustrates the projected evolution of electricity generation in the UNECE region through 2050 across the three scenarios: REF, NDC, and P2C. In all scenarios, natural gas continues to play a critical role, but significant transformation occurs only in the P2C scenario, which emphasizes rapid uptake of low-carbon technologies post-2030. Renewable energy sources, particularly wind and solar PV, expand rapidly beginning in 2025, supported by investments in underdeveloped subregions. Concurrently, retrofitted coal and gas plants with carbon capture and storage (CCS) begin contributing to emissions reduction. Notably, electricity demand under the P2C scenario increases by 30% by 2050, requiring a substantial scale-up in generation capacity. These changes are mirrored in the structural shifts of installed capacity presented in Figure 10.9, which shows nearly six gigawatts of new low-carbon capacity added for each gigawatt of fossil fuel capacity retired.

Figure 10.10 shows the projected evolution of CO_2 Emissions in the UNECE region through 2100 by policy scenario. In the REF scenario, a temperature increase of 4.2°C is expected by 2100. Cumulative regional emissions are expected to amount to 1,250 Gt CO_2 (2020–2100). Under this scenario, the impacts of climate change are expected to be severe and will probably trigger unrecoverable changes in the climate system, which the world probably does not have the resources to control. In the NDC scenario, a temperature increase of 3.0°C is anticipated by 2100. Cumulative

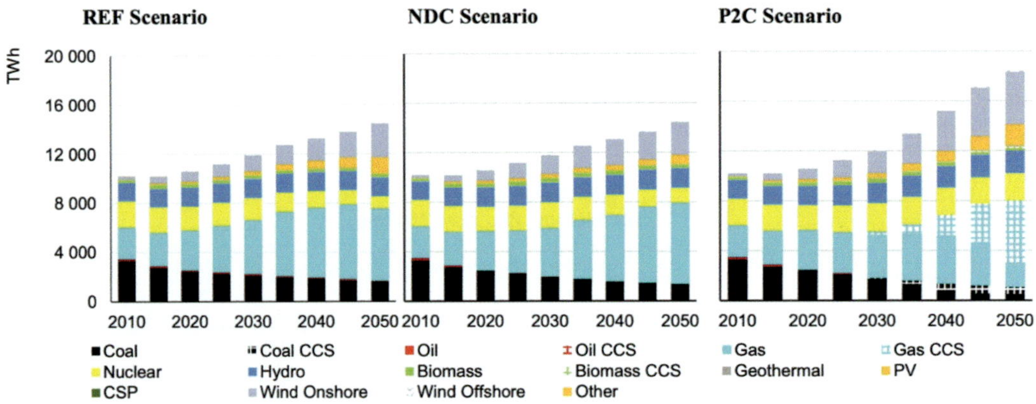

FIGURE 10.9 Electricity generation in the UNECE region by policy scenario.

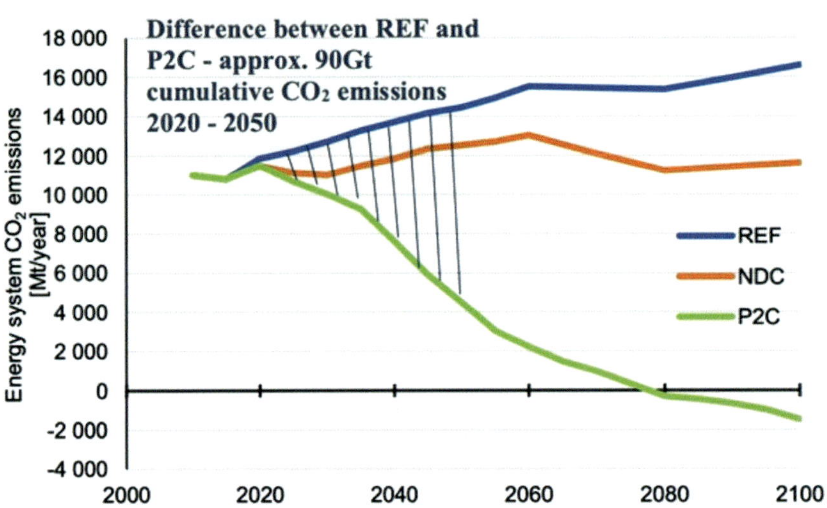

FIGURE 10.10 CO_2 emissions in the UNECE region by policy scenario.

regional emissions are expected to be 18% (225 Gt CO_2) lower compared to the REF scenario (2020–2100). Under this scenario, the impacts of climate change are expected to be severe and may trigger unrecoverable changes in the climate system, with serious consequences for the economy and society as a whole. In the P2C scenario, a temperature increase of 2.1°C is expected by 2100. Under this scenario, emissions peak by 2020. To meet the target, negative emissions are mandatory post-2070. Climate change would be limited to, for example, more extreme weather patterns, major damage to coral reefs and marine systems, and significant shifts in agriculture.

The three pillars of energy for sustainable development encapsulate the intricate balance required to foster economic growth, ensure quality of life for all, and preserve the environment for future generations. The first pillar, Energy Security, underscores the imperative of securing the energy necessary for robust economic development while minimizing dependence on volatile energy sources. This entails diversifying energy portfolios, investing in renewable energy technologies, and enhancing energy efficiency measures. Secondly, the Energy for Quality of Life pillar underscores the importance of providing affordable, reliable energy access to all, encompassing not only electricity but also heating, cooling, and transport. Achieving this requires not just physical infrastructure but also addressing affordability and sustainability concerns, particularly in the context of bio-energy and its impact on food systems. Finally, the Energy and Environment pillar emphasizes the trade-offs inherent in meeting energy demand while safeguarding the environment and combating climate change. Reducing the carbon footprint across the energy supply chain is paramount, with energy emissions contributing significantly to GHG emissions. This pillar extends beyond climate change mitigation, addressing broader environmental concerns such as water usage, transport emissions, and air pollution. In navigating these pillars, it is essential to adopt holistic approaches that prioritize sustainability, equity, and resilience to ensure a prosperous and harmonious future (Figure 10.11).

1. Pursue Energy Efficiency as the Basis for Systemic Efficiencies

   Efficiency is key to the future energy system, with buildings playing a significant role, consuming a majority of power and emitting a substantial portion of CO_2. The High Performance Buildings Initiative (HPBI) by UNECE promotes energy efficiency through global dissemination of guidelines, supported by International Centers of Excellence. Industrial efficiency, especially in energy-intensive sectors like cement and steel, requires

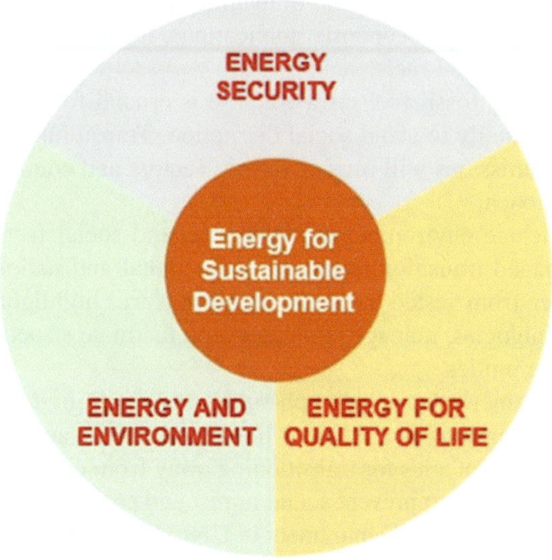

FIGURE 10.11 Energy for sustainable development.

innovation and collaboration with the private sector. Compulsory fuel economy standards are crucial for enhancing vehicle efficiency, while the transport sector can decarbonize through technological advancements like electric vehicles. Urban mobility can be improved with proper infrastructure planning, but long-distance freight transport remains challenging.

2. Address GHG Emissions from the Fossil Fuels

   Reducing GHG emissions from coal mining, particularly methane, is crucial for mitigating environmental impact and addressing climate change. Efforts must encompass the entire coal mining value chain, from exploration to post-mining activities, to convert fugitive emissions into assets. Methane's high GWP underscores the importance of managing emissions from well to burner tip. Additionally, carbon capture and storage (CCS) presents a viable decarbonization option, especially in regions like Western Europe, North America, and the Russian Federation. Investing in cleaner coal technologies such as CCS and high-efficiency, low-emission (HELE) plants is essential for maintaining coal in the energy mix while addressing climate concerns. This approach is vital for all fossil fuel sources to achieve sustainable energy goals.

3. Accelerate the Interplay between Renewable Energy and Renewable/Decarbonized and Low Carbon Gases

   UNECE emphasizes the synergies between traditional and emerging energy sectors to accelerate the integration of renewable energy and low-carbon gases. Flexible systems are crucial for decarbonizing power generation, with natural gas playing a pivotal role in supporting renewable energy introduction due to its rapid response and operational flexibility. However, methane emissions must be managed carefully. The interplay between renewable energy and natural gas, particularly in producing decarbonized gases like hydrogen, is essential for deep decarbonization. Projects such as power-to-gas and renewable hydrogen production can reduce the energy sector's environmental impact, especially in hard-to-electrify sectors like transportation. Despite the potential of the hydrogen economy, legislative and structural challenges persist, including varying standards for hydrogen transmission through natural gas pipelines. Establishing consistent legislative frameworks is crucial to fostering the emergence of disruptive technologies in the energy sector.

4. Address the Social and Economic Impact of the Energy Transition

   The energy transition must address social and economic impacts to ensure a just and sustainable shift away from fossil fuels. Managing reductions in fossil fuel use requires careful consideration of socio-economic implications, including the dependence of communities and sectors on fossil fuels.

   Modernizing existing fossil fuel infrastructure is crucial for sustainable development but must be done gradually to avoid social disruption. Transitioning to renewable energy and reducing GHG emissions will impact various sectors and communities, necessitating a just transition approach.

   Key concerns include environmental, economic, and social factors. Economic realities demand a managed transition to accommodate local and national economies while addressing resistance from vested interests. Social concerns highlight the need to promote carbon-neutral technologies, manage job losses, and facilitate structural changes in fossil fuel-dependent communities.

   A just transition framework can aid in planning, ensuring long-term governmental policies support affected communities. Regions like Upper Silesia and Lausitz, still reliant on coal, require proactive strategies for transitioning away from coal, involving multiple sectors and long-term planning to prevent social unrest and economic decline.

5. Accelerating Renewable Energy Deployment in UNECE Subregions

   Renewable energy competitiveness is rising, yet many UNECE subregions remain untapped. In 2015, UNECE boasted 869 GW of renewable electricity capacity, half of global capacity, but Eastern subregions lag. Hydropower dominates the East, with solar PV rising and wind potential untapped.

However, investment in these regions is minimal, accounting for only 0.2% of global investments in 2015. New business models are emerging, emphasizing bottom-up approaches and consumer control, yet international support remains modest.

To address this, UNECE is advocating for innovative business models to accelerate the energy transition, empowering consumers and leveraging digitalization. This effort will be spearheaded by the Group of Experts on energy efficiency, cleaner electricity systems, and renewable energy.

By focusing on innovative financing, consumer empowerment, and digitalization, UNECE aims to unlock the vast renewable energy potential in Eastern subregions, contributing to global energy transition efforts.

6. Embrace Water-Energy-Food Nexus Approach Across the Whole Region

   Embracing a Water-Energy-Food Nexus approach across the region can optimize resource management, with renewable energy playing a pivotal role. While hydropower dominates renewable energy in the UNECE region, expanding wind and solar capacity must be done carefully to avoid adverse effects on agriculture and food security. Integrated management of natural resources, facilitated by renewable energy technologies, enhances efficiency and sustainability while reducing environmental impact. Transboundary cooperation, advocated by UNECE, is essential for maximizing the benefits of renewable energy while addressing water and land management challenges. Increasing renewable energy's share can alleviate water requirements for power generation, enhance water security, and contribute to food security objectives. However, coordination gaps at national and subregional levels pose challenges.

   The Food-Water-Energy nexus, aligned with the 2030 Agenda, underscores the importance of integrated resource management. The United Nations Framework Classification of Resources (UNFC) provides a framework for seamless policy implementation by harmonizing data on energy and water resources. Facilitating integrated management fosters sustainable development, ensuring the security, accessibility, and affordability of essential resources.

7. Sustainable Resource Management for Energy Storage Solutions

   Ensuring a sustainable energy future poses challenges in managing resources for energy storage solutions. Critical elements like lithium, cobalt, and nickel are essential for batteries and renewable technologies, but their availability is limited, leading to geopolitical shifts. As battery efficiency improves, demand is projected to soar, with prices expected to decrease significantly. However, the finite lifespan of batteries necessitates solutions for prolonging their use, such as repurposing EV batteries for the power sector. Addressing environmental impacts requires innovation, circular economy practices, and adoption of international standards. Further analysis by UNECE's Expert Group on Resource Management and Cleaner Electricity Systems is crucial for navigating this transition toward widespread electrification.

10.3.3 Comparing the Energy Performance of Different Countries

In 2022, France was named the most energy-efficient country with a score of 74.5. France has one of the highest spendings on energy efficiency. Pannier-Runacher, France's minister for energy transition, stated that $787 million is set aside by the government to support their citizens in reducing their energy use in their households. In addition, the government has set up tax credits and loan programs to assist the citizens and is making a big difference. France aims to reduce its energy consumption by 20% and its GHG emissions by 37% by 2030. This country has strict compliance building codes and mandatory energy usage reports for residential and commercial sectors. There is also an ambitious retrofit program, and the country is working to renovate 500,000 dwellings per year. France is also working to reduce its transportation emissions by 29% in 2028. The French government announced that the country plans to invest $100 billion in new railway infrastructure. France has already invested around $2.5 billion in producing electric and hybrid vehicles. Furthermore, the country has invested $12 billion in R&D activities with the aim of developing a low-carbon airplane. By 2050, the country plans to be completely carbon-neutral.

The United Kingdom, with a score of 72.5, was second behind France and is widely known for its efficient power plants and high R&D investment in energy-efficient technologies. The country works on improving the efficiency of buildings with demanding codes, appliance efficiency standards, and a building energy labeling program. They also plan to decarbonize the industrial sector and have a net-zero carbon cluster by 2040. As for transportation, the UK has high fuel economy requirements for vehicles, targeting 64.4 MPG by 2025, and electric vehicles represent over 11% of new sales. The UK plans to have 100% clean electricity by 2035.

Germany, with a score of 71.5, is working to decarbonize heating and cooling systems, which represent a major share of building consumption, and there are energy efficiency loan programs and tax credits. There are mandatory energy performance requirements for new constructions and major renovations, and standards for 42 product categories. There is an electricity tax for the industrial sector, which can be reduced by up to 90% by having an ISO 50,000 energy management system. In the transportation sector, electric vehicles represent 13.5% of new sales. Germany plans to reduce its primary energy consumption by 32.5% by 2030 and 50% by 2050. Germany plans to be completely carbon-neutral in 2045, five years earlier than the EU target.

The Netherlands, with a score of 71.5, is tied with Germany for third place. The government motivates its citizens to use energy sparingly and efficiently by taxing the use of electricity and natural gas. This country is widely known for its strict building codes and retrofit policies, which also include specific requirements for office buildings. Energy use reports are mandatory for residential and commercial buildings, which are subject to a rating system. There are also performance standards for 42 appliance groups. In the transportation sector, the Netherlands offers tax incentives for electric vehicles. The country is targeting a 32.5% improvement in energy efficiency by 2030, with specific goals and measures for each sector, and is increasing its R&D investment. The Netherlands is targeting a 49% reduction of GHGs by 2030 and aims for a 95% reduction by 2050.

Italy, with a score of 68.5, comes in fifth place. This country has the most energy-efficient power plants among all the other countries in the ACEEE scorecard, and there are energy loan and tax credit programs for multiple sectors. The Conto Termico program offers incentives for energy efficiency measures and retrofits in residential and commercial buildings, along with mandatory performance requirements for renovations in residential and commercial buildings. Italy has energy efficiency targets for the industrial sector, which are enforced with regular audits, and there is an energy efficiency certificate trading program for industrial consumers. Combined heat and power (CHP) capacity is high in Italy, and there are subsidies for low-carbon machinery. Italy is improving its performance in the transportation sector with a 64.4 MPG target at the fleet level by 2025, combined with major investments in rail infrastructure. The country also plans to reduce primary energy consumption by 43% and GHG emissions by 33% in 2030.

Spain is ranked among the top three countries in the buildings category. Residential and commercial buildings are subject to strong building codes, which also include renovation requirements. There is also a mandatory labeling and energy disclosure program. Spain is characterized by its high energy efficiency investment per capita compared to many other countries on the scorecard. The country is placed fifth in transportation, achieving maximum scores in fuel economy standards and average fuel economy for passenger vehicles. It is also compliant with the EU emissions reduction target for new cars and has high investment in rail infrastructure. Spain is targeting a 23% reduction of GHG emissions and a 39.5% improvement in energy efficiency by 2030.

Japan has high R&D investment in energy efficiency, combined with multi-sector loan and tax incentives for energy efficiency projects. Japan has great energy-efficient policies, which is how the country has greatly reduced its energy intensity. Japan is the top-ranking country in industrial energy efficiency; this has been achieved through its efforts of combining regulations, voluntary efficiency programs, and financial incentives. The Act Concerning the Rational Use of Energy has energy efficiency requirements per industry sector, and companies are required to have an energy manager. Companies are also required to report annual energy consumption, and there is a benchmarking system with mandatory efficiency targets. Building owners and developers are required to submit a mandatory energy savings plan before any major renovation project. There is also a building energy labeling program, but participation is voluntary. In the transportation sector, Japan has achieved an average fuel economy of 37.94 MPG and is targeting 48.5 MPG by 2025. The country has created the Green Logistic Partnership to improve freight efficiency and public transportation.

Taiwan has some of the most efficient power plants in the world, and the energy service company (ESCO) market is very large as a share of GDP. The country's buildings are characterized by their low-intensity energy usage and minimum energy performance standards for 15 appliance groups. The country has many policies with the specific goal of improving industrial energy efficiency, and local companies have achieved low energy intensity. Companies are subject to mandatory energy audits and must have an energy manager. The country is also widely known for its widespread use of CHP in the industrial sector. Taiwan has low energy intensity overall and has a continuous efficiency improvement goal of 2% per year.

China has the largest ESCO market in the world, which is widely supported by the government. The country ranks 5th in building energy efficiency policy, with firm building codes for commercial and residential buildings, energy performance standards for 40 product categories, and mandatory energy labeling for specific building types. China also has high transportation efficiency, ranking 6th. The country has a fuel economy target of 57.9 MPG for passenger vehicles by 2025, and a standard for heavy-duty vehicles targeting 28% fuel savings by 2025. China also has the world's highest percentage of trips taken in public transit; the annual miles traveled per passenger vehicle per capita is very low. The country has continued reducing its energy intensity over recent years, and there are mandatory efficiency targets for energy-intensive organizations.

The United States is making major investments in energy efficiency programs and related R&D, while offering tax incentives and low-interest loans across multiple sectors. The USA has also taken the lead in energy data reporting, thanks to the help of the Energy Information Administration (EIA), which provides detailed information about energy generation and consumption by state and by sector. Energy use is a U.S. commercial office building's single largest operating expense. It represents about one-third of an enterprise's typical operating budget and accounts for almost 20% of the nation's annual GHG emissions. Office buildings waste up to one-third of the energy they consume. Due to this, the U.S. is ranked 8th in building efficiency. There are mandatory appliance and equipment standards that cover more than 50 product categories and building retrofit codes for the residential and commercial sectors. Many states have withstanding building codes that are updated regularly, providing technical resources and training to ensure proper implementation.

10.4 CONCLUSION

The energy landscape is at a crossroads. Our dependence on fossil fuels and nuclear power has driven progress, but at a steep environmental and social cost, contributing to climate change, pollution, and resource depletion. This chapter has demonstrated that relying on conventional energy sources is no longer viable for a sustainable future.

Fortunately, a new era of energy is emerging. Advances in renewables, energy efficiency, smart grids, and demand-side management are transforming how we generate and use power. From smart thermostats to decentralized grids, technology is making energy cleaner, smarter, and more accessible. International collaboration and bold policy action are accelerating this shift, but challenges remain—transitioning infrastructure, managing costs, and ensuring a just transition for all.

The next chapter takes us from strategy to action. In Chapter 10, we will examine how real-world projects and innovative models are overcoming barriers, leveraging finance and technology, and developing resilient, low-carbon energy systems. The journey ahead is complex, but the path is clear: rethink, innovate, and collaborate for a sustainable energy future.

REVIEW QUESTIONS

1. What are conventional energy resources?
2. Discuss in detail adverse environmental impact of fossil fuels threating the viability of human civilization.
3. Will nuclear energy help reduce GHG emissions relative to the emissions generated by fossil fuels?
4. Discuss the alternative approaches to conventional energy resources. Include use of energy management software, technologies, renewable energy sources.
5. How energy efficiency across various sectors significantly reduces energy consumption, thus minimizing the GHG emissions impact on the climate change? Provide examples?
6. How do smart grids offer efficient delivery network, reliability utilizing AI?
7. How does the DSM offer a multitude of benefits for optimizing energy consumption?
8. What are the challenges and opportunities existing toward the path to sustainability?
9. Briefly discuss the solutions to attain sustainable energy in the UNECE region.
10. Compare and discuss in detail the energy performance of different countries.

REFERENCES

Adewnmi, A., Olu-lawal, K.A., Okoli, C.E., Usman, F.O., Usiagu, G.S. (2023). Sustainable energy solutions and climate change: A policy review of emerging trends and global responses. *World Journal of Advanced Research and Reviews*. https://doi.org/10.30574/wjarr.2024.21.2.0474

Ahmad, S.Y., Hafeez, G., Aurangzeb, K., Rehman, K., Khan, T.A., Alhussein, M. (2023). A sustainable approach for demand side management considering demand response and renewable energy in smart grids. *Frontiers in Energy Research*. https://doi.org/10.3389/fenrg.2023.1212304

Ai, Y., Huang, T., Duan, C., Huang, D., Gong, Y., Cheng, H. (2022). Knowledge domain of greenhouse gas emissions from hydropower reservoirs: Hotspots, frontiers and future perspectives. *Frontiers in Environmental Science*. https://doi.org/10.3389/fenvs.2022.1055891

Alfianto, D., Gayo, S. (2021). Government *Cooperation with Business Entities in the Procurement of Infrastructure based on the Presidential Regulation Number 38 Year 2015 Study Drinking Water Supply System (Spam) in Umbulan. Budapest International Research and Critics Institute (BIRCI) Journal*, 4(4), 8078–8090. https://doi.org/10.33258/birci.v4i4.2724

Amanatidou, E., Samiotis, G., Trikoilidou, E., Tsikritzis, L.I., & Taousanidis, N. (2023). Centennial assessment of greenhouse gases emissions of young and old hydroelectric reservoir in Mediterranean mainland. *Journal of Environmental Informatics*. https://doi.org/10.3808/jei.202300485

Anarene, B. (2024). Revolutionizing energy efficiency in commercial and institutional buildings: A complete analysis. *International Journal of Scientific Research and Management*. https://doi.org/10.18535/ijsrm/v12i09.em12

Bess, M.D. (2022). *Fossil Fuels and Climate Change*. Cambridge University Press. https://doi.org/10.1017/9781009160315.002

Bréchet, Y. (2023). The rôle of Nuclear energy for fighting climate change: assets and weaknesses in a global perspective. *Comptes Rendus Geoscience*. https://doi.org/10.5802/crgeos.21-en

Burns, D., Grubert, E. (2021). Attribution of production-stage methane emissions to assess spatial variability in the climate intensity of US natural gas consumption. *Environmental Research Letters*. https://doi.org/10.1088/1748-9326/ABEF33

Calì, M., & Acri, A. (2023). Photovoltaic panels end-of-life recycling. In H. Bekkay, A. Mellit, A. Gagliano, A. Rabhi, & M. A. Koulali (Eds.), *Proceedings of the 3rd International Conference on Electronic Engineering and Renewable Energy Systems. ICEERE 2022* (Vol. 954). Springer. https://doi.org/10.1007/978-981-19-6223-3_109

Coccia, M. (2024). *Sustainable Technologies for Ecological Societies Directed to Energy* Transition. Academic Press. https://doi.org/10.20944/preprints202407.0806.v1

Cotana, F., Coccia, V., Cavalaglio, G., Barbanera, M., & Petrozzi, A. (2022). Biomass-based systems. In F. Calise, M. Dentice D'Accadia, L. Vanoli, & M. Vicidomini (Eds.), *Polygeneration systems* (pp. 137–192). Academic Press. https://doi.org/10.1016/B978-0-12-820625-6.00009-8

D'Agosto, M. de A. (2019). Air pollutant and greenhouse gas emissions (GHG). In M. de A. D'Agosto (Ed.), *Transportation, energy use and environmental impacts* (pp. 227–257). Elsevier. https://doi.org/10.1016/B978-0-12-813454-2.00006-4

Fu, J., Liu, Y., Sun, F. (2021). Identifying and regulating the environmental risks in the development and utilization of natural gas as a low-carbon energy source. *Frontiers in Energy Research*. https://doi.org/10.3389/FENRG.2021.638105

Gabidullina, E., Gavrilenko, I., Rastegaev, S. (2020). Model of effective partnership between government and business in the course of the implementation of regional infrastructure projects. *E3S Web of Conferences*. https://doi.org/10.1051/E3SCONF/202020808030

Ganea, T. (2022). Benefits, drawbacks, and the path forward for nuclear fission. *Technium*. https://doi.org/10.47577/technium.v4i9.7289

Gemechu, E., Kumar, A. (2022). A review of how life cycle assessment has been used to assess the environmental impacts of hydropower energy. *Renewable & Sustainable Energy Reviews*. https://doi.org/10.1016/j.rser.2022.112684

Gordon, D.M., Reuland, F., Jacob, D.J., Worden, J., Shindell, D., Dyson, M.E.H. (2023). Evaluating net life-cycle greenhouse gas emissions intensities from gas and coal at varying methane leakage rates. *Environmental Research Letters*. https://doi.org/10.1088/1748-9326/ace3db

Hale, G., Belanger, Y.D. (2015). Hale, G., & Belanger, Y. D. (2015). *From "social licence" to "social partnership": Promoting shared interests for resource and infrastructure development* (Commentary No. 440). C.D. Howe Institute. https://doi.org/10.2139/ssrn.2699104

Haque, R., Nurunnahar, S., & Zaman, S. (2024). Eco-friendly waste management for renewable energy generation. In M. R. Rahimpour (Ed.), *Encyclopedia of renewable energy, sustainability and the environment* (1st ed., pp. 885–898). Elsevier. https://doi.org/10.1016/B978-0-323-93940-9.00191-2

Hidy, G.M. (2002). Multiscale impact of fuel consumption on air quality. *Energy & Fuels*. https://doi.org/10.1021/EF0101659

Ikevuje, A.H., Anaba, D.C., Iheanyichukwu, U.T. (2024). Exploring sustainable finance mechanisms for green energy transition: A comprehensive review and analysis. *International Journal of Applied Research in Social Sciences*. https://doi.org/10.51594/farj.v6i7.1314

Imam, S.A., Madkaikar, A. (2024). The costs and concerns of energy transition. *Advances in Logistics, Operations, and Management Science Book Series*. https://doi.org/10.4018/979-8-3693-3880-3.ch016

Ishola, A. (2024). Global renewable energy transition in fossil fuel dependent regions. *World Journal of Advanced Research and Reviews*. https://doi.org/10.30574/wjarr.2024.24.1.3046

Jasińska, E., Jasiński, M., Leonowicz, Z., Martirano, L., Goňo, R., Jasiński, M. (2023). *Various Aspects of Energy Transition: Technologies, Economy, and Social Synergies to Sustainable Future*. https://doi.org/10.1109/eeeic/icpseurope57605.2023.10194631

Khan, M. A. (2024). Advancing sustainable power: A comprehensive review of technologies and strategies for green electricity production. *Collective Journal of Chemical Engineering*, 1, Article ART0039. https://doi.org/10.70107/collectjchemicaleng-art0039

Khattra, S.K., Singh, D., Dogra, R. (2024). A review of energy efficient technology and carbon trading for reducing carbon emissions. *Archives of Current Research International*. https://doi.org/10.9734/acri/2024/v24i6779

Khaturia, S., Singh, H., & Choudhary, G. (2024). Industrial energy efficiency: An interdisciplinary approach. In *Futuristic trends in renewable & sustainable energy* (Vol. 3, Book 5, pp. 261–268). IIP Series. https://doi.org/10.58532/v3bars5p5ch5

Kiranmai, N., Kumari, N.A., Kadari, S.R. (2024). Improved energy efficiency technologies and its applications. *International Journal of Advanced Research in Science, Communication and Technology*. https://doi.org/10.48175/ijarsct-19104

Kraft, M.E. (2024). *Nuclear Energy*. In *Political science and public policy 2024* (pp. 392–396). Edward Elgar Publishing. https://doi.org/10.4337/9781802209204.ch75

Kumar, A., Chaturvedi, M., Joshi, N., Mondal, R., Malyan, S.K. (2023). Greenhouse gas emissions from hydroelectric reservoirs: Mechanistic understanding of influencing factors and future prospect. *Environmental Science and Pollution Research*. https://doi.org/10.1007/s11356-023-25717-y

Kumar, A., Kumar, A, Chaturvedi, A.K., Joshi, N., Mondal, R., Malyan, S.K. (2022). Greenhouse gas emissions from hydroelectric reservoirs: Mechanistic understanding of influencing factors and future prospect. *Research Square*. https://doi.org/10.21203/rs.3.rs-2028526/v1

Leruth, L. (2012). Public-private cooperation in infrastructure development: A principal-agent story of contingent liabilities, fiscal risks, and other (un)pleasant surprises. *Networks and Spatial Economics*. https://doi.org/10.1007/S11067-009-9112-0

Levitt, R.E., Scott, W.R., Garvin, M.J. (2019). *Public–Private Partnerships for Infrastructure Development*. https://doi.org/10.4337/9781788973182

Lopez, G., Aghahosseini, A., Bogdanov, D., Satymov, R., Oyewo, A.S., Breyer, C. (2024). *Sustainable Energy-Industry Systems in the United States and Canada Demonstrating the Value of Solar-to-X*. In *Proceedings of the 2024 IEEE 52nd Photovoltaic Specialist Conference (PVSC)* (p. 0134). IEEE. https://doi.org/10.1109/PVSC57443.2024.10749483

Moudene, K., Ridouane, E.H. (2024). Energy and environmental transition: Perspectives and Challenges. *International Uni-Scientific Research Journal /International Uni-Scientific Research Journal*. https://doi.org/10.59271/s45450.024.2234.15

Nielsen, J.S., Weel, M., Falkenberg, M. (2022). Large scale PEMS for tracking GHG and criteria pollutants. In *Paper Presented at the ADIPEC*, Abu Dhabi, UAE, October 2022. https://doi.org/10.2118/210901-ms

Nusriadi, L., Avianti, I., Tanzil, N.D., Parikesit, D. (2023). The collaborative governance elements contributing to implementing public-private partnerships: A systematic literature review. Journal *of Namibian Studies: History Politics Culture*. https://doi.org/10.59670/jns.v33i.1155

Patra, B. R., Mohapatro, R. N., Routray, S., Swain, R., Nanda, S., & Dalai, A. K. (2022). Thermochemical conversion of organic waste: New horizons for production of green energy. In S. Nanda & D.-V. Vo (Eds.), *Innovations in thermochemical technologies for biofuel processing* (pp. 1–21). Elsevier. https://doi.org/10.1016/B978-0-323-85586-0.00005-6

Peters, I. M. (2024). How *Cradle-to-Cradle Recycling Supports the Transition to 100% Renewable Energies*. In *Proceedings of the 2024 IEEE 52nd Photovoltaic Specialist Conference (PVSC)* (pp. 247–249). IEEE. https://doi.org/10.1109/PVSC57443.2024.10749378

Rahman, A., Murad, S.M.W., Mohsin, A.K.M., Wang, X. (2024). Does renewable energy proactively contribute to mitigating carbon emissions in major fossil fuels consuming countries? *Journal of Cleaner Production*. https://doi.org/10.1016/j.jclepro.2024.142113

Rao, A., Lucey, B., Kumar, S. (2023). Climate risk and carbon emissions: Examining their impact on key energy markets through asymmetric spillovers. *Energy Economics*. https://doi.org/10.1016/j.eneco.2023.106970

Sadekin, S.M.S., Zaman, S., Sarkar, M.A.R., Hossain, A. (2020). Role of nuclear electricity in low carbon economy. *Journal of the Institute of Engineering*. https://doi.org/10.3126/JIE.V15I3.32209

Salooja, N., Butler, C. (2023). Climate change, natural resources, and economic growth. *Reference Module in Social Sciences*. https://doi.org/10.1016/b978-0-44-313776-1.00011-8

Shah, S.H.A., Khan, I.U., Hafeez, G., Murawwat, S. (2024). Load *Scheduling Using Demand Response for Demand-Side Management in Smart Grid*. In *Proceedings of the 2024 7th International Conference on Energy Conservation and Efficiency (ICECE)* (pp. 1–6). IEEE. https://doi.org/10.1109/ICECE61222.2024.10505274

Sun, P., Young, B., Elgowainy, A., Lu, Z., Wang, M., Morelli, B., Hawkins, T.R. (2019). Criteria air pollutant and greenhouse gases emissions from U.S. refineries allocated to refinery products. *Environmental Science & Technology*. https://doi.org/10.1021/ACS.EST.8B05870

SwornaKokila, M.L., Venkatarathinam, R., Rose Bindu Joseph, P., Manivasagam, M.A., Kishore, K.H. (2023). Optimizing energy consumption in smart grids using demand response techniques. *Distributed Generation & Alternative Energy Journal*. https://doi.org/10.13052/dgaej2156-3306.3915

Thakur, N., Samanta, P., Mishra, S. (2023). Methods and technologies for recycling energy storage materials and devices. In R. K. Gupta (Ed.), *Recent advancements in polymeric materials for electrochemical energy storage*. Springer. https://doi.org/10.1007/978-981-99-4193-3_28

Trương, H. (2023). Fossil fuel resources, environment, and climate change. *Contributions to Economics*. https://doi.org/10.1007/978-3-031-27982-9_4

Udayakumar, R., Preethi, R., Al-Fatlawy, R.R., Omirbayevna, T.G., Kavitha, T., Sharma, K., Rao, H.C.-H. (2023). Smart *Grids and Demand Responses Stratergies for Efficient Energy Management in IoT Based Urban Environments*. In *Proceedings of the 2023 International Conference for Technological Engineering and its Applications in Sustainable Development (ICTEASD)* (pp. 325–331). IEEE. https://doi.org/10.1109/ICTEASD57136.2023.10585021

Usman, F.O., Ani, E.C., Ebirim, W., Montero, D.J.P., Olu-lawal, K.A., Ninduwezuor-Ehiobu, N. (2024). Integrating renewable energy solutions in the manufacturing industry: challenges and opportunities: A review. *Engineering Science & Technology*. https://doi.org/10.51594/estj.v5i3.865

Xu, S., Wang, J., Sun, H., Huang, L., Xu, N., Liang, Y. (2022). Life cycle assessment of carbon emission from natural gas pipelines. *Chemical Engineering Research & Design*. https://doi.org/10.1016/j.cherd.2022.07.018

Zhang, F. (2023). Biofuels for transportation. *Journal of Physics: Conference Series*. https://doi.org/10.1088/1742-6596/2608/1/012049

Zhang, M., Gao, L., Wang, Q., Xie, D., Gao, J., Wang, S., Lu, X. (2022). Methane leakage measurement of natural gas heating boilers and greenhouse gas emissions accounting of "coal-to-gas" transition for residential heating in rural Beijing. *Environmental Science & Technology Letters*. https://doi.org/10.1021/acs.estlett.2c00751

11 Decarbonization
Crafting Net-Zero Emissions Strategies

Decarbonization is the cornerstone of global efforts to combat climate change, requiring a fundamental transformation in how societies produce and consume energy. As atmospheric carbon dioxide levels continue to rise due to human activities, the urgency of transitioning to a net-zero emissions economy has never been greater. This chapter explores the multifaceted concept of decarbonization, detailing the strategies, challenges, and opportunities involved in achieving a low-carbon future. It delves into the carbon cycle, emphasizing the role of human activities in disrupting natural carbon balances. This chapter further examines decarbonization's economic and technological dimensions, including carbon offsetting, sequestration, renewable energy investments, and emerging low-emission technologies. Additionally, it highlights the contributions of key sectors—such as transportation, aviation, and private equity—driving decarbonization efforts. Through a comprehensive lens, this chapter underscores the importance of international collaboration, policy innovation, and public-private partnerships in achieving a sustainable and equitable future.

11.1 THE CARBON CYCLE

Carbon is the chemical backbone of all life on Earth, essential for forming key molecules like proteins and DNA. While the amount of carbon dioxide in the atmosphere remained relatively constant for thousands of years, it began to rise in the late 18th century, reaching unprecedented levels in the 20th century due to human activities. The global carbon cycle involves carbon exchange among the atmosphere, oceans, and land surface. Human perturbation of the carbon cycle is recognized as a main factor driving climate change.

Most carbon is stored in rocks and sediments, while significant amounts are also found in the ocean, atmosphere, and living organisms. The sea acts as a massive carbon sink, absorbing and exchanging carbon between surface waters and the atmosphere or storing it in ocean depths for long periods. Marine organisms contribute to this cycle by producing carbon throughout their lives and after they die. Over millions of years, dead organisms can transform into fossil fuels. When humans burn these fuels for energy, large amounts of GHG and CO_2 are released into the atmosphere, leading to climate change by increasing global temperatures, causing ocean acidification, and disrupting ecosystems.

Decarbonization refers to reducing carbon emissions and capturing and storing carbon in soil and vegetation. It requires a fundamental shift in our economic model, prioritizing growth at all costs. Transforming how we generate energy, the sources of energy we use, how we build and move, and how we manage land resources is crucial. To achieve decarbonization, we must drastically reduce our consumption of carbon-intensive products and adopt low-emission technologies and renewable energy alternatives. There are also nature-based solutions for decarbonization. For instance, blue carbon refers to the carbon sequestered and stored in salt marshes, mangroves, and seagrass ecosystems, highlighting their significance in carbon sequestration and storage. Coastal blue carbon ecosystems are highly efficient at storing carbon, with a greater capacity per unit area than the terrestrial carbon pool.

DOI: 10.1201/9781003479666-11

Individuals and corporations can contribute to decarbonization by reducing their carbon footprints by purchasing carbon offsets, which are investments in carbon-reducing activities or technologies to compensate for their emissions and become effectively carbon neutral. Lifestyle choices that reduce carbon footprints include consuming less meat and choosing products that require fewer carbon emissions to produce and transport. Through these combined efforts, individuals and corporations can play a crucial role in the global strategy to achieve net-zero emissions and mitigate climate change.

The pressing need to achieve net-zero emissions arises from the urgent imperative to mitigate the escalating impacts of climate change. Several critical factors underscore this imperative. Firstly, the Earth is experiencing unprecedented warming due to human-induced greenhouse gas emissions, particularly carbon dioxide from fossil fuel combustion. This warming triggers extreme weather events, rising sea levels, disrupted ecosystems, and threats to global food and water security, endangering lives and livelihoods worldwide. Additionally, scientists warn of tipping points beyond which climate change could spiral out of control, leading to irreversible and catastrophic consequences.

To avoid these tipping points and limit global warming to 1.5°C above pre-industrial levels, global greenhouse gas emissions must reach net-zero by mid-century, with substantial reductions beginning immediately. Each year of delay complicates this task, making it more challenging and costly. Moreover, climate change exacerbates existing social, economic, and environmental challenges, heightening poverty, inequality, and conflict. Addressing climate change is therefore essential for building resilience and ensuring the stability and sustainability of societies and economies worldwide. As a global challenge, tackling climate change requires a coordinated and concerted response from all nations. Fulfilling international commitments under the Paris Agreement and safeguarding the well-being of current and future generations hinge on achieving net-zero emissions swiftly and effectively. Failure to do so risks irreparable harm to the planet and its inhabitants.

Carbon neutrality means achieving a balance between emitting carbon and absorbing carbon from the atmosphere in carbon sinks. Companies, governments, and individuals can achieve carbon neutrality by reducing their carbon emissions and compensating for the remaining emissions through carbon offsetting and sequestration initiatives. They often follow a strategy that includes measuring their carbon footprint, reducing emissions through energy efficiency and renewable energy projects, and then balancing the residual emissions with carbon offsets to reach a net-zero carbon output.

Carbon offsetting involves compensating for carbon dioxide emissions produced by one activity by investing in projects that reduce or absorb an equivalent amount of CO_2 elsewhere. For example, a company may fund reforestation projects, renewable energy installations, or methane capture initiatives to offset the emissions generated from their operations. By purchasing carbon credits or investing directly in these projects, organizations and individuals can claim to have offset their emissions, thereby contributing to global carbon reduction efforts. Carbon sequestration is the process of capturing and storing atmospheric carbon dioxide to prevent it from contributing to global warming. Natural processes, such as photosynthesis in forests and oceans, naturally sequester carbon. Additionally, technological methods like carbon capture and storage (CCS) involve capturing CO_2 emissions from sources like power plants and storing it underground in geological formations. Through these methods, carbon sequestration helps reduce the concentration of CO_2 in the atmosphere, mitigating the impact of climate change.

11.2 INVESTMENT IN DECARBONIZATION

Investment in decarbonization represents a significant financial opportunity, with the potential to add \$5–11 trillion in additional investment value by 2030. This opportunity spans across mature solutions such as electricity networks, renewables, and sustainable vehicles, which are currently ready for investment. Decarbonization requires systemic changes across various sectors, including energy generation, industrial processes, and land management. Transitioning from fossil fuels to renewable energy sources is a critical component, as highlighted by the study on major economies like the USA, China, and India, which explores scenarios for achieving a high share of renewable energy by 2030 and 2050, demonstrating significant potential for emissions reduction. Industrial decarbonization also plays a crucial role, with efficiency improvements in energy, materials, and resources offering substantial benefits, though adoption remains hindered by economic and organizational barriers. The macroeconomic perspective emphasizes the need for a historic transformation of economies to achieve climate neutrality, necessitating policy interventions to overcome roadblocks. In the built environment, which accounts for a significant portion of global emissions, sustainable construction practices such as circular construction and the use of eco-efficient materials are essential for reducing energy demands and incorporating low-carbon technologies. Furthermore, the global industry must decarbonize to stabilize the climate, with policies addressing both demand and supply sides of energy use being crucial, especially in developing countries experiencing rapid growth.

Collectively, these efforts, complemented by nature-based solutions like reforestation, are vital for enhancing carbon storage and ecosystem resilience, ultimately contributing to the overarching goal of decarbonization. The decarbonization value chain is projected to grow substantially, driven by increasing capital expenditure investments and the need for raw materials and low-emission fuels. Despite the high initial costs and risks associated with some emerging technologies, the overall trajectory indicates robust growth potential, particularly in Asia-Pacific and for technologies like biofuels and hydrogen, making it an attractive prospect for institutional investors aiming to align with net-zero targets.

11.2.1 Investment in Decarbonization for Private Equity

The growing awareness of environmental, social, and governance (ESG) issues and the urgency of decarbonization are reshaping private equity (PE) investment strategies. Institutional investors, regulators, and the public increasingly demand that PE firms address their environmental impacts. Global decarbonization requires substantial investments, presenting an opportunity for PE to finance the transition to a net-zero economy. The unique characteristics of PE, such as full ownership, operational control, and long-term investment horizons, position it well to lead this transformation.

Events like the COVID-19 pandemic and the war in Ukraine have underscored the importance of sustainable investments. These crises have highlighted systemic vulnerabilities and the necessity for climate resilience, driving better performance in ESG-rated funds compared to their benchmarks. The war in Ukraine has particularly emphasized the need for energy security, accelerating Europe's shift toward renewable energy and decarbonization. Investment strategies are evolving as climate change becomes a critical issue. Nearly half of institutional investors are committed to net-zero emissions by 2050. Limited partners (LPs) are increasingly using ESG criteria to select general partners (GPs), and many PE firms are willing to abandon transactions due to negative ESG findings. However, only a small percentage of climate-related funds are on track to meet the Paris Agreement goals, indicating a significant gap in achieving net zero.

11.2.2 Drivers and Barriers of Decarbonization in PE

The main drivers of decarbonization include regulatory pressure, stakeholder expectations, and intrinsic motivations to create a positive societal impact. New regulations, such as the Sustainable Finance Disclosure Regulation (SFDR), are pushing for greater transparency in ESG-related data. Decarbonization also mitigates risks and can create value, attracting investors.

However, barriers include the lack of high-quality, comparable data, capability gaps within PE firms, and insufficient integration of decarbonization into investment processes. Only a minority of GPs receive ESG reports from their portfolio companies, and few firms have dedicated ESG teams.

11.2.3 Strategies to Overcome Barriers

To successfully incorporate decarbonization into investment strategies, PE firms should:

1. **Develop a Holistic Investment Strategy**: Align portfolios with net-zero goals and integrate climate risks and opportunities into governance, due diligence, and risk management processes.
2. **Build Capabilities and Culture**: Invest in internal knowledge, training, and resources to embed climate considerations across the organization.
3. **Integrate Decarbonization in due Diligence**: Include ESG questions in due diligence to evaluate the climate risks and opportunities of potential investments.
4. **Set Measurable Milestones**: Establish clear, measurable objectives and transparently communicate progress to stakeholders.
5. **Increase Reporting and Transparency**: Improve climate-related reporting at the firm and portfolio company levels, potentially adopting voluntary compliance with regulations like the SFDR.

11.3 ECONOMICS OF DECARBONIZATION

A decarbonized economy, also known as a low-carbon economy, is an economy based on energy sources that produce low levels of greenhouse gas emissions. This global shift aims to combat climate change, promote renewable energy adoption, and enhance sustainability for a greener future. Effective decarbonization actions include shifting the energy mix away from fossil fuels and toward zero-emissions electricity and other low-emissions energy carriers such as hydrogen; adapting industrial and agricultural processes; increasing energy efficiency and managing demand for energy; utilizing the circular economy; consuming fewer emissions-intensive goods; deploying carbon capture, utilization, and storage (CCS) technology; and enhancing sinks of both long-lived and short-lived greenhouse gases.

11.3.1 Capital Allocation

The transition to a net-zero economy by 2050, as outlined in the NGFS Net Zero 2050 scenario, will require approximately $275 trillion in cumulative spending on physical assets between 2021 and 2050, averaging around $9.2 trillion annually. This represents a significant increase of about $3.5 trillion more per year compared to current levels, equating to a 60% rise in annual spending. Most of this increased expenditure will be directed towards low-emissions assets. Currently, $3.7 trillion is spent annually on high-emissions assets, but under the net-zero scenario, approximately $1 trillion of this spending would need to be redirected to low-emissions alternatives. Three major sectors—mobility, power, and buildings—will account for about 75% of this spending.

Economically, this spending is equivalent to about 7.5% of global GDP from 2021 to 2050, with a peak of around 9% of GDP between 2026 and 2030 before declining. Comparatively, under the NGFS Current Policies scenario, which considers expected income and population growth alongside current legislative policies, the additional annual spending required for a net-zero transition is estimated to be around $0.9 trillion. This scenario already anticipates a shift toward low-emissions spending due to existing trends and policies.

The transition also poses the risk of asset stranding, where high-emissions assets, such as fossil fuel–based power plants, may become underutilized or prematurely retired, leading to significant financial losses. For example, it is estimated that $2.1 trillion worth of assets in the power sector could be stranded by 2050. However, a coordinated and orderly economic adjustment toward net zero could mitigate the physical risks and additional costs associated with a more chaotic transition. While the upfront capital expenditures are substantial, they can result in long-term operating savings through improved efficiency and lower maintenance costs, underscoring that such investments are not merely costs but can also yield significant returns.

11.3.2 Economic Impact of the Transition to Low-Carbon Industries

The transition to low-carbon industries will bring significant cost implications across various sectors. In particular, steel and cement production costs are projected to rise by about 30% and 45%, respectively. The power sector will also experience a cost increase, with the global average delivered cost of electricity expected to rise by 25% by 2040 due to investments in renewables, grid, and storage capacity. These investments will create capital costs and depreciation charges, while some fossil fuel–based assets will incur costs even if underutilized or retired prematurely.

By 2050, operating costs for power generation are expected to drop by over 60% from 2020 levels as the energy mix shifts to renewables. However, this decrease will be partly offset by increased costs related to grid flexibility, transmission, and distribution. Thus, the delivered cost of electricity will still be about 20% higher in 2050 compared to 2020. Long-term cost trends are uncertain and will depend on technological innovations and power system evolution. Conversely, the total cost of ownership for electric vehicles is expected to become cheaper than that for internal combustion engine vehicles in most regions by 2025, potentially even sooner in some areas. This shift highlights the diverse financial implications of the transition to a low-carbon economy, affecting production and consumer costs across different sectors.

11.3.3 Job Shifts in Green Industries

The NGFS Net Zero 2050 scenario projects significant shifts in employment across various sectors due to the net-zero transition. This scenario estimates an increase in demand for about 162 million jobs and a decrease of about 152 million jobs in operations and maintenance by 2050. Additionally, 41 million jobs could be gained and 35 million lost related to spending on physical assets. Overall, this results in a net gain of 202 million jobs and a net loss of 187 million jobs by 2050, highlighting a substantial, uneven reallocation of jobs. The job gains are primarily linked to the transition to low-emission production methods, such as renewable power, while job losses are concentrated in fossil fuel–intensive and emissions-intensive sectors. Specifically, the fossil fuel extraction and production sector could see a reduction of nine million jobs, and the fossil fuel–based power sector could lose about four million jobs. The agriculture and food sectors might also experience significant changes, with 34 million direct jobs potentially lost in livestock and feed-related areas, partially offset by gains in poultry farming.

In contrast, low-emissions sectors are expected to see substantial job increases, such as an estimated six million new jobs in renewable power operations and maintenance. These shifts underscore the need for targeted policies to manage the transition's impact on workers and regions heavily reliant on high-emission industries. Compared to other trends like automation and remote work, which could result in job losses of 270–340 million across eight countries by 2030, the net-zero transition's employment impact is considerable but not unprecedented. The transition to a net-zero economy will require a strategic approach to mitigate job dislocations and support affected workers.

11.3.4 Ensuring an Orderly Transition

The transition to net-zero emissions by 2050 is critical, with the NGFS Net Zero 2050 scenario outlining a relatively orderly path. However, the complexity of this transition may lead to a more disorderly reality, making decisive action crucial. Three key risks must be managed: the choice of pathway, stakeholder measures to ease adjustments, and various constraints even under a smooth transition. Pathways vary; some envision immediate, gradual emission reductions, while others assume delayed but rapid reductions, potentially causing abrupt policy changes, high carbon prices, socioeconomic effects, and job transition challenges.

Unmanaged disruptions could disproportionately affect vulnerable communities, with energy cost increases impacting low-income households and displaced workers needing reskilling support. Supply-demand imbalances, shortages, and price volatility may occur if supply cannot scale up to meet the rapid demand for low-emission assets. Investment and adaptation costs are significant. Reskilling workers and economic diversification require funding. Adaptation measures, such as seawalls, system resilience, and relocating assets, are essential to manage increasing physical climate risks. Two NGFS scenarios illustrate the difference in transition pathways. In the below 2°C scenario with immediate reductions, \$150 billion would be added in coal power capacity, with \$100 billion prematurely retired. A delayed reduction scenario would see \$600 billion invested, with \$400 billion prematurely retired. Delaying emissions reductions increases physical climate risk by using up the remaining carbon budget, making it harder to limit warming to 1.5°C or 2.0°C.

11.3.5 Growth Opportunities in the Net-Zero Transition

Economic adjustments driven by the NGFS Net Zero 2050 scenario, which includes \$3.5 trillion in annual spending on physical assets, present substantial growth opportunities for companies. These opportunities fall into three main areas: decarbonizing legacy products and processes to reduce emissions and potentially lower costs, replacing high-emission options with low-emission alternatives such as electric vehicles (EVs) and renewable energy, and developing necessary inputs and services like lithium for batteries, solar panels, and EV charging infrastructure. This transition, costing about 3% of GDP annually through 2050, also mitigates physical climate risks and reduces the likelihood of catastrophic climate impacts.

11.4 SECTOR-WISE DECARBONIZATION PRACTICES

The problem extends to various sectors, such as healthcare, where green competencies positively influence the willingness to mitigate CFP, contributing to sustainable performance in healthcare organizations. The household sector, particularly in Japan, underscores the significance of energy consumption, emphasizing the need for targeted policies to address environmental challenges and promote sustainable consumption practices. The farming sector also comes into focus, as irrigation practices significantly affect the carbon footprint, with potential emission reduction through the adoption of more efficient technologies.

CFP are linked to socioeconomic factors and consumption patterns. Individuals with higher carbon engagement tend to have smaller footprints, suggesting a correlation between awareness, behavior, and emissions. However, despite efforts to enhance knowledge, the overall willingness to pay for offsetting remains insufficient to meet ambitious climate targets. High emissions from energy-intensive practices like irrigation and tillage underscore the need for technological advancements and sustainable practices to reduce carbon intensity. Additionally, the variability in emissions from different activities highlights opportunities for optimization and standardization to minimize environmental burdens.

Addressing CFP requires systemic changes in policy, governance, and international cooperation. While individual efforts are crucial, collective action at national and global levels is essential to achieve meaningful reductions in emissions. Initiatives like carbon pricing mechanisms, clean energy investments, and sustainable production practices are pivotal in transitioning toward low-carbon economies.

11.4.1 Power

Decarbonization of the power sector involves transitioning from fossil fuels to renewable energy sources, implementing carbon trading schemes, and leveraging digital technologies. In India, the power sector's decarbonization is crucial due to its significant contribution to greenhouse gas emissions. The country aims to reduce carbon emissions to zero over a 30-year horizon by integrating more renewable energy, despite current reliance on thermal power plants (Bisht et al., 2024). Similarly, Indonesia faces challenges in phasing out coal, a major energy source, while balancing economic impacts on its coal-producing sectors. The transition involves adopting high-efficiency, low-emission technologies and CCS to mitigate the economic losses associated with stranded assets and untapped coal reserves (Reyseliani et al., 2024). In Türkiye, a comprehensive modeling framework suggests that a coal phase-out by the early 2030s is feasible, with a significant reduction in carbon emissions while maintaining economic stability (Kat et al., 2024). China's approach includes a carbon emissions trading scheme (ETS) that has effectively reduced CO_2 emissions in pilot provinces, promoting technological advancements in power generation (Niu, 2024). The integration of digital technologies is also pivotal, as it enhances the efficiency of renewable energy integration and reduces carbon intensity, particularly in countries with substantial digital investments (Huang & Lin, 2023). Moreover, energy policies enacted globally, especially post-2008 financial crisis, have shown long-term effects in reducing CO_2 intensity when coupled with strong governance capacity (Xu et al., 2024). In China, a combination of carbon constraints, renewable energy development, and carbon trading policies is crucial for transitioning the power sector toward a cleaner energy mix (Liu et al., 2024b). Finally, a holistic approach that includes improvements in power generation, grid infrastructure, and consumer practices, supported by policy and technological innovations, is essential for achieving a low-carbon power system (Wang, 2024b). These diverse strategies highlight the complexity and necessity of tailored approaches to decarbonize the power sector across different regions and economic contexts.

11.4.2 Oil and Gas

The decarbonization of the oil and gas sector requires a combination of technological, financial, and strategic approaches to reduce greenhouse gas emissions and transition toward a sustainable energy future. One significant pathway is reducing methane emissions, primarily released through leaks, venting, and flaring during oil and gas operations. Methane, although short-lived in the atmosphere, has a potent greenhouse effect, contributing significantly to global warming (Elmasry, 2024). To address this, the industry is increasingly adopting technologies such as carbon capture, utilization, and storage (CCUS) and integrating renewable energy sources like wind, solar, and hydrogen into their operations (Adebayo et al., 2024a). The use of lithium-ion batteries is also gaining traction, providing reliable energy storage solutions that enhance operational efficiency and facilitate the integration of renewables (Ahsan, 2024). Additionally, Power Purchase Agreements (PPAs) and Renewable Energy Certificates (RECs) are being utilized, particularly by European companies, to expand renewable energy capacities and align with stringent climate policies (Chrysikopoulos et al., 2024). In the Middle East and North Africa, a screening methodology has been developed to prioritize decarbonization technologies based on factors such as scalability and cost-effectiveness, helping to identify viable pathways for reducing emissions in upstream operations (Ravishankar, 2024). Green financing mechanisms, including green bonds and sustainability-linked loans, are crucial in unlocking investments for projects aimed at reducing emissions and adopting cleaner technologies (Adebayo et al., 2024). The role of digital technologies, such as artificial intelligence and the Internet of Things (IoT), is also pivotal in enhancing operational efficiency and reducing emissions through improved monitoring and data analytics (Ershaghi et al., 2024). In the Arctic region, decarbonization scenarios include the adoption of renewable energy and technological upgrades to existing operations, highlighting the need for tailored strategies in different geographical contexts (Dmitrieva & Skobelev, 2023). Overall, the oil and gas sector's decarbonization efforts are supported by a combination of innovative business models, regulatory frameworks, and strategic investments, all aimed at achieving carbon neutrality and ensuring the industry's relevance in a low-carbon economy (Dodukh, 2023).

11.4.3 Transportation

The transportation sector, responsible for over 20% of global greenhouse gas emissions, is undergoing a significant transformation through electrification and the adoption of alternative fuels, such as hydrogen, to reduce emissions and enhance mobility. EVs are at the forefront of this transition, with substantial progress driven by favorable government policies, subsidies, and technological advancements, particularly in countries like China, Norway, and the United States (Patil et al., 2024; Rajashekara, 2023). China, for instance, leads the global market in EV adoption and manufacturing, with half of the world's EVs on its roads (Rajashekara, 2023). The shift to EVs is crucial for achieving the Paris Agreement's climate goals, as they produce no tailpipe emissions, thus reducing local air pollution and contributing to environmental sustainability (Erickson, 2024; Li & Zhang, 2023). In parallel, hydrogen fuel cell vehicles (HFCVs) are emerging as a promising alternative, offering clean and efficient energy solutions with water as the only byproduct, which significantly reduces carbon emissions compared to traditional vehicles (Li & Wu, 2024).

Although currently, HFCVs may produce more emissions than battery electric vehicles (BEVs), projections suggest that by 2050, HFCVs could become more environmentally beneficial due to advancements in hydrogen production technologies (Hassouna & Shin, 2024). The European Union's efforts in sustainable mobility have already resulted in significant reductions in crude oil imports and greenhouse gas emissions, highlighting the potential of zero-emission transport units and sustainable fuels (Khaustova et al., 2024). Moreover, the integration of renewable energy sources into transportation electrification is crucial for long-term sustainability, as it reduces reliance on fossil fuels and mitigates urban congestion and pollution (Kashem et al., 2024). Despite these advancements, challenges remain, such as the need for substantial infrastructure investments and overcoming market fragmentation (Li & Zhang, 2023).

The combined efforts in EV and HFCV adoption, supported by technological innovations and policy frameworks, are pivotal in steering the transportation sector toward a low-carbon future.

The electrification of the transportation sector presents numerous potential economic benefits, primarily through enhanced energy efficiency and reduced reliance on fossil fuels. EVs are significantly more efficient than internal combustion engine (ICE) vehicles, which only convert 10%–20% of fuel energy into motion, with the rest dissipated as heat (Sharifzadeh, 2022). This efficiency translates into lower operational costs, as electric engines incur about one-third of the fuel costs compared to ICEs. The declining cost of batteries, which fell below $200/kWh in 2018 and are projected to drop further, enhances the economic appeal of EVs, making them a cost-effective transportation option (Erickson & Brase, 2019). This cost reduction, coupled with the potential for EVs to store energy and support grid operations, further underscores their economic viability (Erickson & Brase, 2019).

Additionally, the shift to electrified transportation can significantly reduce greenhouse gas emissions and improve air quality, leading to health-related economic benefits by mitigating risks associated with pollution, such as cancer and cardiovascular diseases. The transition to EVs also promises to decrease dependence on foreign fuels, enhancing energy security and potentially stabilizing energy prices (Brooker et al., 2010). However, the economic benefits are contingent on supporting infrastructure development, such as solar-powered charging stations, which can further reduce emissions and operational costs. The economic case for electrification is particularly strong in countries with cleaner electricity grids, where the environmental benefits of EVs are maximized (Newbery, 2013).

Despite these advantages, the transition requires significant investment in electricity networks, which could impact consumer electricity bills and pose a challenge to economic prosperity if not managed effectively (Alabi et al., 2020). Overall, the economic benefits of transportation electrification are substantial, driven by cost savings, environmental improvements, and enhanced energy security. However, they require careful policy and infrastructure planning to fully realize these advantages (Briceño-Garmendía et al., 2023; Newbery, 2013).

11.4.4 Aviation

Airports are pivotal in decarbonizing aviation, central hubs for implementing new technologies like hydrogen and electric propulsion, and sustainable aviation fuels (SAF). Their extensive infrastructure, large-scale energy needs, and role as convergence points for various stakeholders (airlines, manufacturers, fuel producers, and ground services) position them uniquely to drive the industry's sustainability efforts. The World Economic Forum's "Airports of Tomorrow" initiative emphasizes the importance of developing infrastructure to support the transition to net zero emissions. Airports can leverage their land and resources to integrate solar farms, fuel blending facilities, and hydrogen storage, contributing significantly to decarbonization. The initiative also promotes collaboration among governments, climate experts, and industry leaders to ensure comprehensive support and investment.

SAF are highlighted as a critical immediate step, with regulatory and financial incentives in regions like the E.U. and the U.S. encouraging adoption. The Net Zero Carbon Emission Challenge is that current projections estimate that demand for air passenger journeys in 2050 could exceed 10 billion. The expected 2021–2050 carbon emissions on a 'business as usual' trajectory is approximately 21.2 gigatons of CO_2.

The aviation sector's decarbonization is a complex challenge due to its reliance on energy-dense fuels. However, innovations such as SAF, hydrogen propulsion, and advanced air traffic management systems offer promising pathways to reduce emissions. SAF, particularly synthetic electro fuels (e-fuels), are crucial for long-haul flights, which contribute significantly to emissions. These fuels can achieve economic competitiveness with fossil fuels through advancements in electrolysis and carbon capture technologies, as well as integration with renewable energy sources (Peacock et al., 2023). Hydrogen propulsion, with its high energy density and zero carbon emissions, is another promising alternative, especially for short to medium-haul flights. However, challenges such as fuel storage, infrastructure development, and understanding non-CO_2 climate impacts need to be addressed (Barbosa, 2024; Ivanov, 2023). Hydrogen can be utilized in various propulsion systems, including fuel cells and modified gas turbines, but requires significant technological and industrial collaboration to become viable by 2050 (Tiwari et al., 2024; Cybulsky et al., 2023).

Advanced air traffic management systems can also contribute to emissions reductions by improving operational efficiency, although their impact is relatively modest compared to fuel innovations (Jensen et al., 2023). Collaboration among airports, airlines, and policymakers is essential to scale these solutions, as demonstrated by the need for coordinated efforts in hydrogen production and SAF development (Tait et al., 2024; Degirmenci et al., 2023). Policies such as carbon pricing, mandates, and financial incentives are critical to support the transition to sustainable fuels and technologies (Muelaner, 2023). Overall, achieving net-zero emissions in aviation by 2050 will require a multifaceted approach, leveraging technological advancements, policy support, and industry collaboration to overcome the sector's unique challenges.

SAF offers a straightforward drop-in solution compatible with existing aircraft, allowing for immediate reductions in carbon emissions without significant modifications to aircraft or infrastructure. However, SAF is costly and provides only moderate reductions in climate impact compared to hydrogen, which can achieve a 98% reduction in temperature response, albeit with increased operating costs and infrastructure challenges.

Hydrogen propulsion, particularly for short to medium-haul flights, is promising due to its high energy content and zero carbon emissions. However, it requires significant changes in aircraft design and fuel storage infrastructure (Barbosa, 2024). Hydrogen's potential is further complicated by the need to address non-CO_2 climate impacts, such as contrail formation. Electrification, while beneficial for reducing emissions, is currently limited to smaller aircraft due to battery energy density constraints (Nakano et al., 2022; Babuder et al., 2024). Integrating these technologies could lead to a cost-effective reduction of emissions reductions in line with the Paris Agreement goals, with electric and hydrogen-powered aircraft playing crucial roles in long-term decarbonization pathways (Nakano et al., 2022). However, the adoption of these technologies varies across flight segments, with hydrogen being more suitable for long-range flights and SAF preferred for regional and medium-range categories due to cost considerations (Proesmans & Vos, 2022; Talwar et al., 2023). The transition to these technologies also involves significant operational and infrastructural impacts, necessitating multi-technology airport infrastructures to avoid lock-in effects and facilitate a smooth transition (Babuder et al., 2024). Overall, while each technology offers unique benefits, their successful integration requires addressing economic, technical, and infrastructural challenges to maximize synergies and minimize trade-offs in the aviation sector's decarbonization efforts.

11.4.5 Shipping

The decarbonization of the shipping sector requires joint technological, regulatory, and economic strategies. The International Maritime Organization (IMO) has set ambitious targets to reduce GHG emissions by 50% by 2050, which necessitates the adoption of zero-emission vessel (ZEV) technologies such as battery electric ships, hydrogen fuel cells, and alternative fuels like bio-LNG and ammonia (Ibokette et al., 2024; Shankar et al., 2024). However, the transition to these technologies is hindered by high capital costs, regulatory inconsistencies, and insufficient infrastructure (Ibokette et al., 2024). The use of hydrogen as a carbon-free fuel is promising, but it faces challenges related to storage, infrastructure, and regulatory gaps (Prados, 2024; İnal, 2024). The shipping industry is also exploring the use of digital twin technology to optimize ship design and operational strategies, balancing investment costs with emission reductions (Mavrakos et al., 2024). Fleet modernization efforts are focusing on retrofitting vessels for alternative fuels such as methanol and LNG, which serve as transitional solutions toward more sustainable options like hydrogen (Akgül, 2024).

Economic incentives, such as tax breaks and subsidies, are crucial to overcoming financial barriers and encouraging the adoption of these technologies (Li & Hu, 2024; Baştuğ et al., 2024). Public-private partnerships and international cooperation are essential to address the conflicting interests between economic and environmental goals, ensuring that environmental interests are prioritized in governance frameworks (Li & Hu, 2024). The development of new low-emission fuels and the necessary infrastructure is critical, with hydrogen playing a central role in achieving long-term decarbonization objectives (Prados, 2024). Overall, a coordinated effort involving technological innovation, regulatory alignment, and financial support is necessary to achieve the decarbonization of the maritime sector by 2050 (Ling-Chin et al., 2024).

11.4.6 Steel

Decarbonizing the steel sector is critical due to its significant contribution to global greenhouse gas emissions and its reliance on carbon-intensive processes. The steel industry accounts for approximately 7% of global CO_2 emissions, necessitating urgent and innovative strategies to align with climate objectives such as those outlined in the Paris Agreement (Hishan & Asmawi, 2024; Hariom et al., 2024). Various approaches have been proposed to achieve decarbonization, including the substitution of traditional carbon sources with biochar, which can replace coal in metallurgical processes, potentially reducing emissions by up to 50% in certain applications (Sarker et al., 2024). The use of biomass combined with CCS technologies is another promising strategy, offering the potential for net-negative emissions by compensating for residual emissions (Andrade et al., 2024).

Electrification of processes, such as the use of electric arc furnaces, and the adoption of hydrogen as a reductant in steel production, are also being explored as viable pathways to reduce carbon intensity (Feldmann, 2023; Zabik et al., 2024). Furthermore, energy efficiency improvements and the implementation of energy management systems, such as ISO 50001, are critical for reducing energy consumption and emissions in steel mills (Khalid et al., 2024). The integration of in-situ CO_2 capture and sequestration (ICCS) using waste streams from steel production offers an innovative approach to minimize external resource consumption and enhance sustainability (Zhang et al.,

2024). Despite these advancements, the transition to low-carbon steel production faces significant economic and infrastructural challenges, including the need for substantial investments, policy support, and international collaboration to optimize decarbonization efforts globally (Andrade et al., 2024; Feldmann, 2023). The rapid deployment of these technologies, alongside increased recycling of steel scrap and improvements in material efficiency, is essential to mitigate emissions and avoid stranded assets, thereby supporting the broader goal of limiting global warming to 1.5°C (Speizer et al., 2023). A combination of technological innovation, policy intervention, and strategic investment is required to navigate the complex path toward decarbonizing the steel sector effectively.

11.4.7 Cement

The cement sector has a significant contribution to global CO_2 emissions, accounting for approximately 5%–8% of the total emissions (Pisciotta et al., 2023). The primary sources of emissions in cement production are the calcination process, which releases CO_2 from limestone, and the energy-intensive nature of clinker production, traditionally powered by fossil fuels (Pisciotta et al., 2023; Wang et al., 2024). Various strategies have been proposed to address these emissions, including the use of low-carbon fuels such as renewable natural gas, green hydrogen, and biomass, which can significantly reduce the carbon footprint of the pyroprocessing steps in clinker production (Okeke et al., 2024). CCS technologies are also critical, with studies showing that they can reduce emissions intensity from 571 to 784 kg CO_2eq/metric ton of cement to as low as 166.33–438.66 kg CO_2eq/metric ton (Okeke et al., 2024).

The integration of CCS with other processes, such as the organic Rankine cycle, has been shown to further enhance emission reductions (Wang et al., 2024). Public perception and support are crucial for the successful implementation of these strategies, as evidenced by surveys indicating strong public interest in decarbonization efforts when the benefits are clearly communicated (Prigmore & Olayemi, 2024). In the European context, a mix of alternative fuels and technological improvements could mitigate up to 50% of emissions by 2050, but achieving climate neutrality will require rapid and large-scale CCS implementation (Cavalett et al., 2024). The economic viability of these technologies varies, with some, like hydrogen production, currently not feasible at scale (Domingues et al., 2024). However, novel processes that integrate natural gas with limestone to produce methanol alongside cement show promise for both economic and environmental performance (Wang et al., 2024). Overall, a multifaceted approach that includes material innovations, alternative fuels, and efficiency improvements across the sector is essential to meet net-zero targets by 2050 (Nehdi et al., 2024).

11.4.8 Mining

A significant focus of decarbonization in the mining sector is on electrification and the integration of renewable energy sources, such as solar and wind power, to reduce greenhouse gas emissions. For instance, the use of electric overland conveyor systems in open-pit copper mines has been shown to significantly lower emissions compared to traditional diesel equipment, although the high initial costs pose a barrier to widespread adoption (Aydogdu et al., 2024). Renewable energy systems, including photovoltaic and wind power, are increasingly being implemented in mining operations worldwide, offering both environmental benefits and long-term cost savings (Valikandi & Choi, 2024). The transition to renewable energy is also supported by the potential use of rehabilitated mine lands for utility-scale solar energy generation, which can help mitigate land-use conflicts and support regional economic transformation (Purtill, 2024).

In addition to renewable energy, the adoption of hydrogen as an alternative fuel for mining equipment is gaining traction due to its zero CO_2 emissions, presenting a viable option for reducing the carbon footprint of heavy machinery (Peppas et al., 2023). Digital technologies, such as automation and artificial intelligence, are also playing a crucial role in enhancing energy efficiency and reducing emissions, although their potential is not fully realized due to a focus on other operational benefits (Asa'd & Levesque, 2023). The mining sector's decarbonization efforts are further complicated by regional variations, as seen in Kazakhstan and Canada, where economic and logistical challenges must be addressed to integrate renewable energy and electrification solutions effectively (Аубакирова et al., 2023; Issa et al., 2023). Moreover, the decarbonization of mining truck fleets, which contribute significantly to emissions, requires an optimal pathway that balances cost, technology, and environmental impact, as demonstrated in studies focusing on South African coal mines (Yu et al., 2024). Overall, achieving decarbonization in the mining sector necessitates a comprehensive approach that includes technological innovation, strategic planning, and international cooperation to overcome financial and logistical barriers.

11.4.9 Agriculture

The development of promising technologies in animal feeding, soil carbon sequestration, and crop fertilization is crucial for addressing the environmental challenges posed by traditional agriculture. Cultivated meat, a significant innovation in this realm, is projected to become a $25 billion global industry by 2030, offering a sustainable alternative to conventional meat production. Cultivated meat is produced by culturing animal muscle cells in bioreactors, which significantly reduces greenhouse gas emissions, land use, and water consumption compared to traditional livestock farming (Tuomisto & Roy, 2012). This technology, however, faces several challenges, including optimizing cell culture media and developing stable scaffolds for cell growth, which are essential for producing meat with desirable texture and taste (Puentes, 2022). Despite these hurdles, the potential environmental benefits are substantial, as large-scale adoption could drastically reduce the ecological footprint of meat production (Tuomisto & Roy, 2012). Additionally, recombinant DNA technology and gene transfer are being explored to enhance animal agriculture by improving crop yields and nutritional value, which could complement the shift toward more sustainable meat production (Petters, 1986).

The integration of these biotechnologies with sustainable agricultural practices, such as precision agriculture and integrated nutrient management, can further enhance food security and environmental sustainability (Muhie, 2022). However, the transition to these new systems requires overcoming regulatory, social, and ethical challenges, as well as gaining consumer acceptance (Hocquette et al., 2024; Post et al., 2020). As the global population continues to grow, adopting these innovative technologies will be essential to meet the increasing demand for protein while minimizing environmental impacts (Bonny et al., 2017). Thus, the momentum in developing and implementing these technologies is pivotal for transforming the food system toward greater sustainability and resilience.

11.4.10 Forestry

Forestry and other land-use practices play a crucial role in decarbonization efforts by acting as significant carbon sinks, essential for achieving carbon neutrality goals. The land use, land-use change, and forestry (LULUCF) sector is pivotal in emitting and sequestering greenhouse gases, making it a critical component of climate change mitigation strategies (Liu et al., 2024a; Svensson et al., 2024). In China, for instance, implementing forestry carbon sink transactions could increase forest land area significantly, thereby absorbing substantial amounts of CO_2 annually (Feng & Chen, 2023). Similarly, in countries like Brazil, India, and Indonesia, the LULUCF subsector is vital for medium- and long-term mitigation, with strategies such as reducing deforestation, reforestation, and enhancing soil carbon sequestration being employed (Svensson et al., 2024). However, the economic aspects of these activities, such as the costs associated with carbon sequestration and the challenges of certifying carbon offset credits, present significant hurdles (van Kooten, 2013).

Despite the challenges, forestry activities, particularly in managed forest lands, have shown considerable potential in contributing to the global carbon budget, although they are often underutilized in policy frameworks. Moreover, soil carbon sequestration, which holds more carbon than the atmosphere and

TABLE 11.1
Top 10 Worst and Best Foods for the Environment

Rank	Food Item	Total Emissions ($kg\ CO_2e/kg$)	#1 Source ($kg\ CO_2e$)	#2 Source ($kg\ CO_2e$)
1	Bovine meat (beef herd)	59.57	Farming (39.39)	Land use (16.28)
2	Lamb and mutton	24.42	Farming (19.51)	Feed (2.37)
3	Cheese	21.24	Farming (13.06)	Land use (4.46)
4	Bovine meat (dairy herd)	21.09	Farming (15.69)	Feed (2.51)
5	Dark chocolate	18.68	Land use (14.31)	Farming (3.71)
6	Coffee	16.48	Farming (10.39)	Land use (3.69)
7	Crustaceans (farmed)	11.85	Farming (8.38)	Feed (2.51)
8	Palm oil	7.61	Land use (3.10)	Farming (2.11)
9	Pork	7.28	Feed (2.94)	Farming (1.69)
10	Soybean oil	7.28	Feed (2.94)	Farming (1.69)
		Top 10 Best Foods for the Environment		
1	Nuts	0.28	Farming (2.13)	Packaging (0.08)
2	Citrus fruit	0.32	Farming (0.31)	Transport (0.09)
3	Apples	0.36	Farming (0.23)	Transport (0.10)
4	Root vegetables	0.37	Farming (0.15)	Transport (0.11)
5	Potatoes	0.37	Farming (0.19)	Transport (0.09)
6	Onions and leeks	0.4	Farming (0.21)	Transport (0.10)
7	Brassicas	0.44	Farming (0.28)	Transport (0.10)
8	Bananas	0.68	Transport (0.29)	Farming (0.27)
9	Peas	0.9	Farming (0.72)	Transport (0.10)
10	Soy milk	0.91	Retail (0.27)	Land use (0.18)

vegetation combined, is another critical component, although it is susceptible to losses due to human activities. The integration of forestry and agricultural practices into climate mitigation policies remains contentious, with issues of permanence and the equivalence of sequestration to emission reductions being debated (Conant & Conant, 2011). Nonetheless, the potential of these terrestrial carbon sinks as low-cost mitigation options that support conservation and development is widely recognized, underscoring the need for balanced and effective land use planning and policy implementation (Conant & Conant, 2011).

11.4.11 Food Choices

The food we produce and consume significantly impacts greenhouse gas emissions, making dietary choices essential to decarbonization efforts. Efforts to decarbonize involve shifting dietary patterns toward foods with lower greenhouse gas emissions (Table 11.1). Reducing consumption of high-emission foods like beef, lamb, and cheese can significantly lower one's carbon footprint. Conversely, increasing the intake of low-emission foods such as nuts, fruits, and vegetables can help support more sustainable food systems. These changes, supported by policies and consumer choices, are essential in addressing climate change and promoting environmental sustainability.

11.5 CHALLENGES AND OPPORTUNITIES IN DECARBONIZING

The pursuit of net-zero emissions is fraught with significant technological uncertainties that hinder progress, primarily due to the complexities involved in decarbonizing certain sectors and the nascent state of necessary technologies. One major challenge is the decarbonization of difficult-to-abate sectors such as aviation, long-distance transport, and the production of carbon-intensive materials like steel and cement. These sectors are particularly challenging due to their high energy demands and the current lack of viable low-carbon alternatives, which necessitate substantial technological innovation and integration across energy systems (Davis et al., 2018). Additionally, the deployment of negative emissions technologies (NETs), such as bioenergy with carbon capture and storage (BECCS) and direct air capture (DAC), is critical but remains limited by high costs, regulatory hurdles, and the need for large-scale deployment, which is currently not feasible within the necessary timeframes (Akimoto et al., 2021; Haszeldine et al., 2018). The reliance on CCS is also problematic due to insufficient political and economic incentives to drive its widespread adoption, despite its potential to significantly reduce emissions (Haszeldine et al., 2018). Furthermore, integrating renewable energy sources into the energy grid is essential but complicated by the need for reliable storage solutions and the existing infrastructure's long lifespans, which delay the transition from fossil fuels (Arogyaswamy, 2022; Bandh et al., 2024). The development of green hydrogen and other emerging technologies is promising but requires substantial investment in research and development to achieve the necessary cost reductions and scalability (OECD, 2023). Finally, achieving net-zero emissions also involves overcoming socio-political barriers, such as aligning international policies and financing mechanisms to support technological advancements and ensuring equitable access to these technologies across different regions (Bandh et al., 2024; Xu et al., 2023). These technological uncertainties underscore the need for coordinated global efforts in innovation, policymaking, and infrastructure development to achieve net-zero emissions effectively.

Achieving net-zero emissions presents a complex interplay of challenges and opportunities, as highlighted by various studies. The transition to a net-zero economy is fraught with high capital costs, technological uncertainties, and the need for policy alignment. For instance, the UK faces significant hurdles in modernizing infrastructure and transitioning from fossil fuels to renewable energy, which necessitates substantial investment and innovation in green technologies (Shahzad et al., 2024). Similarly, India's ambitious goals to reduce emission intensity and increase non-fossil fuel electricity generation require an estimated $840 billion investment over 15 years, underscoring the financial challenges involved (Joshi et al., 2024).

The chemical industry, responsible for about 5% of global CO_2 emissions, exemplifies the technological challenges, as it requires high-temperature heat and carbon-rich raw materials, making decarbonization particularly difficult (Gabrielli et al., 2023). Despite these challenges, the transition to net-zero emissions offers significant opportunities for innovation and economic diversification. The development of sustainable energy technologies, such as solar and wind power, and the exploration of new business models like Energy as a Service (EaaS) are promising avenues for economic growth (Kandpal et al., 2024). Moreover, the transition encourages enhanced global cooperation, as seen in the UK's emphasis on international collaboration to address policy and regulatory challenges (Shahzad et al., 2024). Proactive measures to ensure a just transition are crucial, particularly in regions like Latin America and the Caribbean, where the transition could exacerbate inequality if not managed equitably (Alfonso et al., 2023). Addressing equity concerns involves ensuring that the benefits of the transition are distributed fairly and that negative impacts on vulnerable communities are mitigated through targeted policies (Alfonso et al., 2023). Overall, while the path to net-zero emissions is challenging, it also presents a unique opportunity to foster innovation, economic resilience, and global cooperation, provided that equity and inclusivity are prioritized in the transition process.

11.6 CONCLUSION

Decarbonization is now central to the global fight against climate change, necessitating a comprehensive transformation of how we produce, utilize, and manage energy, materials, and land. This chapter outlines the core strategies—carbon offsetting, sequestration, renewable energy, and emerging technologies such as carbon capture and storage—while highlighting the economic opportunities and challenges associated with this transition. We have seen how decarbonization is reshaping industries, driving investment in clean technologies, and creating new job opportunities, even as it requires significant capital shifts, policy reforms, and workforce reskilling.

The chapter also underscored that achieving net-zero emissions is not just a technical challenge but a societal one. Success depends on coordinated action: robust policies, international collaboration, and public-private partnerships are all essential. Sector-specific strategies—from power and transport to steel, cement, and agriculture—show that there is no one-size-fits-all solution but rather a need for tailored approaches that balance economic realities with climate goals. The transition will be disruptive, with risks such as stranded assets and job displacement, but it also presents growth opportunities and long-term savings through efficiency and innovation. Above all, the path to net zero must be equitable, ensuring that the benefits and burdens are shared fairly across regions and communities.

As we move forward, the next chapter turns from the mechanics of decarbonization to the global conversation that makes it possible. Chapter 12 explores how international collaboration, data sharing, and climate diplomacy shape the world's response to climate change. It will examine how agreements like the Paris Agreement are forged, how robust climate data is collected and shared, and how finance and transparency drive global progress. The journey continues from designing net-zero strategies to building the global alliances and trust needed to deliver them.

REVIEW QUESTIONS

1. What constitutes Global Carbon Cycle? Discuss the advantages and disadvantages of the carbon stored in the soil\vegetation, ocean and living organisms toward reducing the increase in the atmospheric temperatures?
2. What is carbon neutrality?
3. Is investing in decarbonatization a profitable venture, especially in industrial as well as PE sectors?
4. What are the drivers and barriers of decarbonization in PE sectors? Discuss the strategies to overcome those barriers.
5. Discuss the economics of decarbonization.
6. The transition to a Net Zero Economy by 2050 has been outlined in the NGFS Net Zero 2050 scenario. Discuss this in detail.
7. Discuss the economic impact of the transition to low-carbon industries.
8. Discuss the job shifts in green industries.
9. Discuss the growth opportunities in the net zero transition. Include sector-wise decarbonization practices such as power, oil and gas industries, transportation aviation, shipping, steel, cement, mining, agriculture, forestry, and food choices.
10. What are the challenges and opportunities involved in decarbonizing?

REFERENCES

Adebayo, Y.A., Ikevuje, A.H., Kwakye, J. M., Esiri, A.E. (2024a). Energy transition in the oil and gas sector: Business models for a sustainable future. *International Journal of Management & Entrepreneurship Research*. https://doi.org/10.51594/ijmer.v6i10.1611

Adebayo, Y.A., Ikevuje, A.H., Kwakye, J.M., Esiri, A.E. (2024b). Green financing in the oil and gas industry: Unlocking investments for energy sustainability. *Finance & Accounting Research Journal*. https://doi.org/10.51594/farj.v6i10.1622

Ahsan, M. (2024). Leveraging lithium-ion batteries for decarbonization in the oil and gas industry. *International Journal of Science and Research Archive*. https://doi.org/10.30574/ijsra.2024.13.1.1584

Akgül, E.F. (2024). Navigating decarbonization: examining shipping companies' fleet modernization strategies worldwide. *Dokuz Eylül Üniversitesi Denizcilik Fakültesi Dergisi*. https://doi.org/10.18613/deudfd.1418186

Akimoto, K., Sano, F., Oda, J., Kanaboshi, H., Nakano, Y. (2021). Climate change mitigation measures for global net-zero emissions and the roles of CO_2 capture and utilization and direct air capture. *Energy and Climate Change*. https://doi.org/10.1016/J.EGYCC.2021.100057

Alabi, O., Turner, K., Calvillo, C., Katris, A., Stewart, J., Brod, C. (2020). *Can the Electrification of Private Transport Lead to Economic Prosperity*. University of Strathclyde. https://doi.org/10.17868/73568

Alfonso, M., Bagolle, A., Baptista, D., Bos, M.S., Fazekas, A., Schwartz, L., Vogth-Schilb, A., Urquidi, M. (2023). *Advancing a just transition in Latin America and the Caribbean* (Policy Brief). Inter-American Development Bank, Climate Change Division. https://doi.org/10.18235/0005216

Andrade, C., Desport, L., Selosse, S. (2024). Net-negative emission opportunities for the iron and steel industry on a global scale. *Applied Energy*. https://doi.org/10.1016/j.apenergy.2023.122566

Arogyaswamy, B. (2022). Climate change mitigation technologies: Prospects and challenges. *Modern Applied Science*. https://doi.org/10.5539/mas.v16n4p1

Asa'd, O., Levesque, M.Y. (2023). Digital technologies for energy efficiency and decarbonization in mining. *CIM Journal*. https://doi.org/10.1080/19236026.2023.2203068

Aydogdu, K., Düzgün, Ş., Yaylacı, E.D., Aranoglu, F. (2024). A systems engineering approach to decarbonizing mining: Analyzing electrification and CO_2 emission reduction scenarios for copper mining haulage systems. *Sustainability*. https://doi.org/10.3390/su16146232

Аубакирова, Г.М., Isataeva, F.M., Мажитова, С. К., Togaibayeva, L. (2023). Decarbonization of mining and steel sectors: experience of Kazakhstan. *"Tūran" Universitetìnìn, Habarsysy*. https://doi.org/10.46914/1562-2959-2023-1-4-54-69

Babuder, D., Lapko, Y., Trucco, P., Taghavi, R. (2024). Impact of emerging sustainable aircraft technologies on the existing operating ecosystem. *Journal of Air Transport Management*. https://doi.org/10.1016/j.jairtraman.2023.102524

Bandh, S.A., Malla, F.A., Hoang, T., Qayoom, I., Mohi-Ud-Din, H., Bashir, S., Betts, R., Le, T. T., Le, D.T.N., Le, N.V.L., Le, H.C., Cao, D.N. (2024). Track to reach net-zero: Progress and pitfalls. *Energy & Environment*. https://doi.org/10.1177/0958305x241260793

Barbosa, F.C. (2024). Zero carbon emission aviation fuel technology review - The hydrogen pathway. *SAE Technical Paper Series*. https://doi.org/10.4271/2023-36-0029

Baştuğ, S., Akgül, E.F., Haralambıdes, H., Notteboom, T. (2024). A decision-making framework for the funding of shipping decarbonization ınitiatives in non-EU countries: insights from Türkiye. *Journal of Shipping and Trade*. https://doi.org/10.1186/s41072-024-00172-1

Bisht, A., Lugovoy, O., & Sharma, T.K. (2024). Decarbonizing the Indian power sector: Assessing the impact of a 30-year plan to net zero. In *2024 IEEE 4th International Conference on Sustainable Energy and Future Electric Transportation (SEFET)*, Hyderabad, India, 2024, pp. 1–5. https://doi.org/10.1109/sefet61574.2024.10717882

Bonny, S.P.F., Gardner, G.E., Pethick, D.W., Hocquette, J.-F. (2017). Artificial meat and the future of the meat industry. *Animal Production Science*. https://doi.org/10.1071/AN17307

Briceño-Garmendía, C., Qiao, W., Foster, V. (2023). *The economics of electric vehicles for passenger transportation* (Sustainable Infrastructure Series). World Bank. http://hdl.handle.net/10986/39513

Brooker, A., Thornton, M., Rugh, J. (2010). Technology *Improvement Pathways to Cost-effective Vehicle Electrification*. SAE Technical Paper 2010-01-0824, 2010, https://doi.org/10.4271/2010-01-0824

Cavalett, O., Watanabe, M.D.B., Voldsund, M., Roussanaly, S., Cherubini, F. (2024). Paving the way for sustainable decarbonization of the European cement industry. *Nature Sustainability*. https://doi.org/10.1038/s41893-024-01320-y

Chrysikopoulos, S.K., Chountalas, P., Géorgakèllos, D.A., Lagodimos, A.G. (2024). Decarbonization in the oil and gas sector: The role of power purchase agreements and renewable energy certificates. *Sustainability*. https://doi.org/10.3390/su16156339

Conant, R.T., Conant, R.T. (2011). Sequestration through forestry and agriculture. *Wiley Interdisciplinary Reviews: Climate Change*. https://doi.org/10.1002/WCC.101

Cybulsky, A., Allroggen, F., Shao-Horn, Y., Mallapragada, D.S. (2023). Decarbonization of aviation via hydrogen propulsion: technology performance targets and energy system impacts. *arXiv.Org*. https://doi.org/10.48550/arxiv.2309.14629

Davis, S.J., Lewis, N.S., Shaner, M.R., Aggarwal, S., Arent, D.J., Azevedo, I., Benson, S.M., Bradley, T.H., Brouwer, J., Chiang, Y.-M., Clack, C.T.M., Cohen, A., Doig, S.J., Edmonds, J., Fennell, P.S., Fennell, P.S., Field, C.B., Hannegan, B., … Caldeira, K. (2018). Net-zero emissions energy systems. *Science*. https://doi.org/10.1126/SCIENCE.AAS9793

Degirmenci, H., Uludag, A., Ekici, S., Karakoc, T.H. (2023). Challenges, prospects and potential future orientation of hydrogen aviation and the airport hydrogen supply network: A state-of-art review. *Progress in Aerospace Sciences*. https://doi.org/10.1016/j.paerosci.2023.100923

Dmitrieva, D.M., & Skobelev, D.O. (2023). Decarbonization of the oil and gas sector in the context of sustainable development: Key directions and possible scnarios for the Arctic region. *Sever i Rynok: Formirovanie Èkonomičeskogo Porâdka*. https://doi.org/10.37614/2220-802x.2.2023.80.001

Dodukh, V.M. (2023). The transformational changes of oil and gas companies in the context of decarbonization as a topical management problem. *Business Inform*. https://doi.org/10.32983/2222-4459-2023-11-319-324

Domingues, B.C., dos Santos, D.M.F., Mateus, M., Cecílio, D.M. (2024). Techno-economic analysis of cement decarbonization techniques: Oxygen enrichment vs. hydrogen fuel. *Hydrogen*. https://doi.org/10.3390/hydrogen5010005

Elmasry, H. (2024). Comprehensive investigation for gas flaring recovery as a potential pathway for oil and gas decarbonization. In *Paper Presented at the Mediterranean Offshore Conference*, Alexandria, Egypt, October 2024. https://doi.org/10.2118/223258-ms

Erickson, L.E. (2024). Electric vehicles for environmental sustainability. In M. N. V. Prasad, L. E. Erickson, F. C. Nunes, & B. S. Ramadan (Eds.), *Decarbonization strategies and drivers to achieve carbon neutrality for sustainability* (pp. 165–177). Elsevier. https://doi.org/10.1016/B978-0-443-13607-8.00012-2

Erickson, L.E., Brase, G.L. (2019). Electrification of transportation. In *Reducing greenhouse gas emissions and improving air quality* (1st ed., pp. 1–12). CRC Press. https://doi.org/10.1201/9781351116589

Ershaghi, I., Paul, D.L., Popa, A. (2024). Digital decarbonization in the oil and gas industry. In *Paper Presented at the SPE Western Regional Meeting*, Palo Alto, CA, April 2024. https://doi.org/10.2118/218905-ms

Feldmann, A. (2023). Circular economy for energy-intensive industries. In *Political science and public policy* 2023 (pp. 228–236). Edward Elgar Publishing. https://doi.org/10.4337/9781800882119.00024

Feng, S., Chen, K. (2023). The impact of forestry carbon sink on land use space based on FLUS model. *Processes*. https://doi.org/10.3390/pr11020608

Gabrielli, P., Rosa, L., Gazzani, M., Meys, R., Bardow, A., Mazzotti, M., Sansavini, G. (2023). Net-zero emissions chemical industry in a world of limited resources. *One Earth*. https://doi.org/10.1016/j.oneear.2023.05.006

Hariom, H., Kumar, Y., Kumar, A. (2024). Decarbonizing Indian steel. *Journal of Mines, Metals and Fuels*. https://doi.org/10.18311/jmmf/2023/43187

Hassouna, F.M.A., Shin, K.H. (2024). The environmental impacts of future global sales of hydrogen fuel cell vehicles. *Energies*. https://doi.org/10.3390/en17194930

Haszeldine, R.S., Flude, S., Johnson, G., Scott, V. (2018). Negative emissions technologies and carbon capture and storage to achieve the Paris Agreement commitments. *Philosophical Transactions of the Royal Society A*. https://doi.org/10.1098/RSTA.2016.0447

Hishan, R.N.D., Asmawi, A. (2024). Breaking the carbon shackles: Navigating the path to decarbonising the Malaysian steel sector. *Energy Research and Social Science*. https://doi.org/10.1016/j.erss.2024.103438

Hocquette, J.-F., Chriki, S., Fournier, D., Ellies-Oury, M.-P. (2024). Review: Will "cultured meat" transform our food system towards more sustainability? *Animal*. https://doi.org/10.1016/j.animal.2024.101145

Huang, C., Lin, B.Q. (2023). Promoting decarbonization in the power sector: How important is digital transformation? *Energy Policy*. https://doi.org/10.1016/j.enpol.2023.113735

İnal, Ö.B. (2024). Decarbonization of shipping: Hydrogen and fuel cells legislation in the maritime industry. *Brodogradnja*. https://doi.org/10.21278/brod75205

Issa, M., Ilinca, A., Rousse, D.R., Boulon, L., Groleau, P. (2023). Renewable energy and decarbonization in the Canadian mining industry: Opportunities and challenges. *Energies*. https://doi.org/10.3390/en16196967

Ivanov, T. (2023). Challenges and opportunities for hydrogen as a viable aviation fuel of the future. *Proceedings Of The International Congress On Process Engineering – Processing*, 36(1), 39–44. https://doi.org/10.24094/ptk.023.039

Jakkaladiki, S.P., Janečková, M., Malý, F. (2024). Similarity analysis study of the carbon emission metrics through fuzzy entropy clustering. *Journal of Infrastructure, Policy and Development*, 8. https://doi.org/10.24294/jipd.v8i5.3926

Jensen, L.L., Bonnefoy, P.A., Hileman, J., Fitzgerald, J.T. (2023). The carbon dioxide challenge facing U.S. aviation and paths to achieve net zero emissions by 2050. *Progress in Aerospace Sciences*. https://doi.org/10.1016/j.paerosci.2023.100921

Joshi, A.M., Anbumaninathan, N., Shah, V.M. (2024). Transformation toward net zero built environment—Indian scenario: Challenges and opportunities. In Bajaj, D., Gajendran, T., Patil, S. (eds) *Sustainable Built Environment*. ICSBE 2023. Lecture Notes in Civil Engineering, vol 451. Springer, Singapore. https://doi.org/10.1007/978-981-99-8842-6_4

Kandpal, V., Jaswal, A., González, E.D.R.S., Agarwal, N. (2024). Challenges and opportunities for sustainable energy transition and circular economy. In *Sustainable Energy Transition. Circular Economy and Sustainability*. Springer, Cham. https://doi.org/10.1007/978-3-031-52943-6_11

Kashem, M.A., Shamsuddoha, M., Nasir, T. (2024). Sustainable transportation solutions for intelligent mobility: A focus on renewable energy and technological advancements for electric vehicles (EVs) and flying cars. *Future Transportation*. https://doi.org/10.3390/futuretransp4030042

Kat, B., Şahin, Ü., Teimourzadeh, S., Tör, O.B., Voyvoda, E., Yeldan, A.E. (2024). A new energy-economy-environment modeling framework: Insights from decarbonization of the Turkish power sector towards net-zero emission targets. *Energy*. https://doi.org/10.1016/j.energy.2024.131760

Khalid, R., Savoia, A., Catania, V., Collarini, N., Anglani, N. (2024). Decarbonization of electric arc furnace steel mills, how to set-up an effective allocation of energy flows, aiming at tracking efficiency and energy savings. *2024 9th International Youth Conference on Energy (IYCE)*, Colmar, France, 2024, pp. 1–6, https://doi.org/10.1109/IYCE60333.2024.10634922

Khaustova, V.Y., Kyzym, M.O., Hubarieva, I., Kotliarov, Y.I., Salashenko, T., Khaustov, M.M. (2024). Tracking the European transition from fuel dependence to sustainable mobility. *Environmental Economics*. https://doi.org/10.21511/ee.15(2).2024.12

van Kooten, G.C. (2013). Avoiding emissions reduction: Terrestrial carbon sinks. In *Climate Change, Climate Science and Economics*. Springer, Dordrecht. https://doi.org/10.1007/978-94-007-4988-7_9

Li, W., Hu, Z. (2024). Pathways in the governance of shipping decarbonization from perspective of balancing the conflicting interests. *Frontiers in Marine Science*. https://doi.org/10.3389/fmars.2024.1479528

Li, C., Wu, D. (2024). Innovation and future development direction of hydrogen fuel cell vehicles. *Highlights in Science Engineering and Technology*. https://doi.org/10.54097/a8n0cr21

Li, W., Zhang, H. (2023). Transport electrification: Opportunities and future challenges. *Highlights in Science, Engineering and Technology*. https://doi.org/10.54097/hset.v46i.7658

Ling-Chin, J., Simpson, R., Cairns, A., Wu, D., Xie, Y., Song, D., Kashkarov, S., Molkov, V., Moutzouris, I., Wright, L., Tricoli, P., Dansoh, C., Panesar, A., Chong, K., Liu, P., Roy, D., Wang, Y., Smallbone, A., Roskilly, A.P. (2024). Research and innovation identified to decarbonise the maritime sector. *Green Energy and Sustainability*. https://doi.org/10.47248/ges2404010001

Liu, L., Qu, J., Gao, F., Maraseni, T.N., Wang, S., Aryal, S., Zhang, Z., Wu, R. (2024a). Land use carbon emissions or sink: research characteristics, hotspots and future perspectives. *Land*. https://doi.org/10.3390/land13030279

Liu, M., Tang, L., Zeng, J., Huang, G., Liu, X., Yao, S., He, G., Shang, N., Tao, H., Ren, S., Wang, P. (2024b). Promoting decarbonization in China: Revealing the impact of various energy policies on the power sector based on a coupled model. *Energies*. https://doi.org/10.3390/en17133234

Mavrakos, A., Tsaousis, T., Faggioni, N., Caviglia, A., Manzo, E. (2024). A digital twin approach for selection and deployment of decarbonization solutions for the maritime sector. In B. Karakostas and T. Katsoulakos (Eds.), *State-of-the-Art Digital Twin Applications for Shipping Sector Decarbonization*. https://doi.org/10.4018/978-1-6684-9848-4.ch002

Muhie, S. (2022). Novel approaches and practices to sustainable agriculture. *Journal of Agriculture and Food Research*. https://doi.org/10.1016/j.jafr.2022.100446

Nakano, Y., Sano, F., Akimoto, K. (2022). Impacts of decarbonization technologies in air transport on the global energy system. *Transportation Research Part D-Transport and Environment*. https://doi.org/10.1016/j.trd.2022.103417

Nehdi, M.L., Marani, A., Zhang, L. (2024). Is net-zero feasible: Systematic review of cement and concrete decarbonization technologies. *Renewable & Sustainable Energy Reviews*. https://doi.org/10.1016/j.rser.2023.114169

Newbery, D. (2013). The Economics of Electric Vehicles [Seminar presentation]. E&E Seminar, Green Emotion – GEM, University of Cambridge and Imperial College London, Cambridge, United Kingdom.

Niu, M. (2024). How does the carbon trading scheme promote the decarbonization of China's power sector? *Journal of Environmental Planning and Management*. https://doi.org/10.1080/09640568.2024.2303735

OECD. (2023). *Driving Low-Carbon Innovations for Climate Neutrality*. OECD.

Okeke, I.J., Kamath, D., Nimbalkar, S., Cresko, J. (2024). The role of low-carbon fuels and carbon capture in decarbonizing the U.S. clinker manufacturing for cement production: CO_2 emissions reduction potentials. *Energies*. https://doi.org/10.3390/en17205233

Patil, G. R., Pode, G., Diouf, B., Pode, R. (2024). Sustainable decarbonization of road transport: Policies, current status, and challenges of electric vehicles. *Sustainability*. https://doi.org/10.3390/su16188058

Peacock, J., Cooper, R., Waller, N., Richardson, G. (2023). Decarbonising aviation at scale through synthesis of sustainable e-fuel: A techno-economic assessment. *International Journal of Hydrogen Energy*. https://doi.org/10.1016/j.ijhydene.2023.09.094

Peppas, A., Kottaridis, S., Politi, C., Oustadakis, P. (2023). Environmental Assessment of Replacing Fossil Fuels with Hydrogen for Motorised Equipment in the Mining Sector. *Materials Proceedings*, 15(1), 43. https://doi.org/10.3390/materproc2023015043

Petters, R.M. (1986). Recombinant DNA, gene transfer and the future of animal agriculture. *Journal of Animal Science*. https://doi.org/10.2527/JAS1986.6261759X

Pisciotta, M., Pilorgé, H., Davids, J., Psarras, P. (2023). Opportunities for cement decarbonization. *Cleaner Engineering and Technology*. https://doi.org/10.1016/j.clet.2023.100667

Post, M.J., Levenberg, S., Kaplan, D.L., Genovese, N., Fu, J., Bryant, C., Negowetti, N., Verzijden, K.E., Moutsatsou, P. (2020). Scientific, sustainability and regulatory challenges of cultured meat. *Nature Food*. https://doi.org/10.1038/S43016-020-0112-Z

Prados, J.M.M. (2024). The decarbonisation of the maritime sector: Horizon 2050. *Brodogradnja*. https://doi.org/10.21278/brod75202

Prigmore, S.M., Olayemi, M. (2024). Public perception of cement industry decarbonization: A case study of Oklahoma. In *Paper presented at 2024 ASEE Midwest Section Conference*, Lawrence, KS. https://doi.org/10.18260/1-2-115-49365

Proesmans, P.-J., Vos, R. (2022). Comparison of future aviation fuels to minimize the climate impact of commercial aircraft. In *AIAA AVIATION 2022 Forum*. Chicago, Il. https://doi.org/10.2514/6.2022-3288

Puentes, J.P. (2022). Sustainable future protein foods: The challenges and the future of cultivated meat. *Foods*. https://doi.org/10.3390/foods11244008

Purtill, J. (2024). How utility-scale solar energy generation on rehabilitated mine lands can contribute to decarbonising the resources sector. *Research Directions. Mine Closure and Transitions*. https://doi.org/10.1017/mcl.2024.3

Rajashekara, K. (2023). Current and future trends in electrification of road and air transportation. *iEnergy*. https://doi.org/10.23919/ien.2023.0027

Ravishankar, K. (2024). Decarbonization technologies screening for enabling net zero in onshore oil and gas operations. In *Paper presented at the SPE International Health, Safety, Environment and Sustainability Conference and Exhibition*, Abu Dhabi, UAE, September 2024. https://doi.org/10.2118/220479-ms

Reyseliani, N., Pratama, Y., Hidayatno, A., Dowell, N.M., Purwanto, W.W. (2024). Power sector decarbonisation in developing and coal-producing countries: A case study of Indonesia. *Journal of Cleaner Production*. https://doi.org/10.1016/j.jclepro.2024.142202

Sarker, T.R., Ethen, D.Z., Nanda, S. (2024). Decarbonization of metallurgy and steelmaking industries using biochar: A review. *Chemical Engineering Technology*. https://doi.org/10.1002/ceat.202400217

Shahzad, S., Faheem, M., ul Muqeet, H.A., Waseem, M. (2024). Charting the UK's path to net zero emissions by 2050: Challenges, strategies, and future directions. *IET Smart Grid*. https://doi.org/10.1049/stg2.12185

Shankar, J., Durga, K.V., Kumar, K., Kumar, T.A.R. (2024). HiGAS marine scrubbers – Revolutionizing shipping industry for MARPOL and decarbonization. In *Paper presented at the ADIPEC*, Abu Dhabi, UAE, November 2024. https://doi.org/10.2118/223043-ms

Sharifzadeh, M. (2022). *Electrification of Transportation*. Wiley. https://doi.org/10.1002/9781119695868.ch10

Speizer, S., Durga, S., Blahut, N., Charles, M., Lehne, J., Edmonds, J., Yu, S. (2023). Rapid implementation of mitigation measures can facilitate decarbonization of the global steel sector in 1.5°C-consistent pathways. *One Earth*. https://doi.org/10.1016/j.oneear.2023.10.016

Svensson, J., Avashia, V., Boer, R., Chaturvedi, R.K., Cotta, M., Dubeux, C.B.S., Immanuel, G.S., Rovere, E.L.L., Patange, O., Rossita, A., Vishwanathan, S.S. (2024). The AFOLU sector's role in national decarbonization: a comparative analysis of low-GHG development pathways in Brazil, India and Indonesia. *Climate Policy*. https://doi.org/10.1080/14693062.2024.2391048

Tait, K.N., Wedemeyer, M., Khan, M.A.H., Lowenberg, M., Shallcross, D.E. (2024). Insights and innovations to mitigate aviation climate impact by 2030. *Communications Engineering*. https://doi.org/10.1038/s44172-024-00290-5

Talwar, C., Joormann, I., Spengler, T. S. (2023). Comparison of adoption rates of hydrogen, hydrogen-electric and SAF in the future air transport system with a system dynamics model. In G. La Rocca & J. T. B. de Hosson (Eds.), *Engineering design applications* (pp. 597–602). Springer International Publishing. https://doi.org/10.1007/978-3-031-24907-5_71

Tiwari, S., Pekris, M.J., Doherty, J.J. (2024). A review of liquid hydrogen aircraft and propulsion technologies. *International Journal of Hydrogen Energy*. https://doi.org/10.1016/j.ijhydene.2023.12.263

Tuomisto, H., & Roy, A. (2012). Could cultured meat reduce environmental impact of agriculture in Europe? In M. S. Corson & H. M. G. van der Werf (Eds.), *Proceedings of the 8th International Conference on Life Cycle Assessment in the Agri-Food Sector (LCA Food 2012)* (pp. 611–615). INRA. https://publications.jrc.ec.europa.eu/repository/handle/JRC71579

Valikandi, E.M., Choi, Y. (2024). Utilization of solar and wind power-generation systems in the mining industry: Recent trends and future prospects. *Journal of Sustainable Mining*. https://doi.org/10.46873/2300-3960.1431

Wang, J. (2024a). Explore the impact of sustainable development strategies on green employment. *Advances in Economics, Management and Political Sciences*. https://doi.org/10.54254/2754-1169/75/20241873

Wang, Y., Yang, M., Shen, F., Zhou, M., Du, W. (2024). Conceptual design and life-cycle environmental and economic assessment of low-carbon cement manufacturing processes. *Journal of Cleaner Production*. https://doi.org/10.1016/j.jclepro.2024.143349

Wang, Z.J. (2024b). Analysis of carbon reduction potential based on carbon flow theory. *Applied and Computational Engineering*. https://doi.org/10.54254/2755-2721/93/20240939

Xu, D., Abbas, S., Rafique, K., Ali, N. (2023). The race to net-zero emissions: Can green technological innovation and environmental regulation be the potential pathway to net-zero emissions? *Technology and Society*. https://doi.org/10.1016/j.techsoc.2023.102364

Xu, P., Sun, L., Feng, K., Zhong, H., Liang, J., Zhang, C., Zhao, D., Chen, H., Long, R., Xing, Z., Hubacek, K. (2024). Extent of global decarbonization of the power sector through energy policies and governance capacity. *Communications Earth & Environment*. https://doi.org/10.1038/s43247-024-01494-5

Yu, G., Ye, X., Ye, Y., Huang, H., Xia, X. (2024). Optimal decarbonisation pathway for mining truck fleets. *Journal of Automation and Intelligence*. https://doi.org/10.1016/j.jai.2024.03.003

Zabik, G., Birkelbach, F., Hofmann, R. (2024). Decarbonizing the steel processing industry: A MILP-based assessment of electrification and hydrogen for hot rolling. https://doi.org/10.1115/es2024-130999

Zhang, Y., Su, Y., Qu, F., Poon, C.-S., Tsang, D.C. (2024). Valorizing inherent resources from waste streams for in-situ CO_2 capture and sequestration in the steel industry. *Journal of Cleaner Production*. https://doi.org/10.1016/j.jclepro.2024.142486

12 The Global Conversation
Data Sharing and Climate Diplomacy

This chapter delves into the critical role of global collaboration in addressing climate change, emphasizing the importance of data sharing and climate diplomacy. It explores the establishment and purpose of the Conference of Parties (COP), highlighting key milestones and outcomes in the fight against climate change. This chapter also discusses the significance of international cooperation in data collection, analysis, and sharing for effective climate action. It examines various climate finance mechanisms and initiatives, showcasing the efforts of organizations like the World Bank Group and UNEP Finance Initiative in mobilizing resources for climate change mitigation and adaptation. Additionally, this chapter explores the concept of open access to climate data and its potential to accelerate research and innovation in addressing the climate crisis.

This chapter investigates the status of global conversations, how data are shared among nations, and the types of climate diplomacies that are in place to face the challenges. Also contained in this chapter are the outcomes of various COP, including international agreements, commitments made by developed countries who are partly responsible for the CO_2 increase, and ensuing progress made to avert the catastrophic climate crisis.

Countries must commit to a decisive decade of climate action. By coming together to set bolder emission reduction targets, bolster adaptation plans, and articulate national roadmaps to achieve those goals, world leaders can help chart a path for a more secure, prosperous, resilient, and sustainable future for all. Global action to reduce short-lived climate pollutants can prevent 0.6°C of warming by 2050. Cutting emissions of carbon dioxide and short-lived climate pollutants is critical to slowing the rate of global warming and achieving the 2°C target set by the Paris Agreement.

12.1 GLOBAL COORDINATION IN TACKLING CLIMATE CHANGE

Emissions anywhere affect people everywhere. It is an issue that requires solutions coordinated at the international level, and it requires international cooperation to help developing countries move toward a low-carbon economy. Hence, international cooperation is crucial for climate change. Addressing climate change as a global problem necessitates international cooperation, as emissions from any country contribute to global warming, impacting communities worldwide. Effective international economic cooperation is crucial for the global energy transition, requiring joint efforts from countries, financial institutions, and the private sector to ensure sufficient investment in renewable energy and technological innovation (Onysiuk, 2024).

The Paris Agreement underscores adaptation as a global challenge, emphasizing the need for international cooperation to enhance adaptive capacity and resilience, particularly between the global North and South (Trivedi & Jolly, 2022). Climate change presents a "free-rider" problem, necessitating transnational negotiations to build a global climate governance system that balances economic development with emission reductions (CICC Research, CICC Global Institute, 2022). Collective responsibility is essential in cooperative governance, transitioning from individual to co-responsibility to effectively manage the climate crisis (Piazza, 2021). Wealthy nations must support vulnerable regions, which suffer disproportionately from climate impacts despite contributing minimally to emissions. This support should focus on resilience and adaptation, as well as

DOI: 10.1201/9781003479666-12

mitigation, to prevent global destabilization (Bellido, 2022). Thus, development cooperation is an investment in shared well-being, enabling countries to tackle interconnected challenges such as climate, migration, and public health through technological innovation, capacity building, and financial investments.

The "free-rider" problem presents significant challenges for global climate governance and international negotiations, as it undermines collective efforts to mitigate climate change. The issue arises from the incentive structure of climate mitigation, which resembles a global public good, encouraging countries to benefit from others' efforts without contributing themselves. This behavior is exacerbated by the Westphalian sovereignty model, where national interests often take precedence over global commitments, leading to governance failures in frameworks like the Kyoto Protocol and the Paris Agreement (Murombedzi & Chikozho, 2023). Although the free-rider hypothesis is prevalent in theoretical discussions, empirical evidence suggests that it may not be the primary driver of lax climate policies, as larger countries do not necessarily set higher carbon prices despite their greater capacity to internalize climate damages (Schmidt et al., 2022). However, the free-rider problem is particularly pronounced in high-growth countries, which have stronger incentives to avoid cooperation due to their economic priorities, contrasting with low-growth countries that are more vulnerable to climate impacts and thus more inclined to cooperate (Wang et al., 2017). This dynamic complicates the development of fair burden-sharing and compensation schemes, as the interests of developed and developing nations often diverge, leading to fragmented negotiations and reduced effectiveness in international climate agreements. Addressing these challenges requires innovative approaches to incentivize cooperation and align national interests with global climate goals.

The divergent interests between developed and developing nations significantly impact the success of climate governance frameworks, as these differences often lead to conflicts in priorities and approaches. Developed countries typically focus on emission mitigation targets, while developing nations emphasize the need for equitable responsibility sharing and addressing socio-economic challenges alongside climate goals (Müller, 2002). The current global climate governance system, which primarily monitors production-based emissions, fails to account for the consumption-based emissions that are more relevant to developing countries, thus exacerbating inequality and hindering effective climate action (Grigoryev et al., 2020). The principle of "common but differentiated responsibilities" is crucial, as it acknowledges nations' varying capabilities and responsibilities, yet its implementation remains contentious (Getu, 2012). Developing countries, which are projected to contribute significantly to future emissions growth, face challenges such as economic development needs and technological limitations, making it difficult to prioritize climate change mitigation without adequate support (Vandenbergh et al., 2009). Lessons from the Montreal Protocol suggest that increased financial assistance, technology transfer, and political commitment could enhance participation from developing nations without compromising their interests (Getu, 2012). Moreover, innovative strategies like expanded corporate carbon reporting and product carbon disclosure could create market pressures that incentivize emissions reductions across global supply chains, potentially aligning the interests of both developed and developing countries (Vandenbergh et al., 2009). Ultimately, bridging the North–South divide requires a comprehensive approach that integrates equity considerations, financial mechanisms, and inclusive policy frameworks to ensure the success of global climate governance (Müller, 2002).

12.1.1 COP Establishments and Its Purpose

The United Nations Climate Change Conference (COP), or Conference of the Parties, is an annual meeting of countries that have joined the United Nations Framework Convention on Climate Change (UNFCCC). It is an international treaty adopted in 1992 to address climate change by limiting global temperature increases and helping countries adapt to its impacts. COPs—meaning those countries that joined and are "party to," in legal terms—are considered crucial in the fight against climate change and are a key decision-making body for the UNFCCC. Despite the adoption of significant agreements like the Kyoto Protocol and the Paris Agreement, the implementation of these commitments has been inconsistent, prompting calls for COP to adopt more proactive measures, such as introducing voting rules and negotiating new legal agreements on specific emission sectors (Depledge, 2024). COP conferences have been effective in aligning global nations toward sustainable growth by facilitating negotiations and agreements that address current and future environmental challenges, such as the greenhouse effect and air pollution (Akhtar et al., 2024). COP28 emphasized the intersection of health, climate, and food, highlighting the need for SMART climate actions that prioritize health and enhance the resilience of food and public health systems (Varyvoda, 2024). However, the carbon footprint of COP events, particularly the use of private jets, has been criticized for undermining the core mission of these discussions, calling for more sustainable travel choices to align with climate commitments (Blenkinsop, 2024). Despite these efforts, greenhouse gas emissions continue to rise, indicating a need for increased ambition and policy intensity in future COP negotiations (Papanikolaki & Fokkas, 2024). The evolving focus on equity and inclusivity at COP underscores the importance of addressing the diverse needs of both developed and developing nations, ensuring that climate actions are fair and effective across different contexts.

The failure of the COP to achieve its ambitious targets in limiting global temperature rise could have profound and multifaceted consequences. Firstly, missing the targets set by the Paris Agreement, which aims to limit global warming to 1.5°C–2°C above pre-industrial levels, could lead to significant and potentially irreversible impacts on Earth's systems. Current warming of 1.2°C is already affecting these systems, and further increases could push them beyond critical tipping points, resulting in severe ecological and physical climate disruptions (Taylor & Vink, 2021). The political and economic challenges, such as the demand for developed countries to achieve net-negative emissions, complicate the equitable distribution of the remaining carbon budget, potentially leading to further ambiguity and conflict in international negotiations (Mohan et al., 2021). Additionally, the prospect of developed economies, like the UK, planning to miss their targets highlights the need for more actionable and realistic policy pathways, as current efforts may not suffice to meet the steep emission reduction requirements (Marsden & Schwanen, 2024). The ongoing failure to peak global emissions exacerbates disparities in emissions both between and within countries, increasing scrutiny and pressure on governments and businesses to avoid over-reliance on carbon dioxide removal technologies (Oxford Analytica, 2022). This situation underscores the urgent need for comprehensive research into climate overshoot risks and the development of effective mitigation and adaptation strategies to manage these risks and ensure a stable climate (Taylor & Vink, 2021).

12.1.2 From the Kyoto Protocol to the Paris Agreement

The Kyoto Protocol was a pioneering effort in the fight against climate change, and while it did not achieve its primary goal of substantial emission reductions, it laid the groundwork for future international climate agreements. The Protocol's mixed legacy offers valuable lessons in crafting more effective and inclusive policies that align global environmental goals with national economic interests. By learning from the Kyoto experience, the international community can enhance its strategies to confront the climate crisis and work toward a sustainable future.

The transition from the Kyoto Protocol to the Paris Agreement reflects significant lessons learned in international climate negotiations, particularly regarding flexibility and inclusivity. While pioneering, the Kyoto Protocol was criticized for its rigid structure and limited participation, primarily involving Annex I countries, which led to challenges in enforcement and effectiveness (Atici, 2021). This rigidity highlighted the need for a more adaptable framework, which the Paris Agreement addressed by allowing countries to set their own nationally determined contributions (NDCs), thus accommodating diverse national circumstances and promoting broader participation (del Castillo, 2018).

The Paris Agreement's emphasis on self-determined contributions and regular reviews represents a shift toward a more inclusive and flexible approach, encouraging both developed and developing countries to engage in climate action. This framework also includes mechanisms for transparency and compliance, aiming to enhance accountability and ambition in climate commitments. Furthermore, the Paris Agreement's integration of social dimensions, such as human rights and gender equality, underscores its comprehensive approach to climate governance. Despite these advancements, challenges remain, as current pledges are insufficient to meet the Agreement's ambitious temperature goals, necessitating further international cooperation and policy innovation (Maslin et al., 2023). The lessons from Kyoto have thus informed a more dynamic and equitable climate regime under the Paris Agreement, fostering a global commitment to sustainable transitions (Doukas et al., 2018).

12.1.3 Key Outcomes and Challenges of COP Milestones

12.1.3.1 COP29 Key Outcomes

COP29, held in Baku, Azerbaijan, concluded with notable developments in climate finance and carbon markets, though opinions on its overall effectiveness were divided. A pivotal outcome was the establishment of the New Collective Quantified Goal (NCQG) for climate finance, committing developed nations to provide funding. In the domain of carbon markets, a decade of negotiations culminated in agreements on regulatory frameworks under Article 6 of the Paris Agreement. These included the formalization of bilateral carbon credit trading (Article 6.2) and standards for a centralized UN-governed carbon market (Article 6.4), which are expected to enable international trading mechanisms and incentivize emissions reductions.

Progress was also made in adaptation efforts, with the establishment of a program to support National Adaptation Plans for Least Developed Countries and the extension of the Lima Work Program on Gender and Climate Change. However, discussions on loss and damage remained limited, with key issues deferred to future negotiations. Marginal advancements were observed in mitigation, and the conference failed to reach a consensus on acknowledging outcomes from COP28 advocating for a shift away from fossil fuels. Additionally, "continuity coalitions" were formed to strengthen sectoral commitments over time. While COP29 marked steps forward, particularly in finance and carbon markets, many participants deemed the results insufficient to address the urgency of climate action. Attention now turns to COP30 in Belém, Brazil, where nations are expected to present more ambitious NDCs to confront the escalating climate crisis.

12.1.3.2 COP28 Key Outcomes

COP28, held in Dubai in 2023, marked a transformative moment in climate diplomacy by achieving significant milestones despite challenges. The conference saw a broad consensus among major nations, including the EU, AOSIS, and the United States, on the necessity of transitioning away from fossil fuels, although this faced resistance from oil-producing countries. A key feature of COP28 was the first Global Stocktake (GST) of the Paris Agreement, which highlighted a significant gap in greenhouse gas emissions reductions needed to limit global warming to 1.5°C, yet it set ambitious goals to triple renewable energy capacity and double energy efficiency by 2030 (Darby et al., 2024). The conference also emphasized the critical role of water in climate action, integrating it into the GST and Global Goal on Adaptation (GGA), recognizing its importance in achieving global mitigation and adaptation objectives (Ransom, 2024).

Additionally, the COP28 UAE Declaration on Climate and Health was a historic milestone, explicitly addressing the intersection of climate change and public health, endorsed by 144 nations, and setting a strategic framework for integrating these issues into global policy (Cai et al., 2024). Despite some disappointments, such as insufficient funding for adaptation and carbon market provisions, the conference underscored the evolving role of civil society and the importance of methane reduction (Stavins, 2024). Overall, COP28 expanded the climate agenda to include interconnected issues like food, health, and water, setting the stage for future climate action and policy integration (Ransom, 2024).

12.2 DATA SHARING AND CLIMATE DIPLOMACY

Reliable and comprehensive data are the cornerstone of effective climate action. It empowers policymakers, scientists, and communities to understand the complex nuances of climate change, identify vulnerabilities, and design targeted interventions. Robust data on greenhouse gas emissions, deforestation rates, sea-level rise, and extreme weather events enable evidence-based decision-making and facilitate the tracking of progress toward climate goals. The advent of advanced technologies such as satellite monitoring, remote sensing, and artificial intelligence has revolutionized data collection and analysis. These tools provide a wealth of real-time information on environmental parameters, enabling a more nuanced understanding of climate patterns and trends.

Open access to climate data is pivotal for informed climate action, as it facilitates collaboration among researchers, policymakers, and communities, thereby addressing complex challenges such as food security and disaster risk management. The transition of Brazil's National Institute for Space Research to an open data model exemplifies how open data can support public debate and inform evidence-based governmental decisions, despite challenges in government transparency and science-policy dialogue (Sá & Grieco, 2016). The availability of climate data is crucial for understanding the climate system, developing Earth system models, and attributing extreme weather events to their causes, which underscores the importance of user-friendly data access for a diverse range of stakeholders (Schneider et al., 2013).

Platforms like the Global Climate Action Portal play a significant role in tracking progress, fostering collaboration, and enhancing transparency in climate governance, thereby enabling a coordinated global response to climate challenges. Collaborative infrastructures like the Collaborative Climate Community Data and Processing Grid (C3Grid) are essential for managing the growing volume of climate data, enabling users to find, access, and share data effectively, thus supporting complex multi-model analysis tasks (Kindermann et al., 2012). Furthermore, open access to primary data is not only vital for development and policy design but also for ensuring government transparency and accountability, which are fundamental to democratic governance (Osorio, 2008). In the context of food security, climate change exacerbates vulnerabilities, making open data crucial for identifying solutions to food insecurity and managing disaster risks (Habiba et al., 2016).

Open data models, as implemented by national institutes, play a crucial role in supporting evidence-based decision-making in climate policy by enhancing transparency, accessibility, and collaboration. These models facilitate the aggregation and analysis of diverse data sources, which is essential for comprehensive policy-making in Science, Technology, and Innovation (STI) ecosystems, as demonstrated in the mapping of Climate Action research in Denmark using open access data sources like OpenAire and CORDIS (Bovenzi et al., 2022). By providing high-quality, relevant data that is easy to understand, open data models empower cities worldwide to manage emissions and risks across various sectors, thereby granting voice and agency to communities affected by climate change (Szasz et al., 2021). The transition of Brazil's National Institute for Space Research to an open data model exemplifies how such systems can inform public debate and governmental decisions by making critical environmental data, like forest cover, widely available (Sá & Grieco, 2016). Furthermore, open development methodologies, such as the ROMA platform, enable stakeholders to engage in "what if" analyses through modular simulations, thus supporting climate policy deliberations and fostering a community focused on model development (Introne et al., 2011). Ultimately, the legitimacy of political systems increasingly relies on their ability to deliver outcomes based on transparent and reliable data, underscoring the importance of open data in making political decisions understandable and sustainable. These initiatives collectively highlight the transformative potential of open data models in advancing evidence-based climate policy.

12.2.1 Orchestrating Global Climate Governance Through Data

The Global Climate Action Portal (GCA Portal) serves as a prime example of how data sharing can be a powerful tool in climate diplomacy. It has evolved from a platform showcasing climate commitments to a tracking tool for implementation progress. The GCA Portal is a prime example of how data sharing can play a crucial role in climate diplomacy. It has helped leverage the power of non-state actors, facilitate collaboration, and track implementation progress. However, for data-driven climate governance to reach its full potential, challenges related to data quality, inclusivity, and neutrality must be addressed.

Key Insights

1. **Leveraging Non-State Actors**: Before the Paris Agreement (COP21), the GCA Portal (formerly NAZCA) was instrumental in showcasing the climate actions of non-state actors, such as cities and businesses. This data helped put pressure on national governments to adopt ambitious climate targets by demonstrating the willingness and capacity for climate action beyond the state level.
2. **Shifting to Implementation Tracking**: After COP21, the focus shifted to tracking the actual implementation of climate pledges. The GCA Portal evolved to incorporate more granular data, including contextual information, to assess the progress of both state and non-state actors in meeting their commitments.
3. **Fostering Collaboration**: Data sharing through the GCA Portal facilitates collaboration between states and non-state actors. It provides a platform for knowledge exchange, learning, and the identification of opportunities for joint action. By demonstrating the effectiveness of climate policies and programs, it can inspire further action and increase confidence in climate action efforts.
4. **Data as a Diplomatic Tool**: The GCA Portal serves as a tool for climate diplomacy. It allows countries to showcase their climate leadership, provides evidence for policy effectiveness, and can be used to encourage other countries to step up their efforts.

Challenges and Considerations

1. **Data Quality and Comparability**: Ensuring the quality, accuracy, and comparability of data is essential for effective data-driven climate governance. This requires the development of standardized metrics, methodologies, and reporting frameworks, as well as addressing data gaps in developing countries.
2. **Inclusivity**: Data-driven climate governance must be inclusive and accessible to all stakeholders. This means addressing the dominance of data from Western countries and ensuring that developing countries have the necessary resources and technical expertise to participate.
3. **Data Neutrality**: The GCA Portal, like any data-driven platform, is not neutral. The data it collects and presents reflect inherent biases and assumptions about what constitutes relevant climate action. It is crucial to critically examine these biases and ensure that diverse perspectives are considered.

12.2.2 Open Access of Data for Climate Justice

The concept of open access, central to the dissemination of scientific research, plays a crucial role in climate diplomacy. Open Access, as highlighted by Royal Society Publishing, promotes the free and unrestricted availability of research data, facilitating collaboration and innovation in addressing climate change.

Data Sharing for Climate Action

1. **Facilitating Interdisciplinary Research**: Open access to climate data enables researchers from diverse fields, like epidemiology and climate science, to collaborate on complex issues such as the impact of climate change on food security.
2. **Informing Policy Decisions**: Long-term climate data sets made accessible through open access initiatives help build accurate models and predictions. These insights are vital for policymakers in formulating effective climate strategies.
3. **Validating Theories**: Open data policies allow researchers to test and validate critical theories related to biodiversity and climate change, leading to a deeper understanding of our planet's complex systems.
4. **Fostering Collaboration**: Open access initiatives encourage collaboration between researchers from different disciplines and geographic locations, fostering a global effort to tackle climate change.

12.2.3 Overcoming and Accessing Data Gaps for Developing Countries

Despite advancements in data collection, significant gaps persist in developing countries, particularly regarding climate indicators and emissions data. The IMF's Data Gaps Initiative aims to address these deficits by improving the granularity and timeliness of climate indicators, which are crucial for crafting equitable and impactful policies. Comprehensive datasets on emissions, household income, and climate risks are essential for understanding the trade-offs faced by developing countries in reducing carbon emissions while maintaining economic growth (Cheng, 2023). The challenge is compounded by the unequal distribution of emissions among households, where energy-intensive consumption, such as fuel and transportation, significantly contributes to carbon emissions. Initiatives like the creation of unified datasets that integrate climate indicators with economic activity proxies—such as population density and night-time light intensity—are vital for providing spatially aggregated data to inform policy decisions (Gortan et al., 2023). Furthermore, the potential diffusion of renewable energy in developing countries offers a pathway to decarbonized development, although this requires technological advances, skilled human capital, and sound governance. However, the global north's control over resources and knowledge continues to perpetuate an unjust north/south narrative, emphasizing the need for equitable access to climate data and resources (Allam et al., 2022). Addressing these data gaps and ensuring equitable access to climate information are critical steps toward achieving global climate justice and effective climate mitigation strategies.

12.2.4 Bridging Data Gaps Can Help Tackle the Climate Crisis

Policymakers need better data to address climate change mitigation and adaptation effectively. The new Data Gaps Initiative, supported by G20 leaders, aims to enhance official statistics by making them more detailed and timely. This is especially crucial in the energy sector, the largest contributor to greenhouse gas emissions. Comprehensive data is necessary to monitor the progress of the energy transition and devise effective measures to achieve net-zero emissions. While economies must expand renewable energy sources and curb fossil fuel use, the pace of this shift remains insufficient. Beyond a lack of policy ambition in many cases, there is also a shortage of comprehensive, internationally comparable data to accurately track progress.

The initiative also emphasizes the need for improved data on climate indicators, household income, and wealth distribution to help policymakers better understand the wide-ranging impacts of climate change and associated risks. By expanding national statistics beyond GDP, policymakers can more effectively assess the distributional consequences of climate policies. The IMF, working alongside other international agencies, will coordinate efforts to fill these data gaps by leveraging collective expertise and existing initiatives. This work aligns closely with other IMF projects, such as the Climate Indicators Dashboard and the joint initiative on sustainable finance alignment. The Data Gaps Initiative represents a crucial step toward addressing data deficiencies that currently hinder effective climate action, recognizing the importance of comprehensive and internationally comparable data to guide policy decisions and track progress in combating the climate crisis.

12.3 CLIMATE FINANCE AND CLIMATE DIPLOMACY

Climate finance is pivotal in supporting mitigation and adaptation efforts, especially in developing nations, as it facilitates the transition to low-carbon economies and enhances resilience against climate impacts. Mechanisms such as the Green Climate Fund (GCF) and sovereign green bonds play crucial roles in mobilizing resources for renewable energy and sustainable infrastructure projects. The GCF, for instance, is designed to accelerate low-emission and climate-resilient development by balancing oversight with funding accessibility, thus fostering international collaboration between developed and developing countries (The Green Climate Fund, 2022). Despite these efforts, developed countries have not yet fulfilled their commitment to provide $100 billion annually, a pledge made at the Copenhagen conference in 2009, highlighting ongoing challenges in equity and accountability within international climate finance. Green bonds are another significant tool, offering a pathway to bridge the climate finance gap by encouraging investments in environmentally friendly projects, although their development is hindered by the lack of a universal policy framework (Banerjee, 2019). Furthermore, integrating environmental, social, and governance (ESG) factors into investment decisions is essential for advancing green finance and achieving sustainable development goals (Fu et al., 2023). The Paris Agreement underscores the importance of international finance, recognizing the differentiated responsibilities of nations and the need for substantial investment flows to decarbonize the global energy system (Steckel et al., 2017). Overall, aligning financial incentives with sustainable outcomes and fostering transparency are critical for advancing climate finance and promoting a greener future.

Sovereign green bonds (SGBs) play a significant role in the global effort to address climate change by providing a financial mechanism that aligns investment with ecological priorities. These bonds, issued by governments with sovereign guarantees, are designed to fund projects that mitigate climate change impacts, such as renewable energy initiatives and greenhouse gas emission reductions (Yadav et al., 2024). The issuance of SGBs has gained momentum since the Paris Agreement in 2015, with cumulative volumes reaching $1.6 trillion by June 2021, primarily driven by the United States, China, and Europe (Bohner, 2022). In developing countries, sovereign climate-aligned bonds (SCBs) have been instrumental in mobilizing capital for climate investments, despite challenges such as higher transaction costs and structural barriers in private markets (Chavarría-Flores & Warren, 2024). In Europe, SGBs signal a long-term governmental commitment to sustainability, encouraging similar actions in the private sector (Chesini, 2023). The global green bond market, although still in its nascent stage, is crucial for achieving UN sustainable development goals, with China emerging as a leading issuer (Basak & Sana, 2024). Overall, SGBs provide an unconventional yet effective tool for linking fiscal policy with science-based solutions to address pressing global sustainability issues (Yadav et al., 2024).

12.3.1 Examples of Climate Finance

Developing countries can access some of their climate finance in the form of grants from the GEF, the GCF, and the AF. Governments and the private sector can also access market-based and concessional loans from financial institutions such as the World Bank, the African Development Bank, and the Inter-American Development Bank, among others. These grants and loans can be used to invest in projects that reduce, absorb, or prevent greenhouse gas (GHG) emissions, such as renewable energy power plants, electric buses, and forest conservation, or projects that build resilience to climate change, such as establishing early warning systems, increasing coastal protection, enhancing the resilience of agriculture and food systems, and building infrastructure that can withstand storms and flooding.

Moreover, governments, through their budgeting processes, can allocate funding to priority climate actions, such as those set out in their national climate pledges (referred to as NDCs under the Paris Agreement), or issue SGBs to fund those projects. Sovereign bonds are loans that governments take from a pool of investors in exchange for regular interest payments over a certain number of years. At the end of this period, when the bond reaches maturity, the government returns the initial investment to the investors.

Governments can also mobilize climate finance from carbon trading and carbon taxes. Through carbon trading, GHG emissions are quantified into carbon credits that can be bought and sold. One tradable carbon credit equals one ton of carbon dioxide, or the equivalent amount of a different GHG reduced, sequestered, or avoided. Carbon credits can be bought by countries or private companies that want to enhance their GHG emissions reduction efforts. Carbon taxes are typically applied to discourage the use of products and services with big carbon footprints. For instance, a tax can be applied on gasoline at the pump or on electricity generated from fossil fuels. Proceeds from these taxes can be used to invest in renewable energy, forest conservation, and other forms of climate action.

12.3.2 World Bank Group's Climate Finance

1. The World Bank Group delivered a record $31.7 billion in fiscal year 2022 (FY22) to help countries address climate change. This is a 19% increase from the $26.6 billion all-time high in financing reached in the previous fiscal year.
2. The Bank Group continues to be the largest multilateral financier of climate action in developing countries.
3. Financing for climate action in FY22 exceeded the target set in the Group's Climate Change Action Plan for 2021–2025 to deploy an average of 35% of the institution's financing in support of climate action: it reached 36% of total Bank Group financing.
4. IBRD and IDA together delivered $26.2 billion in FY22 in climate finance.
5. Building resilience to climate shocks is a priority. Nearly half of the Bank's finance—$12.9 billion—supported investments in adaptation and resilience.
6. IFC, the private sector arm of the World Bank Group, delivered an unprecedented $4.4 billion in climate finance and mobilized an additional $3.3 billion from other sources.
7. MIGA, the World Bank Group's political risk insurance and credit enhancement arm, delivered $1.1 billion in climate finance in FY22.
8. The Bank's first Country Climate and Development Reports (CCDRs) are examples of our ongoing effort to help countries integrate climate and development objectives and prioritize the most impactful actions that can reduce GHG emissions and boost adaptation,
9. The Bank Group's climate finance is calculated based on the agreed joint Multilateral Development Bank methodology: we count only the share of financing directly tied to climate action across all Bank Group projects.
10. In FY22, the Bank is also making its first disclosure aligned with the framework developed by the Task Force on Climate-related Financial Disclosures (TCFD) in its annual report, joining IFC and MIGA, which have produced TCFD-aligned disclosures since Fiscal Year 2018 and Fiscal Year 2021, respectively.

12.3.3 UNEP Finance Initiative

The United Nations Environment Program Finance Initiative (UNEP FI) is a global partnership established in 1992 between the UN and over 500 financial institutions. This collaboration aims to drive the integration of sustainability into the global financial system. UNEP FI has played a pioneering role in developing frameworks and tools that help banks, insurers, and investors address ESG challenges. Through voluntary participation, financial institutions can access resources and guidance to align their practices with sustainable and inclusive economic principles.

With a secretariat based in Geneva, UNEP FI's influence spans across continents. Their global programs are tailored to specific regions, including Africa & the Middle East, Asia Pacific, Europe, Latin America & the Caribbean, and North America. Notably, UNEP FI was the first organization to engage the finance sector on sustainability and fostered the development of the Principles for Responsible Investment, which has since become a leading advocate for responsible investment worldwide.

UNEP FI convenes financial institutions to advance knowledge and practice in areas including climate, nature, pollution & circular economy, the SDGs & impact, and social issues. UNEP FI has a long-standing program on climate change, which has accelerated action to contribute to achieving the goals of the UN Paris Agreement. Aligning all public and private financing with pathways to net-zero by 2050 is critical to limit the average global temperature rise to 1.5°C above pre-industrial levels by 2100.

UNEP FI convenes three finance sector 'net-zero' alliances and is working with key stakeholders to help members align portfolios with science-based GHG emissions pathways through interim targets:

1. Net-Zero Asset Owner Alliance ('AOA'), launched in 2019, with a membership of more than 80 institutional asset owners (pension funds, insurers, sovereign wealth funds) with over USD 10 trillion in assets, about 7% of global investment.
2. Net Zero Banking Alliance ('NZBA'), launched in 2021, with over 130 banks and USD 65 trillion, about half of global banking industry assets. The majority of NZBA members are PRB signatories who are setting ambitious climate targets.
3. Net-Zero Export Credit Agencies Alliance ('NZECA') was launched in 2023 by five founding members and three affiliate members that together supported an estimated $120 billion in international trade in 2022.

12.4 IMPACTS OF CLIMATE FINANCE

Climate finance has been instrumental in supporting various impactful projects aimed at mitigating climate change and promoting sustainable development. In India, climate finance has facilitated the growth of micro, small, and medium enterprises (MSMEs) through projects focused on solar power plants, renewable energy, green machinery, waste management, electric vehicles, clean energy, recycling, poverty alleviation, and energy conservation. These initiatives are part of India's broader commitment to achieving net-zero emissions by 2070 and are supported by financial instruments such as climate funds, green bonds, and impact financing. The Reserve Bank of India (RBI) has prioritized eco-friendly initiatives for priority sector lending, thereby fostering the green economy and aligning with sustainable development goals (SDGs) and ESG criteria (Ranjan & Chaturvedi, 2023). In Bangladesh, the Bangladesh Climate Change Trust Fund (BCCTF) has funded 236 projects by 2014, focusing primarily on infrastructure development, which is one of the six pillars of the Bangladesh Climate Change Strategy and Action Plan (BCCSAP) established in 2009. These projects, managed by various government organizations, aim to mitigate the impacts of climate change by enhancing the country's resilience to natural hazards. However, the emphasis on infrastructure development has overshadowed other critical areas of the BCCSAP, indicating a need for a more balanced approach in addressing climate change impacts (Das, 2017). Together, these examples illustrate the diverse applications of climate finance in supporting projects that address both mitigation and adaptation to climate change across different regions.

12.5 CHALLENGES AND OPPORTUNITIES IN CLIMATE DIPLOMACY

Climate diplomacy faces significant challenges, primarily due to the complex interplay of economic disparities, political inertia, and mutual recrimination between developed and developing nations. One of the core issues is the malignant incentive structure of climate mitigation, which encourages free-riding and discourages deep commitments to GHG emissions reduction, thereby impeding effective international cooperation (Daniel, 2023). The North–South divide further complicates negotiations, as developing countries fear economic constraints while industrialized nations are reluctant to make sacrifices without reciprocal actions from the developing world (Roberts & Parks, 2006). This divide is exacerbated by the inequitable distribution of emissions rights under frameworks like the Kyoto Protocol, which fail to account for the disproportionate contributions of wealthier individuals globally to climate change.

To overcome these barriers, trust-building and equitable frameworks are essential. Proposals include shifting the focus from cap-and-trade systems to development pathways that align climate policies with domestic development goals (Méjean et al., 2015). Additionally, a per-capita emissions allowance system could offer a more equitable solution, allowing flexibility through national trading of unused allowances while maintaining a constant global reduction target. Enforcement mechanisms are also crucial; however, they must be credible and legitimate, potentially through sector-specific agreements that ensure compliance without over-reliance on trade restrictions (Barrett, 2008). Addressing these challenges requires a concerted effort to recognize and mitigate global inequalities, fostering a cooperative environment for effective climate diplomacy (Roberts & Parks, 2006).

Designing robust enforcement mechanisms to ensure compliance with global climate commitments involves a multifaceted approach that integrates mechanism design principles, legal frameworks, and innovative technologies. A mechanism design perspective suggests that optimal environmental agreements must balance the free-rider problem with the need to incentivize countries to ratify and comply with the agreements. This can be achieved through a simple menu of options that are both attractive and robust, although enforcement and commitment limitations may still pose challenges, potentially leading to a "business as usual" scenario (Martimort & Sand-Zantman, 2016a). The Paris Agreement exemplifies a multilateral effort to incorporate compliance mechanisms, establishing a committee under Article 15 to facilitate implementation and promote adherence to its provisions. However, this committee's exact competencies and functions are still under negotiation, highlighting the ongoing need for detailed procedural development (Voigt, 2016).

Additionally, blockchain technology offers a novel approach to enforcement, where smart contracts can serve as self-enforcing mechanisms without relying on traditional legal systems. These digital courts can identify and penalize non-compliance, ensuring that agreements are upheld as long as there is a sufficient presence of honest agents within the system (Matsushima & Noda, 2020). By combining these strategies, a comprehensive enforcement mechanism can be developed that not only encourages compliance but also adapts to the evolving landscape of international climate agreements.

Emerging technologies such as satellite monitoring, artificial intelligence (AI), and advanced analytics are pivotal in enhancing climate action by improving data accuracy, enabling real-time tracking, and supporting adaptive decision-making. AI, in particular, offers revolutionary solutions for environmental sustainability by optimizing natural resource management and monitoring environmental changes, such as deforestation and pollution levels (Jain & Mitra, 2024). These technologies are integral to achieving the SDGs, particularly SDG-9, which emphasizes building resilient infrastructure and fostering innovation (Colombage, 2019). Initiatives like the Alliance of

Excellence for Research and Innovation on Aephoria (AE4RIA) and the UN SDSN Global Climate Hub (GCH) exemplify how innovative research can guide policymaking and drive global efforts toward a sustainable future (Koundouri, 2024). However, the deployment of AI in climate action is not without challenges, including data privacy concerns, algorithmic bias, and the carbon footprint associated with AI research (Jain & Mitra, 2024; Cowls et al., 2021). Despite these challenges, AI's potential to enhance understanding and combat climate change is significant, provided that effective governance and evidence-based policies are implemented to mitigate its environmental impact (Cowls et al., 2021). The 28th Conference of the Parties (COP28) further underscores the importance of transitioning away from fossil fuels, with AI playing a crucial role in crafting comprehensive climate strategies across various sectors. Thus, harnessing these technologies through collaborative efforts among policymakers, researchers, and practitioners is essential for strengthening climate diplomacy and achieving SDGs.

12.6 CONCLUSION

International collaboration and open data are now central pillars in the global response to climate change. This chapter has demonstrated how mechanisms such as the Conference of the Parties (COP), the Kyoto Protocol, and the Paris Agreement have shaped climate diplomacy—setting targets, mobilizing finance, and driving both mitigation and adaptation efforts worldwide. Robust, transparent data collection and sharing—powered by advanced technologies—are enabling more accurate tracking of emissions, progress, and the effectiveness of policies. Climate finance tools, such as the Green Climate Fund and green bonds, are unlocking resources for developing countries, although challenges remain in meeting commitments and ensuring equitable access.

The chapter also highlighted persistent obstacles, including the free-rider problem, divergent national interests, and the North-South divide, which complicate negotiations and the fair distribution of climate responsibilities. However, lessons from past agreements and recent COPs point to the need for more flexible, inclusive, and enforceable frameworks. Open access to climate data, capacity building for developing nations, and innovative enforcement mechanisms are essential for building trust and accelerating progress. Ultimately, effective climate diplomacy depends on global solidarity, transparent governance, and the willingness of all nations to align their interests with a sustainable future.

As we move to the next chapter, the focus shifts from the architecture of global cooperation to the tangible benefits of compliance and climate action. Chapter 13 examines how robust governance, economic incentives, and technological innovation not only drive emissions reductions but also yield economic, social, and environmental benefits. The journey continues from negotiating commitments to realizing the rewards of collective climate action.

REVIEW QUESTIONS

1. How do global carbon emissions vary by country and sector?
2. Which countries are the largest emitters of GHGs, and why?
3. How can emissions be reduced in high-emitting cities worldwide?
4. What role does the carbon disclosure project (CDP) play in promoting environmental transparency?
5. How can data-sharing efforts support global carbon reduction initiatives?
6. What are some effective strategies for decarbonization at a national level?
7. How does the COP-28 conference set goals for future climate actions globally?
8. How do developed and developing countries differ in their approaches to carbon reduction?
9. What are the implications of carbon neutrality pledges made by various nations?
10. How does urbanization affect global carbon emissions, and what solutions exist?

REFERENCES

Akhtar, S., Shaima, S., Rita, G., Rashid, A., Rashed, A. (2024). Navigating the global environmental agenda: A comprehensive analysis of COP conferences, with a spotlight on COP28 and key environmental challenges. *Nature Environment and Pollution Technology*. https://doi.org/10.46488/nept.2024.v23i03.057

Allam, Z., Jones, D., Roös, P. (2022). Addressing knowledge gaps for global climate justice. *Geographies*. https://doi.org/10.3390/geographies2020014

Atici, C. (2021). Reconciling the flexibility mechanisms of climate policies towards the inclusiveness of developing countries: Commitments and prospects. *Environment, Development and Sustainability*. https://doi.org/10.1007/S10668-021-01834-X

Banerjee, A. (2019). Financing the climate change through green (climate sustainable) bonds. In *2019 Advances in Science and Engineering Technology International Conferences (ASET)* (pp. 1–4). IEEE. https://doi.org/10.1109/ICASET.2019.8714408

Barrett, S. (2008). Climate treaties and the imperative of enforcement. *Oxford Review of Economic Policy*. https://doi.org/10.1093/OXREP/GRN015

Basak, S.K., Sana, A.K. (2024). How Does Green Finance Reduce the Adverse Effect of Global Climate Change Issues: An Investigation. In B. Paul & S. Poddar (Eds.), *Sustainable Partnership and Investment Strategies for Startups and SMEs* (pp. 30-48). IGI Global Scientific Publishing. https://doi.org/10.4018/979-8-3693-2197-3.ch003

Bellido, F.J. (2022). COP27 Climate Change Conference: Urgent action needed for Africa and the world. *Lancet Oncology*. https://doi.org/10.1016/s1470-2045(22)00645-3

Blenkinsop, S. (2024). *Review: Navigating the climate conferences: Comparing the carbon footprint of private jet travel and other modes of transport to COP28* [Preprint]. UCL Open: Environment Preprints. https://doi.org/10.14324/111.444/ucloe.3036.r393

Bohner, S. (2022). Green bonds: Quick insight into one of the most implemented sustainable debt instruments to tackle climate change. *Ciencias Administrativas,* 21, Article e117. https://doi.org/10.24215/23143738e117

Bovenzi, N., Duran-Silva, N., Massucci, F. A., Multari, F., Parra-Rojas, C., Pujol-Llatse, J. (2022). *Mapping Climate Change Research via Open Repositories & AI: advantages and limitations for an evidence-based R&D policy-making*. Cornell University. https://doi.org/10.48550/arxiv.2209.09246

Cai, W., Zhang, S., Zhang, C., Dai, F., Lu, B., Luo, Y., Gong, P. (2024). 《cop28气候与健康阿联酋宣言》解读. *Chinese Science Bulletin*. https://doi.org/10.1360/tb-2023-1355

del Castillo, T. F. (2018). El acuerdo de París sobre el cambio climático : sus aportaciones al desarrollo progresivo del Derecho internacional y las consecuencias de la retirada de los Estados Unidos. *Redia-Giornale Di Zoologia*. https://doi.org/10.17103/REDI.70.1.2018.1.01

Chavarría-Flores, A., Warren, P. (2024). Sovereign climate bonds: Policy innovation for a just transition in developing countries. *Journal of Climate Finance*. https://doi.org/10.1016/j.jclimf.2024.100041

Cheng, X. (2023). Exploring CO2 emission in developing countries by PCA and time series analysis. In *Proceedings of the Second International Conference on Statistics, Applied Mathematics, and Computing Science (CSAMCS 2022), Proc. SPIE* 12597, 1259718. https://doi.org/10.1117/12.2672670

Chesini, G. (2023). *Sovereign Green Bonds in Europe: Are They Effective in Supporting the Green Transition?* In S. Carbó-Valverde & P. J. Cuadros-Solas (Eds.), *New challenges for the banking industry* (pp. 151–171). Palgrave Macmillan. https://doi.org/10.1007/978-3-031-32931-9_8

CICC Research, CICC Global Institute. (2022). Tackling climate change: Global cooperation and China's commitment In *Guidebook to Carbon Neutrality in China* (pp. 317–335). Springer. https://doi.org/10.1007/978-981-16-9024-2_13

Colombage, S. (2019). Harnessing science, technology and innovation for sustainable development. *Sri Lanka Journal of Social Sciences*. https://doi.org/10.4038/SLJSS.V42I2.7967

Cowls, J., Cowls, J., Tsamados, A., Taddeo, M., Taddeo, M., Floridi, L., Floridi, L. (2021). The AI gambit: Leveraging artificial intelligence to combat climate change-opportunities, challenges, and recommendations. *Ai & Society*. https://doi.org/10.1007/S00146-021-01294-X

Darby, S., Hammond, G.P., Wu, J. (2024). Stocktaking global warming – the outcomes of the 2023 Dubai climate summit (COP28). *Proceedings of the Institution of Civil Engineers*. https://doi.org/10.1680/jener.24.00005

Das, A. (2017). Appraising climate change impact mitigation standards to ground realities: The lessons from bangladesh climate change trust funded projects. *Social Science Research Network*. https://doi.org/10.2139/SSRN.3641765

Depledge, J. (2024). The future of negotiations under the climate change COP (Conference of the Parties): Implementation is not enough. *Dialogues on Climate Change*, 1(1), 12–17. https://doi.org/10.1177/29768659241293212

Doukas, H., Nikas, A., González-Eguino, M., Arto, I., Anger-Kraavi, A. (2018). From integrated to integrative: Delivering on the Paris Agreement. *Sustainability*. 10(7), 2299. https://doi.org/10.3390/SU10072299

Fu, C. T., Lu, L., Pirabi, M. (2023). Advancing green finance: A review of sustainable development. *DESD*. https://doi.org/10.1007/s44265-023-00020-3

Getu, M. (2012). Accommodating the interests of developing countries in the climate change regime: Lessons from the ozone layer regime. *Mizan Law Review*. https://doi.org/10.4314/MLR.V6I1.1

Gortan, M., Testa, L., Fagiolo, G., Lamperti, F. (2023). A unified dataset for pre-processed climate indicators weighted by gridded economic activity. *Scientific Data*. https://doi.org/10.1038/s41597-024-03304-1

Grigoryev, L., Makarov, I., Sokolova, A., Pavlyushina, V., Stepanov, I. (2020). Climate change and inequality: How to solve these problems jointly? *International Organisations Research Journal*. https://doi.org/10.17323/1996-7845-2020-01-01

Habiba, U., Abedin, Md. A., Shaw, R. (2016). Food security, climate change adaptation, and disaster risk. In Uitto, J., Shaw, R. (Eds.), *Sustainable Development and Disaster Risk Reduction. Disaster Risk Reduction*. Springer, Tokyo. https://doi.org/10.1007/978-4-431-55078-5_6

Introne, J., Laubacher, R., Malone, T.W. (2011). Enabling open development methodologies in climate change assessment modeling. *IEEE Software*. https://doi.org/10.1109/MS.2011.115

Jain, V., Mitra, A. (2024). Integrative innovations. *Practice, Progress, and Proficiency in Sustainability*. https://doi.org/10.4018/979-8-3693-6336-2.ch014

Kindermann, S., Schintke, F., Fritzsch, B. (2012). *A Collaborative Data Management Infrastructure for Climate Data Analysis*. Geophysical Research Abstracts, 14 (EGU201), p. 10569.

Koundouri, P. (2024). Innovative climate research for guiding policy-making. *Open Access Government*. https://doi.org/10.56367/oag-042-11363

Leys, D., & Anderson, R. (2022). The Green Climate Fund: A unique financing vehicle among international organizations. In *Funding international development organizations* (Chapter 9). Brill | Nijhoff. https://doi.org/10.1163/9789004460010_010

Marsden, G., Schwanen, T. (2024). Planning to fail? How science can respond to reduced climate mitigation ambition. *npj Sustainable Mobility and Transport*. https://doi.org/10.1038/s44333-024-00002-8

Martimort, D., Sand-Zantman, W. (2016a). A mechanism design approach to climate-change agreements. *Journal of the European Economic Association*. https://doi.org/10.1111/JEEA.12150

Maslin, M.A., Lang, J.W., Harvey, F. (2023). A short history of the successes and failures of the international climate change negotiations. *UCL Open Environment*. https://doi.org/10.14324/111.444/ucloe.000059

Matsushima, H., Noda, S. (2020). *Mechanism design with blockchain enforcement* (CARF F-Series No. CARF-F-474). Center for Advanced Research in Finance, Faculty of Economics, The University of Tokyo.

Méjean, A., Lecocq, F., Mulugetta, Y. (2015). Equity, burden sharing and development pathways: Reframing international climate negotiations. *International Environmental Agreements: Politics, Law and Economics, 15*(4), 387–402. https://doi.org/10.1007/s10784-015-9284-x.

Mohan, A., Geden, O., Fridahl, M., Buck, H.J., Peters, G.P. (2021). UNFCCC must confront the political economy of net-negative emissions. *One Earth*. https://doi.org/10.1016/J.ONEEAR.2021.10.001

Müller, B. (2002). *Equity in Climate Change: The Great Divide* (EV31; Publisher's version). Oxford Institute for Energy Studies.

Murombedzi, J.C., Chikozho, C. (2023). Westphalian sovereignty and the free-rider conundrum in the atmospheric commons: Examining global governance regimes for addressing climate change adaptation. *The International Journal of the Commons*. https://doi.org/10.5334/ijc.1159

Onysiuk, S. (2024). Mechanisms of international economic cooperation to support the global energy transition: Ways to improve. Вчені Записки. https://doi.org/10.33111/vz_kneu.36.24.03.22.152.158

Osorio, R.G. (2008). *Free access to primary data should be a right* (One Pager No. 72). International Policy Centre for Inclusive Growth. https://ideas.repec.org/p/ipc/opager/72.html.

Oxford Analytica. (2022). Global climate goals are on course to be missed. *Emerald Expert Briefings*. https://doi.org/10.1108/oxan-db268638

Papanikolaki, E., Fokkas, E. (2024). Climate ambition of the conference of parties, a content analysis. *Journal of Student Research*. https://doi.org/10.47611/jsr.v13i2.2496

Piazza, A. (2021). Collective responsibility in the cooperative governance of climate change. *Sustainability*. https://doi.org/10.3390/SU13084363

Ranjan, R., Chaturvedi, P. (2023). An analysis of impact of climate finance on MSMEs and entrepreneurial growth. *Adhyayan: A Journal of Management Sciences*. https://doi.org/10.21567/adhyayan.v12i2.05

Ransom, S. (2024). Guest editorial: Returning from Dubai: reflections on water in the climate conference. *World Water Policy*. https://doi.org/10.1002/wwp2.12170

Roberts, J.T., Parks, B.C. (2006). *A Climate of Injustice: Global Inequality, North-South Politics, and Climate Policy* (Global Environmental Accord: Strategies for Sustainability and Institutional Innovation). MIT Press.

Sá, C.M., Grieco, J. (2016). Open data for science, policy, and the public good. *Review of Policy Research*. https://doi.org/10.1111/ROPR.12188

Schmidt, R., Drupp, M.A., Nesje, F., Hoegen, H. (2022). *Testing the Free-Rider Hypothesis in Climate Policy* [Preprint]. arXiv. https://arxiv.org/abs/2211.06209.

Schneider, D.P., Deser, C., Fasullo, J.T., Trenberth, K.E. (2013). Climate data guide spurs discovery and understanding. *Eos, Transactions American Geophysical Union*. https://doi.org/10.1002/2013EO130001

Stavins, R.N. (2024). What really happened at COP-28 in Dubai: A personal view. *Economics and Policy of Energy and the Environment*. https://doi.org/10.3280/efe2024-001001

Steckel, J.C., Jakob, M., Flachsland, C., Kornek, U., Lessmann, K., Edenhofer, O. (2017). From climate finance toward sustainable development finance. *Wiley Interdisciplinary Reviews: Climate Change*. https://doi.org/10.1002/WCC.437

Szasz, D., Neuner, J., George, S., Eichholzer, A. (2021). *Accelerating Progress on Tackling the Climate Crisis through Data Collaboration*. Open Data Institute. https://doi.org/10.61557/qiun1896

Taylor, G., Vink, S. (2021). Managing the risks of missing international climate targets. *Climate Risk Management*. https://doi.org/10.1016/J.CRM.2021.100379

Trivedi, A., Jolly, S. (2022). Construing climate change adaptation as global public good under international law: Problems and prospects. *The Liverpool Law Review*. https://doi.org/10.1007/s10991-022-09317-3

Varyvoda, Y. (2024). Conference of the parties (COP) 28 takeaways at the intersection of health, climate, and food. *Nutrition Reviews*. https://doi.org/10.1093/nutrit/nuae083

Voigt, C. (2016). The compliance and implementation mechanism of the Paris Agreement. *Review of European, Comparative and International Environmental Law*. https://doi.org/10.1111/REEL.12155

Wang, C., Jin, Z., Cai, W. (2017). Incentive to freeride in international climate cooperation. *Environmental Economics*, Volume 4, Issue 2, 2013. pp 88–92.

Yadav, A., Manda, V.K., Shivani, S., Sangwan, V., Vambol, S. (2024). Sovereign green bonds as an unconventional tool to address climate change. *Ecological Questions*. https://doi.org/10.12775/eq.2024.028

13 The Benefit of Compliance and Climate Action

Compliance and climate action offer numerous benefits that span economic, environmental, and social spheres, contributing to a sustainable and resilient future. Economically, companies can achieve significant cost savings through energy-efficient practices and technologies, which reduce energy consumption and operational costs, and by minimizing waste through efficient resource use. Staying ahead of regulations helps avoid fines and penalties associated with non-compliance, while robust climate policies can secure supply chains against disruptions caused by extreme weather, leading to better insurance and financing terms. Additionally, climate action drives innovation by developing green products and services, opening new markets, and enhancing competitive advantage. An environmentally responsible reputation also boosts sales and market share, as consumers increasingly favor such companies.

Environmentally, climate action leads to a reduction in greenhouse gas emissions, mitigating the impacts of climate change, such as global warming, sea-level rise, and extreme weather events. It also promotes the conservation of natural habitats and biodiversity, and reduces pollution, contributing to cleaner air, water, and soil. Socially, improved air quality from reduced pollutants benefits public health by reducing respiratory and cardiovascular issues. Climate measures also protect communities from climate-related hazards, fostering resilience. Investments in renewable energy and other green industries create jobs and offer workforce training in new technologies and sustainable practices, enhancing employability and economic stability.

Furthermore, inclusive climate policies help protect vulnerable populations from climate impacts and ensure fair access to resources and opportunities, promoting social equity. In the long term, economies that integrate climate action are better prepared to adapt to future conditions and shocks, fostering sustainable growth and stability. These actions ensure that future generations inherit a stable climate and a healthy environment, supporting intergenerational equity. In summary, compliance and proactive climate action provide extensive economic, environmental, and social benefits, paving the way for a sustainable and resilient future.

13.1 ECONOMIC INCENTIVES

Economic incentives for compliance and climate action are essential tools to drive the transformation toward a sustainable and resilient global economy. By aligning financial interests with environmental goals, these incentives encourage businesses and governments to adopt practices that reduce emissions, protect natural ecosystems, and promote sustainable development. One way is through nature performance bonds. These financial instruments reschedule debt based on the achievement of specific environmental targets, such as reforestation or emission reductions. They provide countries, especially those with high debt burdens, a way to invest in sustainability while managing their financial obligations. The world needs to shift its spending priorities toward a healthier planet. Right now, $5.2 trillion is **spent** each year subsidizing fossil fuels. Investing in solutions is much cheaper, such as reversing nature loss and reaching net-zero emissions, which would cost just $700 billion a year—less than 1% of global income. Governments should move beyond using just GDP as an indicator of economic success and also adopt more inclusive approaches that measure well-being and prosperity across generations. Implementing frameworks like the Task Force on Climate-related Financial Disclosures (TCFD) and the Taskforce on

DOI: 10.1201/9781003479666-13

Nature-related Financial Disclosures (TNFD) ensures that companies and financial institutions are transparent about their environmental impacts. This can attract investors looking for sustainable opportunities and push businesses toward greener practices.

13.1.1 Nature Performance Bonds

Nature performance bonds, such as debt-for-nature swaps, represent a strategic financial tool that aligns fiscal obligations with environmental objectives, enabling countries to invest in sustainability while managing their debt burdens. These bonds incentivize nations to achieve specific environmental targets, such as reforestation or emissions reductions, by integrating financial and ecological goals. The concept of environmental bonds has evolved to address the complex issue of environmental debt, which includes both spatial and temporal dimensions, highlighting the need for sustainable financing mechanisms to mitigate the environmental impacts of past and present generations (Kamel & Khalaf, 2024). The issuance of eco-friendly bonds has shown a positive correlation with economic growth, although challenges remain in ensuring that these bonds consistently drive sustainable economic expansion (Bajra & Wagner, 2024). In Ukraine, for instance, debt-for-nature swaps have been proposed as a means to reduce the national debt burden while expanding investments in environmental programs, demonstrating the potential of these instruments to facilitate green public finance management (Naumenkova et al., 2023). Despite the challenges, the global trend toward green finance, including climate bonds, underscores the growing recognition of the need for innovative financial solutions to support environmental protection and sustainability goals (Kashif et al., 2023). However, the substantial investment required and the lengthy payback periods associated with environmental projects necessitate a unique approach to financing, which nature performance bonds can provide by aligning investment strategies with both economic and environmental objectives (Kashif et al., 2023).

13.1.2 Reforming Subsidies and Metrics

Redirecting subsidies from fossil fuels to renewable energy presents a significant opportunity for global economic and environmental benefits. Fossil fuel subsidies, estimated at around USD 425 billion, represent a substantial financial resource that could be reallocated to support renewable energy and energy efficiency initiatives, potentially saving trillions of dollars annually (Merrill et al., 2017). The transition to a low-carbon energy industry is critical, with investment needs estimated between 5 trillion and 7 trillion per year, highlighting the importance of capital markets and private sector involvement to bridge this gap (de Putte et al., 2020). The TCFD and TNFD play crucial roles in promoting transparency and attracting investment in sustainable practices, aligning with the growing investor demand for positive environmental, social, and governance impacts (de Putte et al., 2020). Technological innovations in renewable energy, such as advancements in solar photovoltaics and wind turbines, have improved efficiency and affordability, further supporting the shift toward sustainable energy systems (Uzondu & Lele, 2024). Policy frameworks, including feed-in tariffs and tax incentives, have been instrumental in driving renewable energy adoption, while international collaborations, such as those facilitated by IRENA and the Paris Agreement, are essential for global deployment (Uzondu & Lele, 2024). Reforming fossil fuel subsidies could lead to an average reduction in national greenhouse gas emissions by 11%, with additional benefits from modest resource recycling into renewables (Merrill et al., 2015). Overall, a multifaceted approach involving policy reform, technological innovation, and international cooperation is necessary to accelerate the transition to a sustainable and resilient energy system.

13.1.3 Market-Based Instruments

Market-based instruments such as carbon taxes and cap-and-trade systems are pivotal in creating economic incentives for emissions reductions, fostering both environmental sustainability and economic development. Carbon taxes penalize carbon-intensive activities, generating revenue that can be reinvested into green initiatives, while cap-and-trade systems set emission limits and allow entities to trade allowances, promoting cost-effective compliance. The effectiveness of these instruments is evident in their ability to drive green innovation, as demonstrated by China's carbon emission trading scheme (ETS), which significantly increases green innovation through invention and utility model patents (Feng et al., 2024). Furthermore, carbon pricing mechanisms, including both carbon taxes and cap-and-trade systems, are recognized for their potential to mitigate climate change while supporting economic growth and structural transformation (Simpa et al., 2024). However, the proliferation of diverse carbon pricing schemes can lead to a complex patchwork, necessitating greater international cooperation to harmonize efforts and address issues such as cross-border carbon leakage (World Trade Organization, 2023). The European and Chinese ETS have shown varying degrees of success in reducing pollution and impacting GDP, highlighting the importance of integrating carbon markets with other policies and ensuring environmental integrity and cost-effectiveness (Ratra, 2023). Despite these successes, challenges such as greenwashing and the continued reliance on non-renewable sources in some systems underscore the need for robust verification and ethical considerations in the design and implementation of carbon pricing strategies. Overall, when effectively designed and equitably implemented, market-based instruments can align environmental sustainability with economic development goals, serving as a cornerstone of global climate governance (Simpa et al., 2024).

13.2 ENVIRONMENTAL HEALTH AND SAFETY

Environmental compliance starts with measurement; without the measuring of environmental stressors and the role that they play on not only the environmental regulations but the health of the environment itself, compliance would never be able to begin. Measurement can be used in a multitude of different ways. Another way to ensure environmental compliance is being met worldwide is by performing audits. Environmental audits are a type of performance test that entities can either do internally or hire an outside company to do. An audit will give a corporation/company an examination of their environmental impact and, oftentimes, lay down a baseline for them to assess what steps are next/necessary. Another way that companies can measure and keep track of what it is they are releasing into the environment/ecosystem is the scope of emissions.

Finding a base level of a healthy environment, or better known as Baseline Monitoring, is the key to understanding when an environment is in danger. Baseline monitoring is the initial state of environmental indicators before any compliance issue is identified. For instance, if there is a factory located on a river, the initial measurements taken are the baseline of a healthy ecosystem. This ultimately allows us to figure out what these elevated numbers must fall back to before the incident of stressors was introduced into the environment.

Another reason measurement is so key to environmental compliance is to see if the government action taking place has the intended impact. Often, however, there are times when other variables might be influencing these outcomes, other than the agency's impacts, and this is where evaluation methods come into play. One measurement method used to do so is regression analysis; this is a method that is used for estimating a relationship of a specified functional form between a response variable and one or more explanatory variables. In recent years, the EPA has supported the thought of regression analysis to understand factors affecting compliance assurance and ways to measure the impact of compliance assurance.

Positive outliers are another way to measure the positive impact of regulatory compliance. Regulatory units, parties, or locations that are trending in the correct "positive" direction are a suitable place to start measurements. When looking at these positive trends, it can be asked, "What is going so well here? Why is this site trending upwards when others are not? And what can be done at other locations to make them all trend positively like this site?" On the opposite end of the spectrum, negative outliers are a terrific way to decide what is not going correctly for other sites. Focusing on these negative outliers might shed light on patterns of problems, issues that need attention, and other key characteristics that might not be trending positively.

13.2.1 Real-Time Environmental Monitoring

Real-time environmental monitoring systems, integrating sensor networks, geographic information systems (GIS), and artificial intelligence, are pivotal in tracking emissions and environmental parameters continuously, thereby enabling early detection of regulatory breaches and facilitating adaptive management strategies. These systems leverage advanced technologies such as the Internet of Things (IoT), AI, and big data analytics to enhance the accuracy, transparency, and efficiency of environmental compliance, particularly in industries like oil and gas, where regulatory challenges are significant (Audu & Umana, 2024). In industrial environments, such as the chrome plating industry, real-time IoT-powered AI systems are employed to monitor and forecast air pollution, addressing health risks associated with hazardous pollutants like hexavalent chromium and volatile organic compounds (Ramadan et al., 2024). The integration of AI in ecological environment monitoring improves the effectiveness of supervision and protection, addressing issues like resource constraints and job pressure, thus enhancing decision-making quality in environmental governance (Li, 2023). The Industrial Internet of Things (IIoT) further supports environmental monitoring by enabling proactive approaches to emission and air pollution management, with applications in both stationary and dynamic settings (Lou et al., 2022). Additionally, IoT-generated open government data can be utilized to develop automated compliance systems, such as those monitoring river flow rates to ensure regulatory compliance, demonstrating the scalability and cost-effectiveness of these technologies in real-world applications (Miasayedava et al., 2020). Collectively, these advancements underscore the transformative potential of real-time monitoring technologies in achieving sustainable environmental management and compliance.

13.2.2 Environmental Audits

Environmental audits play a crucial role in ensuring compliance with regulations and internal sustainability standards, serving as both internal and external evaluations to identify gaps, recommend improvements, and ensure adherence to environmental policies. These audits can be categorized into compliance audits, management system audits, and functional audits, each targeting specific issues such as waste management or air quality. Compliance audits, for instance, are essential tools in modern environmental management, systematically assessing an organization's processes to verify adherence to legal requirements, as highlighted by Manfredini and Manfredini, who discuss the strategic importance of such audits in relation to standards like ISO 14001 and CONAMA Resolution 306/2002 (Manfredini & Manfredini, 2024). In Nairobi, environmental audits have been shown to enhance industrial performance by improving organizational effectiveness and ensuring regulatory compliance, despite challenges like corruption and high costs (Kariuki & Nzuki, 2024). Third-party environmental audits contribute significantly to business sustainability by promoting accountability and continuous improvement, leading to better resource efficiency and reduced emissions (Thoti et al., 2024). However, the effectiveness of these audits can be compromised by superficial compliance and the lack of uniform audit frameworks (Thoti et al., 2024). In Russia, the lack of detailed legal regulation hinders the implementation of environmental audits, despite their potential benefits in reducing external costs and maintaining environmental quality (Gumarova, 2024). Green auditing further emphasizes the role of audits in achieving environmental sustainability by improving performance and controlling environmental risks (Yousef & Al-Fatlawy, 2024). Overall, while environmental audits are pivotal in promoting sustainability and compliance, their success depends on overcoming regulatory and procedural challenges.

13.2.3 Environmental Impact Assessments

Environmental impact assessments (EIAs) are critical tools for evaluating the potential environmental, socio-economic, and cultural impacts of proposed projects, ensuring alignment with both local and international environmental standards. The EIA process is structured into several key stages, including scoping, impact evaluation, and mitigation planning, which collectively aim to integrate environmental considerations into decision-making and promote sustainable development (Moses et al., 2024; Ravi, 2024). Scoping involves identifying the potential impacts and alternatives, while impact evaluation assesses these effects in detail, considering various environmental, physical, chemical, biological, and socio-economic parameters. Mitigation planning then proposes measures to minimize adverse impacts, which may include redesigning project components or implementing additional studies. Despite its structured approach, the efficacy of EIAs can be limited by political influences, as final decisions often rest with authorities who may prioritize socio-economic interests over environmental concerns (Nadir & Ahmed, 2023). Furthermore, international legal frameworks sometimes lack clarity on EIA activation criteria and the duty of states to collaborate, particularly in cross-border contexts, which can lead to incomplete assessments (Pang, 2024). Nonetheless, EIAs remain a vital process, supported by international agreements and legal obligations, to safeguard the environment from the hazards of development projects (Nadir & Ahmed, 2023). Public participation and adherence to operational principles such as rigor and cost-effectiveness are essential for effective EIAs, ensuring that environmental audits and new technologies are leveraged to achieve sustainable outcomes (Moses et al., 2024).

Additionally, cultural considerations are essential, particularly in regions with diverse cultural landscapes, as projects can affect cultural heritage sites and practices. Public participation is a fundamental principle in the EIA process, ensuring that the voices of affected communities are heard and considered, which can help mitigate adverse socio-cultural impacts (Moses et al., 2024). Furthermore, EIAs should evaluate the socio-economic development prospects and the potential for both direct and indirect impacts on public health and community well-being (Morozov et al., 2022). In the context of water resource projects, EIAs should aim to maximize social and economic benefits while minimizing environmental impacts, thereby balancing development with environmental sustainability (Vyas et al., 2023). Overall, integrating socio-economic and cultural considerations into EIAs not only enhances the environmental assessment process but also promotes sustainable development by aligning projects with the needs and values of local communities (Ravi, 2024)

13.3 TECHNOLOGICAL INNOVATIONS IN COMPLIANCE

13.3.1 Advanced Emission Tracking

Technologies such as blockchain and IoT devices significantly enhance data accuracy and transparency in emissions reporting, streamlining compliance tracking and ensuring accountability across supply chains. Blockchain's decentralized and immutable nature provides a robust platform for secure and transparent data management, reducing fraud and enhancing accountability in carbon credit exchanges (Boumaiza & Maher, 2024). IoT devices play a crucial role in real-time monitoring and verification of carbon emissions, facilitating detailed tracking of environmental impacts across various stages of supply chains, such as in the meat industry (Ali et al., 2024). This integration allows for the capture of comprehensive data streams, enabling fine-grained emission tracking and promoting collaborative efforts to reduce carbon footprints. In the oil and gas industry, the adoption of blockchain and AI technologies has improved real-time monitoring, data accuracy, and regulatory transparency, offering predictive insights and automating compliance processes (Audu & Umana, 2024). Furthermore, the synergy between IoT and blockchain fosters environmentally responsible practices by ensuring traceability and enhancing operational efficiency in supply chains (Mofatteh et al., 2024).

Digital tools and AI further enable real-time adjustments for sustainability improvements, optimizing resource allocation and minimizing environmental footprints (Ameh, 2024). Collectively, these technologies provide a powerful framework for organizations to adapt to changing environmental regulations and consumer expectations, ultimately contributing to a more sustainable future (Ameh, 2024; Mofatteh et al., 2024).

13.3.2 Machine Learning

Machine learning and predictive analytics play a crucial role in identifying and mitigating non-compliance risks, thereby reducing environmental risks and operational inefficiencies. The integration of big data and machine learning within management information systems (MIS) enhances predictive analytics by enabling real-time decision-making and improving data preprocessing techniques, which are essential for maintaining data accuracy and relevance in predictive models (Adewale et al., 2024). The selection of appropriate machine learning algorithms is vital, as different algorithms offer varying levels of prediction accuracy, computational efficiency, and scalability, particularly in big data environments. Algorithms such as random forests and neural networks, while accurate, may require substantial computational resources, posing challenges for real-time processing (Raju & Devi, 2024).

In industrial settings, the use of machine learning-enabled smart sensors has demonstrated significant improvements in machine uptime and fault detection, highlighting the potential for predictive maintenance to enhance operational efficiency and financial returns (Val et al., 2024). Furthermore, machine learning algorithms, including supervised, unsupervised, and reinforcement learning, have broad applications across sectors such as agriculture, where they aid in crop forecasting and pest identification, thereby optimizing resource use and reducing losses (Saini et al., 2024). In financial risk management, machine learning models improve predictive accuracy and compliance with regulations like Basel III and GDPR, underscoring their importance in maintaining financial stability and customer confidence (Sarioguz & Miser, 2024). Collectively, these advancements illustrate the transformative impact of machine learning and predictive analytics in proactively managing risks and enhancing operational efficiencies across diverse sectors.

Machine learning models can be trained to predict operational inefficiencies in real time, leveraging various algorithms and methodologies across different sectors. In industrial operations, predictive maintenance (PdM) is a prominent application where machine learning models such as support vector machines (SVM), random forests, and neural networks are employed to predict equipment failures before they occur, thereby enhancing energy efficiency and reducing downtime (Márquez & Garcia, 2024; Shah et al., 2024). These models analyze historical operational data, including equipment performance metrics and environmental conditions, to forecast maintenance needs, allowing for timely interventions and optimized operational strategies (Márquez & Garcia, 2024).

In the business intelligence domain, machine learning models, particularly regression and neural network algorithms, are used to predict sales volumes and optimize supply chain strategies by identifying hidden patterns and trends within historical sales data (Wang & Aviles, 2023). This predictive capability supports real-time decision-making, enabling enterprises to adjust production and inventory management to meet market demands effectively (Wang & Aviles, 2023). However, the reliability of these predictions is contingent upon the representativeness of the training data to the operational space, as models excel at interpolation but struggle with extrapolation tasks. Techniques such as out-of-distribution (OOD) detection are crucial for assessing the applicability of predictions under varying conditions (Chen et al., 2024). Overall, the integration of machine learning in predictive analytics across different industries demonstrates its potential to enhance operational efficiency by providing real-time insights and facilitating informed decision-making.

13.4 PATHWAYS TO NET-ZERO EMISSIONS

13.4.1 Emission Reduction Strategies

Organizations can significantly reduce their carbon footprints by integrating low-carbon technologies, enhancing energy efficiency, and adopting circular economy principles, as evidenced by various studies. Transitioning to low-carbon technologies, such as electric vehicles and renewable energy sources, is crucial for reducing emissions in sectors like transportation and energy production (Thakur et al., 2024). Energy-efficient practices, including the use of energy-efficient lighting and smart infrastructure, further contribute to emission reductions by optimizing energy use and integrating renewable energy into the grid (Thakur et al., 2024). The adoption of circular economy principles, which emphasize waste reduction, recycling, and resource efficiency, plays a pivotal role in minimizing carbon emissions (Haris et al., 2024). For instance, circular economy-integrated waste management strategies, such as composting and recycling, have shown substantial potential in reducing urban carbon emissions (Hailemariam & Erdiaw-Kwasie, 2022). Additionally, sourcing materials from carbon-neutral suppliers aligns with corporate social responsibility (CSR) strategies, which focus on enhancing supply chain sustainability and reducing carbon intensity (Balukja, 2024).

The role of waste-to-energy (WTE) processes in circular economy policies is also highlighted, as they provide a means to recover energy from non-recyclable materials, thus reducing reliance on fossil fuels and contributing to a low-carbon economy (Bourtsalas, 2023). Overall, the transition to a circular economy not only helps in achieving emission reduction targets but also promotes environmental sustainability by improving resource efficiency and reducing CO_2 emissions. These integrated approaches are essential for organizations aiming to achieve long-term sustainability and align with global climate goals.

Circular economy principles significantly impact the development of low-carbon technologies by promoting resource efficiency, waste minimization, and the sustainable management of critical raw materials (CRMs). These principles facilitate the transition to low-carbon processes by enhancing the reuse and recycling of materials, thereby reducing carbon emissions and supporting carbon-neutral development (Mazzi & Ren, 2021). The circular economy framework, which includes strategies such as end-of-life collection and recycling, can meet a substantial portion of CRM demand for low-carbon technologies, potentially fulfilling 37% to 91% of this demand by 2050, depending on the technology type (Karali & Shah, 2022). This approach not only secures the supply chain for CRMs but also encourages the development of cost-competitive recycling technologies and recycling-friendly product designs (Karali & Shah, 2022). Furthermore, the integration of circular economy principles into business models, such as energy-independent agrarian-industrial complexes, exemplifies how these principles can drive sustainable development projects (Bushuyev et al., 2023). The transition from a linear to a circular economy model, as part of Industry 4.0, emphasizes energy-saving and regenerative production, aligning with low-carbon development goals. Additionally, the circular economy's role in climate change mitigation is underscored by its potential to reduce reliance on primary materials and energy, thereby aiding in achieving climate neutrality by 2050 (Zorpas, 2024). Overall, circular economy principles provide a comprehensive toolkit for advancing low-carbon technologies through enhanced resource management, policy support, and innovative project development.

13.4.2 Sustainable Urbanization

Sustainable urbanization is a critical focus in addressing the environmental impacts of urban areas, which are significant contributors to global emissions. Strategies such as green infrastructure, energy-efficient buildings, and sustainable transportation systems are essential for achieving net-zero emissions in urban settings. In the United States, urban sustainability practices are led by cities like those on the Pacific Coast, Washington D.C., Boston, and New York City, which have implemented various innovations to promote smart, sustainable city concepts (Sargsyan, 2024). Globally, green city initiatives emphasize sustainable infrastructure, green spaces, energy efficiency, waste management, and community engagement, highlighting the importance of holistic planning and collaborative governance to enhance urban sustainability and quality of life (Raghubanshi, 2024). Green building technologies, including energy-efficient designs and renewable energy integration, are pivotal in reducing the environmental impact of urbanization, with innovations like passive design strategies and smart building systems playing a crucial role

(Singla, 2024). Urban sustainability also involves balancing economic vibrancy, social inclusivity, and environmental stewardship, as cities are expected to house nearly 70% of the global population by 2050 (Singh et al., 2024). Sustainable urban planning strategies, such as implementing green building standards and promoting renewable energy, are vital in mitigating environmental pollution and fostering balanced development across economic, social, and environmental dimensions (Li, 2024). These strategies collectively offer a roadmap toward more inclusive, resilient, and environmentally conscious urban futures.

13.5 THE ROLE OF COMPLIANCE IN ENVIRONMENTAL HEALTH

13.5.1 Baseline Monitoring

Compliance plays a crucial role in environmental health baseline monitoring by ensuring that industrial activities adhere to established environmental standards and regulations, thereby maintaining ecosystem integrity. Baseline monitoring serves as a reference point to detect deviations caused by industrial activities, and compliance ensures that these deviations are addressed effectively. The concept of compliance, as discussed in various contexts, involves adherence to rules and norms to maintain order and efficiency, which is essential in environmental health interventions (Utami et al., 2023; Brown et al., 2019). In the environmental sector, compliance is integral to the implementation of programs that minimize harmful impacts, as seen in the case of JSC Naftan, where an environmental compliance program was proposed to reduce environmental harm from oil refining activities (Mikhailova, 2022). Furthermore, compliance in environmental health interventions, such as water and sanitation, is critical for reducing exposure to environmental hazards and achieving health gains (Brown et al., 2019). The preventive nature of compliance, as highlighted in environmental governance, can lead to both immediate and long-term benefits, aligning with sustainability goals (Vasconcelos et al., 2019). Additionally, compliance programs in organizations, such as health plan operators, emphasize the importance of managing risks and strengthening internal controls to meet regulatory requirements, which parallels the need for rigorous environmental compliance to protect ecosystems. Overall, compliance ensures that baseline monitoring effectively informs and enforces environmental standards, thereby safeguarding environmental health.

13.5.2 Impact on Public Health

Environmental compliance for climate change, environmental health, and human health is a multifaceted challenge that requires integrated strategies and cross-sector collaboration. The intricate relationship between environmental conditions and human health is increasingly recognized, with climate change exacerbating health risks through altered weather patterns, increased vector-borne diseases, and compromised water quality (Chandar & Ravipragasam, 2024; Kumar, 2024). Compliance involves a complex array of self-monitoring reports and coordination among various professionals to meet environmental standards across different matrices, such as air, water, and soil. The health sector plays a crucial role in advocating for policies that reduce greenhouse gas emissions and build climate-resilient health systems, emphasizing the need for sector-wide emission cuts to protect both ecological and human systems (Neville et al., 2023). The One Health approach, which integrates environmental, veterinary, and public health disciplines, is essential for addressing the interconnected challenges of climate change and antimicrobial resistance (AMR), as these issues impact both ecosystem and human health (Reid & Broadbent, 2023). For instance, the thawing of permafrost in the Canadian Arctic releases ancient carbon and contaminants, posing new health risks (Reid & Broadbent, 2023). Additionally, the concept of "green prescriptions" promotes nature-based interventions to enhance well-being and foster environmental stewardship, highlighting the importance of sustainable resource management (Chandar & Ravipragasam, 2024). Overall, fostering collaboration among stakeholders and adopting holistic perspectives are vital for developing resilient communities capable of navigating the intertwined challenges of environmental compliance, climate change, and health sustainability (Chandar & Ravipragasam, 2024; Reid & Broadbent, 2023).

13.5.3 Impacts on Biodiversity

Biodiversity conservation through compliance measures such as habitat restoration and pollution control is crucial for maintaining ecosystem services and ensuring resilience against climate-related challenges. Habitat restoration, particularly in forested areas, is essential, as forests harbor a significant portion of Earth's terrestrial biodiversity. An integrated framework that incorporates landscape resilience and habitat risk assessment can effectively identify priority areas for diversified management actions, enhancing the planning process and outcomes for biodiversity conservation (Tan et al., 2024). Pollution control is equally important, as pollution and climate change pose significant threats to biodiversity, leading to ecosystem disruptions and species loss. Holistic conservation strategies that consider ecological tolerances and promote sustainable resource use are vital for preserving biodiversity and ensuring long-term ecological balance (Durmaz et al., 2024). Sustainable agricultural practices, such as organic farming and agroforestry, also play a significant role in biodiversity conservation by reducing environmental stresses, increasing species variety, and improving habitat quality. These practices contribute to a more resilient ecosystem by reducing chemical inputs and providing diverse ecological niche (Singh, 2024b). In India, despite being a megadiverse country with significant biodiversity, the implementation of regulatory acts faces barriers such as inadequate funding and bureaucratic challenges. Effective enforcement and adaptive management strategies are necessary to overcome these barriers and enhance biodiversity conservation efforts (Singh, 2024a). Overall, a combination of habitat restoration, pollution control, sustainable agriculture, and effective regulatory frameworks is essential for protecting biodiversity and maintaining ecosystem resilience against climate-related challenges.

13.6 ENVISIONING A SUSTAINABLE FUTURE

13.6.1 Future Temperature Projections

Global temperature projections indicate a significant rise if current high-emission scenarios persist, underscoring the urgent need for achieving net-zero emissions to prevent catastrophic climate impacts. The food sector, responsible for 20%–30% of global human-induced greenhouse gas emissions, plays a critical role in this context. Innovative design processes, such as those promoted by the DC4DM European project, aim to transform food practices toward sustainability, highlighting the sector's potential for reducing emissions (Canina, 2023). The broader economic and political systems also contribute to climate change, with mass production paradigms exacerbating resource depletion and environmental degradation.

The transition to an Information and Communication Technology (ICT) paradigm offers potential for more sustainable production and consumption, though this shift requires careful consideration of its environmental impacts (Schulz, 2023; McEvoy, 2022). Furthermore, sustainable futures must incorporate more-than-human perspectives, recognizing the interdependent relationships between human and non-human entities, as advocated by speculative design techniques (Stead & Coulton, 2022). The integration of specialty rights, diversity, and political coherence into sustainability assessments is crucial for fostering equitable and inclusive outcomes, as emphasized by the current sustainability discourse (Nunes, 2023). Collectively, these insights highlight the multifaceted approach needed to address climate change, involving technological innovation, systemic economic shifts, and inclusive sustainability practices.

13.6.2 Emerging Technologies

Emerging technologies play a crucial role in achieving sustainability goals by enhancing resilience and the effectiveness of climate action. The integration of green technologies is essential for meeting the Paris Agreement's climate objectives, as demonstrated by the Advanced Climate Change Long-term model (ACCL), which shows that a multi-lever strategy combining energy efficiency, carbon sequestration, and price adjustments for clean and polluting energies is necessary to achieve these goals (Alestra et al., 2023). Technological innovation, particularly in green energy, significantly contributes to sustainable development, as evidenced by Türkiye's efforts to align with international environmental agreements, highlighting the positive impact of technological advancements on economic growth and sustainability (Atay Polat, 2023). Real-time energy monitoring systems further optimize energy consumption and foster sustainability by reducing carbon footprints and enhancing resource utilization, with significant socio-economic benefits observed in regions like Canada, the USA, and Africa (Hamdan et al., 2024). However, the development and deployment of new technologies must be carefully monitored to avoid unintended consequences, necessitating reflexive guidance processes and early assessments to ensure that technological advancements align with sustainability objectives. The concept of "green innovation" underscores the importance of connecting governance, practices, and outcomes to address environmental challenges, emphasizing the need for collaboration between academia and practitioners to maximize the impact of sustainability initiatives (Blundel et al., 2018). Collectively, these technologies and strategies underscore the potential for technological advancements to drive significant progress toward global sustainability goals.

13.6.3 Global Collaboration

Achieving a sustainable future necessitates robust international cooperation, which involves joint efforts in data sharing, policy development, and financial mobilization to address global environmental challenges. The conceptual framework for international collaboration, as illustrated by the China–US sponge city exchange, highlights the importance of communication, cultural understanding, and policy alignment in fostering sustainable urban development (Smith, 2024). The UN Summit of the Future emphasized the critical role of multi-stakeholder partnerships in achieving the Sustainable Development Goals (SDGs), particularly SDG 17, which focuses on partnerships for the goals. This requires innovative, bottom-up approaches that connect research, policy, and communities to ensure equitable solutions and measurable impacts (Koundouri et al., 2024). Regional environmental policies also play a crucial role in global sustainability efforts, with international cooperation being essential to address issues like air pollution and climate change. However, the effectiveness of these efforts often depends on a nation's social and economic development level (Öztopçu, 2024).

Global partnerships face challenges such as technological barriers and regional diversification, which need to be addressed to enhance their effectiveness in achieving the SDGs (Ahmed et al., 2024). The Asia-Pacific region exemplifies the significance of collaboration, where geographical proximity and resource sharing underscore the need for a strong communication infrastructure to support sustainable development initiatives (Okawa, 2023). Collectively, these insights underscore the necessity of international cooperation in creating a sustainable future, highlighting the need for integrated efforts across various sectors and regions.

13.7 CONCLUSION

Compliance and proactive climate action deliver wide-ranging economic, environmental, and social benefits. Companies and governments that invest in energy efficiency, clean technologies, and robust compliance systems not only reduce greenhouse gas emissions but also cut operational costs, avoid regulatory penalties, and strengthen supply chain resilience. Market-based tools, such as emission taxes and cap-and-trade, create powerful incentives for innovation, encouraging the development of green products and services while opening new markets and enhancing competitiveness. Transparent reporting frameworks and environmental audits, supported by real-time monitoring and advanced analytics, are making compliance more effective and measurable than ever.

The environmental gains are clear: reduced emissions slow global warming, protect biodiversity and lead to cleaner air, water, and soil. Socially, climate action improves public health, creates jobs in green industries, and builds community resilience to climate hazards. Inclusive policies ensure that vulnerable populations benefit from the transition, supporting social equity and intergenerational fairness. Ultimately, economies that integrate climate action are better equipped to prepare for future shocks and can achieve sustainable growth and stability.

As we move forward, the next chapter brings these benefits to life through real-world examples. Chapter 14 presents case studies from major emission sectors—such as buildings, cement, and chemicals—showing how compliance strategies, innovation, and policy support are driving measurable progress. These stories offer practical lessons and inspiration for organizations and communities seeking to translate climate commitments into tangible results.

REVIEW QUESTIONS

1. What are the benefits of compliance in climate action?
2. Discuss the economic incentives for compliance and climate action.
3. Discuss the "nature performance bonds", how they are issued and how they help in reducing the adverse climate crisis.
4. Discuss reforming "subsidies and metrics" from fossil fuels to renewable energies and how do they present significant opportunities for global economic and environmental benefits.
5. What are the "market-based instruments" and why are they pivotal in creating economic incentives for reducing GHG emissions.
6. Discuss the role of environmental health and safety (EHS) with regard to monitoring, measurement, and how they help toward positive impact of regulatory compliance.
7. How does "real-time environmental monitoring" help track emissions and environmental parameters continuously?
8. Discuss the steps involved in the environmental audits.
9. Discuss the steps involved in the EIAs.
10. Briefly discuss the technological innovations in compliance.

REFERENCES

Adewale, G.T., Alexander, V., Sylvia, A.E., Sonubi, T., Mesogboriwon, A.O. (2024). Integrating big data and machine learning in management information systems for predictive analytics: A focus on data preprocessing and technological advancements. *World Journal of Advanced Research and Reviews*. https://doi.org/10.30574/wjarr.2024.24.2.3427

Ahmed, S., Memon, N.A., Memon, A.H., Jattak, Z.A., Shah, S.K. (2024). Role of global partnerships for sustainable development for a resilient future. *Construction Technologies and Architecture*. https://doi.org/10.4028/p-5e86o7

Alestra, C., Cette, G., Chouard, V., Lecat, R. (2023). How can technology significantly contribute to climate change mitigation? *Applied Economics*. https://doi.org/10.1080/00036846.2023.2227416

Ali, S., Gondro, C., Yan, Q., Banzhaf, W. (2024). End-to-end decentralized tracking of carbon footprint using Internet of Things and distributed databases. *2024 7th Conference on Cloud and Internet of Things (CIoT)*, Montreal, QC, Canada, 2024, pp. 1–8. https://doi.org/10.1109/CIoT63799.2024.10757144

Ameh, B. (2024). Digital tools and AI: Using technology to monitor carbon emissions and waste at each stage of the supply chain, enabling real-time adjustments for sustainability improvements. *International Journal of Science and Research Archive*. https://doi.org/10.30574/ijsra.2024.13.1.1995

Atay Polat, M.A. (2023). Technological innovation and sustainable development. In *Reference module in social sciences*. Elsevier. https://doi.org/10.1016/B978-0-44-313776-1.00140-9

Audu, A.J., Umana, A.U. (2024). Advances in environmental compliance monitoring in the oil and gas industry: Challenges and opportunities. *International Journal of Scientific Research Updates*. https://doi.org/10.53430/ijsru.2024.8.2.0062

Bajra, U., Wagner, N. (2024). Analyzing the impact of eco-friendly bond on economic growth and environmental sustainability. *Borsa Istanbul Review*. https://doi.org/10.1016/j.bir.2024.04.003

Balukja, I. (2024). Carbon reduction indicators and strategies: A cornerstone of corporate social responsibility. *Balkan Journal of Interdisciplinary Research*. https://doi.org/10.2478/bjir-2024-0012

Blundel, R., Smith, D., Ackrill, R., Schaefer, A. (2018). Making 'greener' connections: An introduction to the Special Issue. *The International Journal of Entrepreneurship and Innovation*. https://doi.org/10.1177/1465750317753494

Boumaiza, A., Maher, K. (2024). Harnessing blockchain and IoT for carbon credit exchange to achieve pollution reduction goals. *Energies*. https://doi.org/10.3390/en17194811

Bourtsalas, A.C. (2023). The role of fuel and energy recovery from non-recyclable materials in circular/low carbon economy policies. In M. J. Castaldi & S. S. Ali (Eds.), *Proceedings of the 2022 EEC/WTERT Conference* (Chapter 5). ASME. https://doi.org/10.1115/1.887271_ch5

Brown, J., Hayashi, M.A.L., Eisenberg, J.N.S. (2019). The critical role of compliance in delivering health gains from environmental health interventions. *American Journal of Tropical Medicine and Hygiene*. https://doi.org/10.4269/AJTMH.18-0624

Bushuyev, S., Bushuyeva, N., Ivko, A., Bushuiev, D., Bushuieva, V., Iazykov, D. (2023). Principles of circular economy as a driver of development projects. *2023 IEEE International Conference on Smart Information Systems and Technologies (SIST)*, Astana, Kazakhstan, 2023, pp. 37–41, https://doi.org/10.1109/SIST58284.2023.10223513

Canina, M. (2023). Design futures to support sustainable food practices. In De Sainz Molestina, D., Galluzzo, L., Rizzo, F., Spallazzo, D. (Eds.), *IASDR 2023: Life-Changing Design*, pp. 9–13 October, Milan, Italy. https://doi.org/10.21606/iasdr.2023.521

Chandar, D., Ravipragasam, S. (2024). Environmental health and human health-unifying strategies, conceptual framework and green prescriptions: A review. *International Journal of Community Medicine and Public Health*. https://doi.org/10.18203/2394-6040.ijcmph20243424

Chen, E., Bao, H., Dinh, N. (2024). Evaluating the reliability of machine-learning-based predictions used in nuclear power plant instrumentation and control systems. *Reliability Engineering & System Safety*. https://doi.org/10.1016/j.ress.2024.110266

Durmaz, A., Torunoğlu, E.İ., Aydın, B., Aytar, E.C. (2024). *Need for Biodiversity Conservation and Conservation Strategies*. In A. Rathoure (Ed.), *Biodiversity loss assessment for ecosystem protection* (pp. 167–187). IGI Global Scientific Publishing. https://doi.org/10.4018/979-8-3693-3330-3.ch010

Feng, T., Wang, X., Shi, Y., Tu, Q. (2024). The role of carbon price signal in green innovation: Evidence from China. *Journal of Environmental Management*. https://doi.org/10.1016/j.jenvman.2024.122787

Gumarova, R. (2024). On the legal regulation of environmental audit. *Pravo i Politika*. https://doi.org/10.7256/2454-0706.2024.10.69392

Hailemariam, A., Erdiaw-Kwasie, M.O. (2022). Towards a circular economy: Implications for emission reduction and environmental sustainability. *Business Strategy and the Environment*. https://doi.org/10.1002/bse.3229

Hamdan, A., Sonko, S., Fabuyide, A., Daudu, C.D., Etukudoh, E.A. (2024). Real-time energy monitoring systems: Technological applications in Canada, USA, and Africa. *World Journal of Advanced Research and Reviews*. https://doi.org/10.30574/wjarr.2024.21.1.0255

Haris, N.I.N., Edrus, S.S.O.A., Raof, N.A., Wondi, M.H., Khan, W.R., Sien, L., Ilyas, R.A., Norrrahim, M.N.F., Sawatdeenarunat, C. (2024). Toward low-carbon cities: A review of circular economy integration in urban waste management and its impact on carbon emissions. *WIREs Energy and Environment*, 13(5), e535. https://doi.org/10.1002/wene.535

Kamel, G. M., & Khalaf, H. A. A. (2024). The role of environmental bonds in the settlement of environmental debts. (2024). *Pakistan Journal of Criminology*. 16(4), 1111–1126. https://doi.org/10.62271/pjc.16.4.1111.1126

Karali, N., Shah, N. (2022). Bolstering supplies of critical raw materials for low-carbon technologies through circular economy strategies. *Energy Research and Social Science*. https://doi.org/10.1016/j.erss.2022.102534

Kariuki, P., Nzuki, P.M. (2024). The role of environmental audits in enhancing sustainability in industrial property management. Insights from Nairobi County. *East African Journal of Environment and Natural Resources*. https://doi.org/10.37284/eajenr.7.2.2306

Kashif, M., Shajar, S.N., Singhal, N., Kumar, P. (2023). Achieving sustainable investment practices through green finance: Challenges and opportunities. In S. Taneja, P. Kumar, S. Grima, E. Ozen, & K. Sood (Eds.), *Sustainable investments in green finance* (pp. 234–244). IGI Global Scientific Publishing. https://doi.org/10.4018/979-8-3693-1388-6.ch016

Koundouri, P., Prohic, A., Karouzos, M. (2024). *Uniting for a sustainable future: Connecting research, policy, and communities for real-world impact* [Seminar presentation]. SDG Talks, Springer Nature Sustainable Development Goals Programme. https://doi.org/10.52843/cassyni.xqb1k8

Kumar, V. (2024). Climate change impacts on human health. *Current Research in Medical Sciences*. https://doi.org/10.56397/crms.2024.03.01

Li, N. (2023). Research of ecological environment monitoring based on artificial intelligence. *Theoretical and Natural Science*, 7, 110–115.

Li, G. (2024). Sustainable development strategies in urban planning. *Advances in Economics, Management and Political Sciences*. https://doi.org/10.54254/2754-1169/96/2024mur0118

Lou, H.H., Mukherjee, R., Wang, Z., Olsen, T., Diwekar, U.M., Lin, S. (2022). A new area of utilizing industrial Internet of Things in environmental monitoring. *Frontiers in Chemical Engineering*. https://doi.org/10.3389/fceng.2022.842514

Manfredini, F.N., Manfredini, L.M.V. (2024). Environmental legal compliance audit: Perspectives, challenges and importance. *Biological Sciences,* 28 (138/SET 2024). https://doi.org/10.69849/revistaft/ch10202409261045

Márquez, J. C. de la C., Garcia, A.M. (2024). Machine learning for predictive maintenance to enhance energy efficiency in industrial operations. *ITEj (Information Technology Engineering Journals)*. https://doi.org/10.24235/itej.v9i2.125

Mazzi, A., & Ren, J. (2021). Circular economy in low-carbon transition. *Energies,* 14(23), 8061. https://doi.org/10.3390/en14238061

McEvoy, J.K. (2022). Sustainable futures: An agenda for action. *International Affairs*. https://doi.org/10.1093/ia/iiac093

Merrill, L., Bassi, A.M., Bridle, R., Christensen, L.T. (2015). *Tackling fossil fuel subsidies and climate change: Levelling the energy playing field* (TemaNord 2015:575). Nordic Council of Ministers. https://doi.org/10.6027/TN2015-575

Merrill, L., Bridle, R., Klimscheffskij, M., Tommila, P., Lontoh, L., Sharma, S., Touchette, Y., Gass, P., Gagnon-Lebrun, F., Sanchez, L., Gerasimchuk, I. (2017). *Making the switch: From fossil fuel subsidies to sustainable energy* (TemaNord 2017:537). Nordic Council of Ministers. https://doi.org/10.6027/TN2017-537

Miasayedava, L., McBride, K., Tuhtan, J.A. (2020). Automated environmental compliance monitoring with iot and open government data. *arXiv: Computers and Society.*

Mikhailova, A. (2022). The role of environmental compliance in the environmental activities of a legal entity: legal aspect (on the example of the oil refinering enterprise jsc «Naftan»). *Вестник Полоцкого Государственного Университета*. https://doi.org/10.52928/2070-1632-2022-60-5-120-125

Mofatteh, M.Y., Davallou, R., Ishimwe, C.N., Divekar, S.S., Valilai, O.F. (2024). Developing a blockchain based supply chain CO2 footprint tracking framework enabled by IoT. *International Journal of Advanced Computer Science and Applications*. https://doi.org/10.14569/ijacsa.2024.0151002

Morozov, A., Morozova, T., Ziuziun, V.I. (2022). Environmental impact assessment for construction and reconstruction of motor roads. *Dorogi ì Mosti*. https://doi.org/10.36100/dorogimosti2022.26.285

Moses, T.S., Ajilore, B.G., Moronkola, A.A., Paul, O., Ahmed, A.J. (2024). The operating principles in an EIA process. *International Journal of Science and Technology Research Archive*. https://doi.org/10.53771/ijstra.2024.7.2.0060

Nadir, H., Ahmed, A. (2023). The critical review of the efficacy of the Environmental Impact Assessment (EIA) process as a tool intended to protect the environment from construction development's hazards. *Journal of Civil Engineering Research & Technology,* 5(2), 1–7. https://doi.org/10.47363/JCERT/2023(5)143

Naumenkova, S., Mishchenko, V., Chugunov, I., Mishchenko, S. (2023). Debt-for-nature or climate swaps in public finance management. *Problems and Perspectives in Management*. https://doi.org/10.21511/ppm.21(3).2023.54

Neville, T., Campbell-Lendrum, D., Maiero, M., Wyns, A., Neira, M. (2023). Climate Change and Human Health. In Raviglione, M.C.B., Tediosi, F., Villa, S., Casamitjana, N., Plasència, A. (Eds.), *Global Health Essentials*. Sustainable Development Goals Series. Springer, Cham. https://doi.org/10.1007/978-3-031-33851-9_74

Nunes, M.C. (2023). The future of sustainable impact. *Sustainable Development Goals Series*. https://doi.org/10.1007/978-3-031-36907-0_18

Okawa, K. (2023). Fostering *Sustainable Futures through Collaborative Innovation: Lessons from the AI3/SOI-Asia Project*. In *Proceedings of the 8th International Conference on Sustainable Information Engineering and Technology (SIET '23)* (p. 2). Association for Computing Machinery. https://doi.org/10.1145/3626641.3629026

Öztopçu, A. (2024). Environmental policies in global economies and environmental concern in regional sustainable production. *Advances in Finance, Accounting, and Economics Book Series*. https://doi.org/10.4018/979-8-3693-5508-4.ch018

Pang, S. (2024). Environmental impact assessment system: Case analysis of existing legal disputes. *Lecture Notes in Education Psychology and Public Media*. https://doi.org/10.54254/2753-7048/69/20240160

Raghubanshi, A.S. (2024). Sustainable urban planning: A comparative study of green city initiatives around the world. *Journal of Sustainable Solutions,* 1(3), 18–23. https://doi.org/10.36676/j.sust.sol.v1.i3.16

Raju, N., Devi, P.N. (2024). A comparative analysis of machine learning algorithms for big data applications in predictive analytics. *International Journal of Scientific Research and Management*. https://doi.org/10.18535/ijsrm/v12i10.ec09

Ramadan, M.N.A., Ali, M.A.H., Khoo, S.Y., Alkhedher, M., Alherbawi, M. (2024). Real-time IoT-powered AI system for monitoring and forecasting of air pollution in industrial environment. *Ecotoxicology and Environmental Safety*. https://doi.org/10.1016/j.ecoenv.2024.116856

Ratra, R. (2023). Carbon credits, emissions trading and energy saving. *International Journal of Social Science and Economic Research*. https://doi.org/10.46609/ijsser.2023.v08i07.020

Ravi, S. (2024). Environmental impact assessment processes & attributes. In *Futuristic trends in social sciences* (Vol. 3, Book 7, pp. 91–111). IIP Series. https://doi.org/10.58532/V3BJSO7P1CH8

Reid, T., Broadbent, J. (2023). From climate change to AMR: Understanding environmental-human health issues in a One Health framework. *ARPHA Conference Abstracts*. https://doi.org/10.3897/aca.6.e108165

Saini, A., Dhuriya, G., Jain, A., Mishra, A. (2024). Machine learning algorithms and applications. In *Artificial intelligence for precision agriculture* (1st ed., pp. 1–31). Auerbach Publications. https://doi.org/10.1201/9781003504900

Sargsyan, T. (2024). Sustainable urban development policies and practices in the United States. *Czasopismo Geograficzne*. https://doi.org/10.12657/czageo-95-26

Sarioguz, O., Miser, E. (2024). Integrating AI in financial risk management: Evaluating the effects of machine learning algorithms on predictive accuracy and regulatory compliance. *International Journal of Science and Research Archive*. https://doi.org/10.30574/ijsra.2024.13.2.2206

Schulz, C. (2023). Sustainable futures—An agenda for action. *Economic Geography*. https://doi.org/10.1080/00130095.2023.2168393

Shah, H., Haley, B.J., Abdelfattah, E. (2024). Application of machine learning models for predictive analytics on AI4I 2020. In *2024 IEEE 15th Annual Ubiquitous Computing, Electronics & Mobile Communication Conference (UEMCON)* (pp. 506–511). IEEE. https://doi.org/10.1109/UEMCON62879.2024.10754668

Simpa, P., Solomon, N.O., Adenekan, O.A., Obasi, S.C. (2024). Strategic implications of carbon pricing on global environmental sustainability and economic development: A conceptual framework. *International Journal of Advanced Economics*. https://doi.org/10.51594/ijae.v6i5.1134

Singh, M.P. (2024a). Biodiversity in India: Implementation and barriers for regulatory acts. *Journal of Science Innovations and Nature of Earth,* 4(1), 17–20. https://doi.org/10.59436/jsiane256

Singh, R. (2024b). Assessing *the Impact of Sustainable Agriculture Practices on Biodiversity Conservation*. https://doi.org/10.36676/j.sust.sol.v1.i3.13

Singh, M., Arora, V., Kulshreshta, K. (2024). Towards sustainable cities. *Practice, Progress, and Proficiency in Sustainability*. https://doi.org/10.4018/979-8-3693-6567-0.ch011

Singla, A. (2024). Sustainable urban infrastructure: Innovations in green building technologies and urban planning. *Universal Research Reports*, 11(4), 106–112. https://doi.org/10.36676/urr.v11.i4.1316

Smith, J.S. (2024). International collaboration for sustainable development in the Third Space: A conceptual framework and relevant approaches based on China-US sponge city exchange. *International Journal of Chinese Education*. https://doi.org/10.1177/2212585x241278376

Stead, M., Coulton, P. (2022). Sustainable technological futures: Moving beyond a one-world-world perspective. *Nordic Conference on Human-Computer Interaction*. https://doi.org/10.1145/3546155.3547283

Tan, C., Xu, B., Hong, G., Wu, X. (2024). Integrating habitat risk and landscape resilience in forest protection and restoration planning for biodiversity conservation. *Landscape and Urban Planning*. https://doi.org/10.1016/j.landurbplan.2024.105111

Thakur, S., Swathi, S., Nijhawan, G., Sharma, N., Kumar, A., Mashkour, M.S., Nalli, N. (2024). Reduction of green house gases emission: Towards a sustainable future. *E3S Web of Conferences*. https://doi.org/10.1051/e3sconf/202455201116

Thoti, K.K., Pasha, M.A., Shankar, D., Miranda, P.A., Yaziz, D.M.F.A., Uday, M. (2024). An analysis examining the way supply chain management-focused third-party environmental audits contributed to business sustainability efforts. *Economic Sciences*. https://doi.org/10.69889/2y3pxd38

Utami, N.T., Wolor, C.W., Marsofiyati, M. (2023). Analisis Kepatuhan Kerja Anggota di Polsek Pademangan. *Jurnal Riset Rumpun Ilmu Sosial, Politik Dan Humaniora*. https://doi.org/10.55606/jurrish.v3i1.2267

Uzondu, N.C., Lele, D.D. (2024). Multifaceted impact of renewable energy on achieving global climate targets: Technological innovations, policy frameworks, and international collaborations. *International Journal of Applied Research in Social Sciences*. https://doi.org/10.51594/ijarss.v6i7.1338

Val, O.O., Olaniyi, O.O., Selesi-Aina, O., Gbadebo, M.O., Kolade, T.M. (2024). Machine learning-enabled smart sensors for real-time industrial monitoring: Revolutionizing predictive analytics and decision-making in diverse sector. *Asian Journal of Research in Computer Science*. https://doi.org/10.9734/ajrcos/2024/v17i11522

Van de Putte, A., Campbell-Holt, A., Littlejohn, G. (2020). Financing the Sustainable Energy Transition. In Hafner, M., Tagliapietra, S. (Eds.), *The Geopolitics of the Global Energy Transition*. Lecture Notes in Energy, vol. 73. Springer, Cham. https://doi.org/10.1007/978-3-030-39066-2_11

Vasconcelos, P.E.A., Oliveira, R.C.R., Vasconcelos, P.S. (2019). Environmental compliance: Search for effectiveness in the application of environmental standards. *International Journal of Advanced Engineering Research and Science*. https://doi.org/10.22161/IJAERS.68.34

Vyas, J. N., Nath, S., Kumar, D. N., Deogade, R. B., & Chandra, P. (2023). Environmental impact assessment of water resource projects. In P. V. Timbadiya, P. L. Patel, V. P. Singh, & B. Barman (Eds.), *Fluid mechanics and hydraulics. HYDRO 2021* (Vol. 314). Springer. https://doi.org/10.1007/978-981-19-9151-6_47

Wang, F., Aviles, J.S. (2023). Enhancing operational efficiency: Integrating machine learning predictive capabilities in business intellgence for informed decision-making. *Frontiers in Business, Economics and Management*. https://doi.org/10.54097/fbem.v9i1.8694

World Trade Organization. (2023). Carbon pricing and international trade. In *World trade report 2022* (pp. 78–97). https://doi.org/10.30875/9789287053961c010

Yousef, H.H., Al-Fatlawy, A.K. (2024). The role of green auditing in realizing environmental sustainability. *Jurnal Nuansa: Publikasi Ilmu Manajemen Dan Ekonomi Syariah*, 2(3), 124–132. https://doi.org/10.61132/nuansa.v2i3.1106

Zorpas, A.A. (2024). The hidden concept and the beauty of multiple "R" in the framework of waste strategies development reflecting to circular economy principles. *Science of the Total Environment*. https://doi.org/10.1016/j.scitotenv.2024.175508

14 Case Studies

It has been unequivocally established that global warming is real, and the cause of it is man-made operations such as extreme energy consumption by industries such as cement, glass, chemical manufacturers, and facilities using fossil fuels and higher electricity consumption. Highly populated cities like NYC have their own climate acts enforcing their regulations to reduce greenhouse gas (GHG) emissions. Other industries across the globe are trying their part to comply with the mandates established by the COP of the global partners by the year 2030, and beyond.

In this chapter, what and how major GHG emission contributors (industries) to the climate crisis such as cement, glass, chemical, and healthcare account for their CFP, measures taken to reduce and offset are illustrated. Also narrated in this chapter are the estimated penalties for noncompliance in the industrial sector including recommended mitigation efforts.

14.1 CASE STUDY: COMPLIANCE WITH CLIMATE MOBILIZATION ACT AND LOCAL LAW (LL 97) OF NYC

New York City (NYC) is densely populated, with 8.5 million diverse people, thousands of high-rise buildings, hospitals, and universities with the potential to emit GHG (www.nyc.gov/site/buildings). The NYC Comptroller's office reported that, as of June 30, 2023, its GHG emissions (scopes 1 and 2) were higher in 2022 compared to 2019. Scope 3 emissions increased by 65.4% in 2023, exceeding the benchmark. Buildings account for approximately 2/3rd of GHG emissions emitted. Several compliance efforts are underway in order to be carbon neutral by 2050, one of which is LL 97, the most ambitious plan for reducing GHG emissions. LL 97 was included in Climate Mobilization Act (CMA) as the "Green New Deal" and is vigorously enforced.

The law establishes carbon emission limits for NYC's largest buildings. Covered buildings that exceed the emissions cap will face a penalty of \$268.00 per ton (2024 level) starting in 2025. Emissions are calculated based on tons of CO2eq per ft². Emission caps will become more stringent from 2030. In essence, buildings (>25,000 ft², or 2 or more buildings on the same tax, >50,000 ft², or 2 or more buildings owned by an association and governed by the same board, >50,000 ft²) are forced to reduce their emissions by 40% by 2030 and 80% by 2050 to meet these aggressive sustainability goals (Amended version of Local Law 97 of 2019.

$$\text{LL 97 Penalties Maximum annual penalty} = (\text{actual emissions} - \text{annual emissions limit}) \times \$268.00$$

GHG Calculations and Compliance Measures: For example, one of the hospitals operates 1 boiler rated at 20 mm btu/hr., firing NG, one absorber rated at 5 MMBtu/hr., firing NO_2 oil, and the annual electricity consumption is 4.8 MM KWH. Estimated annual total CO_2 emissions from this facility could be 3,883 tons. The allowable emission limit for this occupancy (healthcare) until the year 2030 is 2,381 tons, and because of this exceedance, this facility is required to pay annually a penalty of \$402,390.74 as shown in Table 14.1. Figure 14.1 presents NYC's GHG gas reduction mandate for various occupancy groups for the years 2024 through 2050 (Kasinathan, 2023).

Some of the CO_2 offset efforts (mitigation measures) this facility is undertaking are:

- Retiring some of the non-working or unnecessary equipment such as compressors, old emergency generators that consume power even if idle.
- Replacing inefficient controls for the HVAC systems.

DOI: 10.1201/9781003479666-14

TABLE 14.1
Estimation of Annual GHG Emissions and Penalty for Non-Compliance with NYC LL 97 for a Facility Located in NYC

Sources	Capacity	EPA CO_2eq Factor	Annual Operation Hour	Estimated CO_2e (ton/yr)
1 Boiler	$20*10^6$ BTU/hr.	0.00005311 T/ KBTU	2,000	2,124
1 Absorber	$5*10^6$ BTU/hr.	0.00007421 T/ KBTU	1,000	371
Electricity	$4.8*10^6$ kwh	0.00028896 T/ KWH	500	1,387
			Total emissions	3,882
Penalty: $268/Ton of CO_2 Exceeded			LL97 allowed cap	2,381
			Emissions exceeded	1,501
			Estimated penalty	$402,390

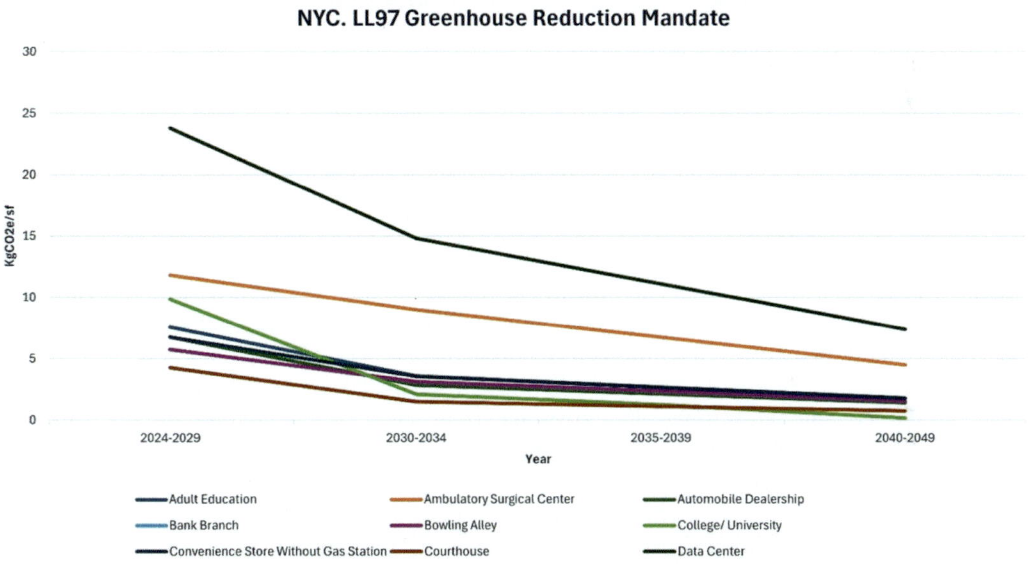

FIGURE 14.1 NYC's GHG reduction mandate for various occupancies (2024–2050).

- Lighting retrofits: Replacing the current CFL (Compact Fluorescent Lights) and incandescent lighting systems throughout the premises with efficient and long-lasting LED bulbs.
- Installing smart sensors integrated with IoT to measure and monitor the energy and emissions, and apply control measures (Kasinathan, 2022).
- On site solar solution.
- On site battery storage systems.
- Clean distributed energy resource.
- Undertaking periodic energy audits by qualified energy managers throughout the campus, and making required repairs and renovations as and when required (Kasinathan, 2006).

14.1.1 Conclusions and Lessons Learned

LL 97 is about estimating facilities' GHG emissions to assess where they are in the year 2024 versus how they will reduce their emissions in 2030 and 2050. NYC has established and requires the facilities, depending upon the building occupancies, to make considerable efforts to offset the increased emissions from the baseline to avoid stiff penalties. For instance, a hospital facility in NYC having multiple emission sources is required to pay thousands of dollars in penalties, which requires it to implement several energy savings programs to offset the penalties. This case study helps to understand how a new project with the potential to emit GHG emissions can opt for energy-saving components within the design.

14.2 CASE STUDY: CEMENT INDUSTRIES AND CFP REDUCTION INITIATIVES

Case Overview: The cement industry is one of the largest industrial sources of GHG emissions due to the chemical reactions involved and the energy-intensive nature of cement production (www.weforum.org/stories/2024). However, the industry has been making significant improvements to reduce its carbon footprint (CFP) through a variety of process modifications and upgrades. Cement industries that operate in countries like India have undertaken several qualitative and quantitative measures to reduce the CFP (www.ultratechcement.com). Incorporation of Waste Heat Recovery Systems (WHRS) reduces a significant amount of CFP. Other measures cement industries are employing are renewable energy integration (REI) and alternative fuels and raw materials (AFRs).

India's cement industries are keen on the application of WHRS. Ultratech Cement Company in India has implemented WHRS at several of its plants. A plant in Rajashree, Karnataka, uses waste heat to generate electricity, thus avoiding external power. Based on the published data, WHRS reduces CO_2 emissions by about 10%–15% per ton of cement produced due to lower energy consumption. AFRs are used in cement industries, such as ACC Cement in Wadi, India (www.cement-equipment.org). They include fuels such as biomass and industrial wastes. ACC has also set up a dedicated AFR plant to process and utilize these materials. The use of AFR can cut CO_2 emissions by up to 20% per ton of cement, depending on the proportion of alternative fuels used.

REI is employed at Ambuja Cements, which has integrated solar power at its plants, including a significant project at its Marwar Mundwa plant in India. The integration of solar energy can reduce reliance on grid power, cutting CO_2 emissions by around 20%–30% at the plant level.

In the United States, CEMEX, located in Houston, Texas, produces low-carbon cement with reduced clinker content and utilizes alternative materials. Their CFP reduction initiatives include the utilization of renewable energy in their production. Investing in solar and wind energy for plant operations may significantly reduce the CFP.

Blended cements production offers a great deal of CFP reduction since fly ash and slag are used in the cement production, which already have the lowest CFP rate. India's Dalmia Bharat Cement is spearheading this effort. Blended cements can also lower CO_2 emissions by up to 30% compared to traditional Portland cement due to reduced clinker content.

Carbon capture and storage (CCS) is another means of CFP reduction. Heidelberg Cement, India, has initiated research into CCS technologies in collaboration with local research institutions. CCS can potentially capture about 90% of the CO_2 emissions from cement production, although this is still in the pilot phase.

Advanced Control Systems and Process Optimization are being utilized in many cement industries. For example, JK Cement has implemented advanced process control systems and automation to optimize kiln operations. These technologies can improve energy efficiency by up to 10% and reduce emissions proportionally.

14.2.1 Qualitative and Quantitative CFP Reductions

Qualitative Reductions involve the development of low-carbon cements, improvements in process efficiency, and innovations in technology. For example, incorporating supplementary cementitious materials (SCMs) can improve the properties of cement and reduce its carbon intensity.

Quantitative Reductions: Cement industries globally have set ambitious targets by aiming to reduce their carbon intensity by 20%–30% by 2030 compared to 1990 levels.

Achievements: Some specific achievements include Lafarge Holcim, which reduced CO_2 emissions per ton of cement by 20% from 1990 to 2020; CEMEX achieved a 22% reduction in CO_2 emissions per ton of cement between 1999 and 2021 (www.cemex.com). Ambuja Cements aimed to reduce CO_2 emissions by 24% by 2030 compared to 2010 levels. ACC Cement reduced CO_2 emissions by 10% between 2016 and 2021 through energy efficiency and SCM use.

14.2.2 Specific Emission Reductions

- **Energy Efficiency Projects**: Calculate the reduction in energy consumption and associated CO_2 emissions. If energy use decreases by a certain percentage, multiply this by the emission factor of the energy source.
- **Clinker Substitution**: If using SCMs reduces the amount of clinker needed, calculate the CO_2 saved by the reduction in clinker production.
- **Carbon Capture**: Determine the percentage of CO_2 captured and use: Captured CO_2 = Total CO_2 Emissions × Capture Rate.

Lifecycle Analysis: This assesses the total CO_2 emissions over the lifecycle of a product, from raw material extraction to disposal. It includes all stages of production and can help in understanding the broader impact of new technologies.

14.2.3 Problems Faced

- **High Costs**: Implementing new technologies like CCS or alternative fuel systems can be expensive and economically challenging, especially for smaller companies.
- **Technical Challenges**: Efficiently integrating new technologies into existing plants and ensuring they operate reliably can be complex.
- **Regulatory and Policy Uncertainty**: Inconsistent regulations and lack of clear policies can hinder investment in new technologies.
- **Supply Chain Issues**: Availability and cost of alternative fuels and raw materials can be variable and unpredictable.

14.2.4 Solutions and Future Directions

- **Innovation and R&D**: Continued investment in research and development to lower the costs and improve the efficiency of low-carbon technologies.
- **Policy Support**: Advocating for clear, consistent policies and incentives that support low-carbon technologies and practices.
- **Industry Collaboration**: Collaborative efforts among companies and stakeholders to share knowledge, develop standards, and scale up successful technologies.
- **Circular Economy Approaches**: Embracing circular economy principles by finding new uses for waste products and integrating them into the production process.
- **Digitalization**: Utilizing digital technologies and data analytics to optimize operations, reduce energy consumption, and improve efficiency.

14.2.5 Conclusions and Lessons Learned

Cement industries are actively working on a range of projects and technologies to mitigate their carbon footprints. The combination of improving energy efficiency, adopting alternative fuels, incorporating SCMs, and investing in carbon capture and utilization technologies is key to these efforts (www.imperial.ac.uk/news). Despite facing significant challenges, ongoing innovation and supportive policies offer pathways to substantial reductions in GHG emissions (www.cement.

org/a-sustainable-future). AI technologies will offer significant CFP reductions in the future, despite initial capital investments.

14.3 CASE STUDY: CHEMICAL INDUSTRIES AND CFP REDUCTION INITIATIVES

Case Overview: The chemical industry is another major contributor to global GHG emissions due to its energy-intensive processes and the production of carbon-intensive products. However, there are several initiatives and technologies being adopted to mitigate the industry's carbon footprint.

14.3.1 Energy Efficiency Improvements

Process Optimization: Implementing advanced process control systems and automation to enhance the efficiency of chemical processes. Techniques like real-time optimization and advanced analytics help in reducing energy consumption and waste. Tata Chemicals Limited has adopted process optimization and control technologies to improve energy efficiency in its manufacturing processes.

Energy Recovery Systems: Using heat recovery systems to capture and reuse waste heat from chemical processes. This can be applied to various processes, such as steam generation and heating. Reliance Industries Limited has installed heat recovery systems in its petrochemical operations to improve energy efficiency.

Alternative Feedstocks and Raw Materials: Developing and using renewable or less carbon-intensive feedstocks to replace traditional fossil-based raw materials. This includes bio-based feedstocks and recycled materials. BASF India is exploring the use of bio-based raw materials and waste recycling in its chemical production processes.

Improved Catalysts: Using advanced catalysts that enhance process efficiency and reduce by-product formation, leading to lower energy consumption and emissions. Hindustan Organic Chemicals Ltd. (HOCL) focuses on developing and using more efficient catalysts in its chemical processes.

Carbon Capture and Storage (CCS): Some companies are exploring carbon capture and storage technologies to capture CO_2 emissions from chemical processes. These projects are often in the pilot or early implementation stages. Grasim Industries is involved in research and pilot projects related to CCS. Chevron and Occidental Petroleum companies are involved in CCS projects that capture CO_2 emissions from chemical production and inject it underground for permanent storage. The Gulf Coast Carbon Center is one such initiative focused on large-scale implementation.

Renewable Energy Integration: Integrating renewable energy sources such as solar and wind into the energy mix of chemical manufacturing facilities to reduce reliance on fossil fuels. Gujarat Alkalies and Chemicals Ltd. (GACL) has invested in renewable energy projects to power its operations and reduce its carbon footprint.

Waste Management and Recycling/Zero-Waste Initiatives: Implementing waste management strategies to reduce waste generation and promote recycling of industrial by-products. This includes using waste as a feedstock for other processes. Shree Cements has a zero-waste policy where industrial by-products are used in other processes, reducing the overall waste and emissions.

Circular Economy Practices: Adopting circular economy principles to close the loop of product life cycles through greater resource efficiency and recycling. Tata Chemicals has embraced circular economy principles, focusing on recycling and reusing waste materials in its production processes.

Sustainable Practices and Green Certifications: Pursuing certifications like ISO 14001 (Environmental Management Systems) and other sustainability certifications to ensure adherence to environmental standards. Reliance Industries holds ISO 14001 certification and other environmental certifications, reflecting its commitment to sustainability. Clariant company is developing sustainable products with lower environmental impacts, such as biodegradable plastics and specialty chemicals derived from renewable resources.

14.3.2 Qualitative and Quantitative Reductions

BASF aims to reduce its CO_2 emissions by 25% by 2030 compared to 2018 levels. Dow targets a 15% reduction in GHG emissions by 2030 from a 2020 baseline (corporate.dow.com/en-us). Clariant reduced GHG emissions by 10% between 2015 and 2020 through energy efficiency measures and alternative feedstocks. Solvay achieved a 22% reduction in CO_2 emissions per ton of product between 2012 and 2020 by improving energy efficiency and using renewable energy. Reliance Industries plans to reduce its carbon intensity by 20% by 2030 from a 2016 baseline (Singh, 2024). Tata Chemicals achieved a 15% reduction in GHG emissions per ton of product between 2010 and 2020 (sustainability.tatacommunications.com) (Statista, n.d.). Asian Paints reduced energy consumption by 7% and GHG emissions by 5% from 2018 to 2020 through energy efficiency measures and process improvements (Asian Paints, BASF report, 2022; sustainability.asianpaints.com/sustainability).

14.3.3 Conclusions and Lessons Learned

By adopting these strategies and addressing the associated challenges, the chemical industry can make significant progress in reducing its carbon footprint and contributing to global GHG reduction goals.

14.4 CONCLUSION

The case studies in this chapter demonstrate how major emission-intensive sectors—such as the building, cement, and chemical industries—are addressing their greenhouse gas footprints through compliance, innovation, and targeted mitigation. In the United States, New York City's Climate Mobilization Act and Local Law 97 demonstrate how ambitious regulatory frameworks can drive emissions reductions in large buildings and hospitals, with strict penalties incentivizing the adoption of energy-efficient technologies, smart monitoring, and renewable energy sources.

The cement industry case studies span both the U.S. and international contexts, with detailed examples from India and the United States. Indian companies such as Ultratech, ACC, Ambuja, and Dalmia Bharat are implementing waste heat recovery, alternative fuels, solar integration, blended cement, and pilot carbon capture projects, resulting in significant reductions in carbon intensity. In the U.S., companies like CEMEX are producing low-carbon cement and investing in renewable energy to further reduce their carbon footprint.

The chemical industry section highlights efforts by global companies to reduce emissions through process optimization, energy recovery, alternative feedstocks, renewable energy sources, waste management, and circular economy practices. Notable achievements include double-digit percentage reductions in emissions per ton of product and industry-wide commitments to ambitious 2030 targets.

Across all sectors, the most effective strategies blend regulatory compliance, operational efficiency, technological innovation, and collaboration. Key lessons include the importance of robust data and monitoring, early investment in low-carbon solutions, and cross-sector partnerships. While challenges remain—such as high upfront costs and policy uncertainty—the direction is clear: proactive climate action not only ensures compliance but also delivers operational savings, resilience, and long-term competitiveness.

These case studies confirm that substantial decarbonization is both possible and practical. The path forward requires continuous learning, adaptation, and a commitment to turning climate goals into measurable results for industries, cities, and the environment.

REVIEW QUESTIONS

1. What are the lessons learned from Case Study 14.1?
2. What are the lessons learned from Case Study 14.2?
3. What are the lessons learned from Case Study 14.3?
4. What conclusions can be drawn across all sectors?

REFERENCES

Amended version of Local Law 97 of 2019, §28–320 and §28–321 of the Administrative Code/Buildings Emissions Limits Under Local Law 97 of 2019.

Asian Paints, BASF report, 2022, Energy and Climate Protection, https://report.basf.com/2022/en/managements-report/sustainability-along-the-value-chain/safe-and-efficient-production/energy-and-climate-protection.html

https://www.nyc.gov/site/buildings/codes/ll97-buildings-emissions-limits.page

https://www.weforum.org/stories/2024/09/cement-production-sustainable-concrete-co2-emissions/

https://www.ultratechcement.com/corporate/media/press-releases/waste-heat-recovery

https://www.cementequipment.org/home/alternative-fuels-and-raw-materials/

https://www.cemex.com/w/cemex-develops-innovative-approach-to-reduce-carbon-emissions

https://www.cement.org/a-sustainable-future/reaching-our-goal/blended-cements/

https://www.imperial.ac.uk/news/221654/best-ways-carbon-emissions-from-cement/

https://sustainability.asianpaints.com/sustainability/shrinking-carbon-footprint.php

https://sustainability.tatacommunications.com/Planet/energy-&-climate-changeenergy&climatechange

https://corporate.dow.com/en-us/news/press-releases/dow-sets-targets-to-reduce-ghg-emissions--stop-plastic-waste--an.html

https://corporate.dow.com/en-us/purpose-in-action/sustainability-targets.html

https://sustainability.tatacommunications.com/Planet/energy-&-climate-change

https://www.spglobal.com/commodity-insights/en/news-research/latest-news/energy-transition/020724-reliance-industries-plans-to-reuse-captured-carbon-for-internal-industrial-processes

Kasinathan, R. (2006). *Environmental Compliance Tracking Software.* US Copyright.

Kasinathan, R. (2022). System of sensors and Controllers integrated with IoT for maintaining environmental health and safety. US Patent.

Kasinathan, R. (2023). *Environmental Compliance Guide for Facility Managers and Engineers*. CRC Press,

Singh, R. (2024, February 7). Reliance Industries plans to reuse captured carbon for internal industrial processes. *S&P Global*. https://www.spglobal.com

Statista. (n.d.). Greenhouse gas emissions of the chemical and petrochemical industry worldwide in 2010 and 2020, with a forecast until 2030, by scenario (in million metric tons of CO_2 equivalent).

Answer Keys

CHAPTER 1: REVIEW QUESTIONS AND ANSWER KEYS

1. **What is a carbon footprint, and why is it important in understanding climate change?**
   A carbon footprint (CFP) is the total amount of carbon dioxide and other greenhouse gases (GHGs) emitted directly or indirectly by human activities. It is crucial for understanding climate change because it quantifies human impact on the planet, especially in terms of emissions that drive global warming.
2. **How do the carbon cycle and greenhouse gas (GHG) emissions impact global warming?**
   The carbon cycle maintains balance in the Earth's systems. Human-induced GHG emissions from fossil fuels, deforestation, and agriculture disrupt this cycle, causing excess CO_2 and methane in the atmosphere, enhancing the greenhouse effect, and contributing to global warming and ocean acidification.
3. **Why are climate feedback loops significant in the study of climate change?**
   Climate feedback loops, such as increased water vapor and melting ice, amplify the effects of climate change. For instance, warmer temperatures cause ice to melt, reducing reflectivity and increasing heat absorption, which leads to further warming. These loops can accelerate global temperature rise beyond predictions.
4. **How does the Paris Agreement aim to address carbon emissions on a global scale?**
   The Paris Agreement aims to limit global warming to well below 2°C, preferably 1.5°C, by requiring countries to set and regularly update nationally determined contributions (NDCs). It also promotes financial aid, technology sharing, and capacity building for developing countries to reduce emissions and adapt to climate change.
5. **What actions can individuals take to reduce their carbon footprints?**
   Individuals can reduce their CFP by:
   - Consuming local and seasonal foods
   - Limiting meat (especially beef) consumption
   - Using reusable bags and minimizing plastic use
   - Reducing waste and only buying what is necessary
   - Conserving energy and opting for sustainable transport
6. **How do different greenhouse gases contribute to global warming potentials?**
   Different GHGs have different global warming potentials (GWPs) based on their heat-trapping ability and atmospheric lifespan. For example:
   - CO_2 (reference gas) = GWP 1
   - Methane (CH_4) = GWP ~30
   - Nitrous oxide (N_2O) = GWP ~273
   - SF_6 = GWP over 25,000

   These metrics allow emissions from various gases to be compared on a common scale (CO_2e).
7. **In what ways does reducing CFP impact human health and ecosystems?**
   Reducing CFP improves air and water quality, reduces exposure to pollution-related illnesses, and lessens the impact of climate change on food systems and biodiversity. It helps preserve ecosystems, reduces climate-induced disasters, and supports public health and food security.

8. **How does international collaboration influence carbon reduction efforts?**
   International collaboration, through agreements like the Paris Accord and COP summits, ensures collective commitment to reducing emissions. It enables resource sharing, technology transfer, and unified tracking systems like the Enhanced Transparency Framework, ensuring equitable action among nations.
9. **What role do natural vs. human-made emissions play in the global CFP?**
   Natural processes (e.g., volcanic eruptions and respiration) emit GHGs but are typically balanced by natural sinks. Human-made emissions—especially from fossil fuel combustion, deforestation, and industrial activities—are the primary drivers of recent increases in atmospheric GHGs and the global carbon footprint.
10. **How is CFP measurement and reporting essential in environmental policies?**
   Accurate CFP measurement using standardized units like CO_2e is vital for setting benchmarks, comparing emissions across sectors and regions, and designing effective policies. It also supports transparency, accountability, and progress tracking in international climate commitments and environmental governance.

CHAPTER 2: REVIEW QUESTIONS AND ANSWER KEYS

1. **What are the primary sectors contributing to GHG emissions?**
   The primary sectors contributing to greenhouse gas emissions are:
   - **Electric power generation** (burning fossil fuels like coal and natural gas)
   - **Industry** (chemical processing, combustion, electricity use)
   - **Transportation** (vehicles, aviation, shipping)
   - **Commercial and residential sectors** (heating, cooling, electricity use)
   - **Agriculture** (soil management, livestock, manure)
   - **Land use, land-use change, and forestry** (both as a source and a carbon sink)
2. **How does carbon dioxide (CO_2) differ from other GHGs like methane (CH_2) in its impact on global warming?**
   CO_2 is the most prevalent GHG and remains in the atmosphere for centuries, causing long-term warming. Methane (CH_4), although less abundant, has a much higher global warming potential (around 30 times that of CO_2 over 100 years) but persists for a shorter period. Thus, CH_4 has a stronger short-term warming effect.
3. **What are the specific sources of emissions in the transportation sector, and how can they be reduced?**
   Sources include fossil fuel combustion by cars, trucks, planes, and ships. Emissions can be reduced by:
   - Adopting **zero-emission vehicles** (e.g., EVs and hydrogen)
   - Switching to **biofuels or CNG**
   - Improving **fuel efficiency and driving practices**
   - Reducing travel demand **by urban planning**
   - Investing in **public transport and infrastructure**
4. **How do agricultural practices contribute to GHG emissions, and what solutions exist?**
   Agriculture emits GHGs through:
   - **Methane** from enteric fermentation in livestock
   - **Nitrous oxide** from fertilizers and soil management
   - **Manure management**

   Solutions include:
   - Precision farming and efficient fertilizer use
   - Improved livestock feeding
   - Manure treatment technologies
   - Promoting organic and sustainable farming

5. **Why is it important to measure emissions in carbon dioxide equivalent (CO_2e) across sectors?**
   CO_2e allows comparison of different GHGs based on their global warming potential (GWP). Standardizing emissions this way enables policymakers to track, report, and compare impacts across gases and sectors effectively, ensuring better mitigation strategies.
6. **What role does industrial activity play in GHG emissions, and what steps can industries take to mitigate this?**
   Industry contributes significantly via combustion, chemical reactions, and electricity use. Mitigation steps include:
   - **Energy efficiency improvements**
   - **Fuel switching** to lower-carbon or renewable sources
   - **Recycling and waste reduction**
   - **Carbon capture and storage (CCS)**
   - **Training and leak prevention**
7. **How do land-use changes, like deforestation, contribute to climate change?**
   Deforestation reduces carbon sinks and releases stored CO_2 into the atmosphere. It disrupts ecosystems, lowers biodiversity, and contributes to rising atmospheric GHG concentrations, accelerating climate change.
8. **How can renewable energy sources replace fossil fuels in the power sector?**
   Renewables like solar, wind, hydro, and geothermal produce electricity without emitting CO_2. Transitioning involves:
   - Expanding renewable infrastructure
   - Upgrading grids (e.g., smart grids)
   - Implementing energy storage solutions
   - Encouraging policy and investment in clean energy
9. **What are carbon sinks, and how do they help offset emissions?**
   Carbon sinks are natural systems like forests, soils, and oceans that absorb more CO_2 than they emit. They offset emissions by removing CO_2 from the atmosphere, playing a vital role in climate regulation.
10. **How can governments and corporations work together to reduce emissions in high-impact sectors?**
    Governments can set regulations, provide subsidies, and fund research. Corporations can adopt clean technologies, report emissions transparently, and invest in sustainability. Collaboration ensures technological advancement, economic viability, and global climate progress.

CHAPTER 3: REVIEW QUESTIONS AND ANSWER KEYS

1. **What are the main health risks associated with increased global temperatures?**
   Increased temperatures elevate risks of heat-related illnesses and mortality, waterborne and vector-borne diseases, foodborne illnesses, and mental health disorders. Vulnerable populations are especially at risk due to limited adaptive capacity.
2. **How does climate change impact food security and nutrition?**
   It disrupts food supply chains, increases food prices, spreads pathogens, and reduces the protein and mineral content of crops due to elevated CO_2. Marginalized communities are most affected.
3. **What are vector-borne diseases, and how does climate change influence their spread?**
   These are illnesses transmitted by vectors like mosquitoes and ticks (e.g., malaria and West Nile virus). Climate change alters habitats and extends the geographic and seasonal range of vectors, increasing disease risk.

4. **How does climate change exacerbate mental health issues, particularly among vulnerable populations?**
   It heightens stress, anxiety, and PTSD—especially after extreme weather events. Children, the elderly, and low-income groups face greater psychological harm due to reduced resources and support systems.
5. **What are the projected impacts of sea-level rise on coastal cities by 2050?**
   Coastal cities will face more flooding, property damage, and population displacement. The Atlantic and Gulf Coasts are especially vulnerable, with billions in property value at risk and tens to hundreds of thousands potentially displaced.
6. **How are extreme weather events linked to climate change, and what are their effects on communities?**
   Climate change increases the frequency and intensity of heatwaves, floods, droughts, and storms. These events cause death, displacement, infrastructure damage, and strain on health systems.
7. **Why is water availability a concern in the context of global warming?**
   Warming reduces freshwater through evaporation and altered precipitation. It threatens drinking water, agriculture, and power generation, especially in arid and overpopulated regions.
8. **How does climate change affect energy demand, particularly in hotter regions?**
   Hotter climates increase energy use for cooling, stressing power grids. Tropical and mid-latitude areas may see major demand spikes, especially under scenarios with rapid economic growth.
9. **What are the socioeconomic impacts of climate change on disadvantaged communities?**
   Disadvantaged groups suffer more from health effects, food and water insecurity, economic instability, and displacement due to limited resources for adaptation and recovery.
10. **How can countries prepare for and adapt to the health challenges posed by climate change?**
    Countries can prepare for and adapt to the health challenges posed by climate change through climate-smart agriculture, resilient infrastructure, early-warning systems, improved healthcare, international cooperation, and policies targeting vulnerable populations.

CHAPTER 4: REVIEW QUESTIONS AND ANSWER KEYS

1. **How has scientific understanding of climate change evolved over recent decades?**
   Scientific understanding of climate change has evolved from early theories about the greenhouse effect in the 19th century to the modern-day integration of complex Earth system models and robust observational data. Key milestones include the identification of greenhouse gases' role in warming, the development of climate models, and the increasing evidence from paleoclimate records and satellite data. By the 21st century, climate change became an established scientific fact, supported by improved climate models (e.g., CMIP6) and global consensus through IPCC reports.
2. **What is the current consensus on the causes of climate change?**
   There is overwhelming consensus that **human activities**—primarily the emission of greenhouse gases from fossil fuel combustion, deforestation, and industrial processes—are the dominant cause of recent climate change. These anthropogenic factors have altered the Earth's radiative balance and are responsible for observed increases in global temperatures, extreme weather events, and sea-level rise.
3. **How do scientists model and predict future climate change impacts?**
   Scientists use **climate models**, particularly Earth system models (ESMs) like those in the CMIP6 framework, to simulate the Earth's climate. These models incorporate physical

processes, greenhouse gas emissions, aerosols, feedback mechanisms, and biogeochemical cycles. Future projections are made under various **Shared Socio-economic Pathways (SSPs)** that represent different emissions scenarios, enabling assessments of potential climate change impacts and risks.

4. **What is the significance of the IPCC's assessment reports in guiding climate policy?**
   IPCC reports synthesize the latest climate science and provide **authoritative, policy-relevant guidance** to governments and organizations. These reports have shaped international agreements like the **Paris Agreement**, helped quantify global warming thresholds (e.g., 1.5°C and 2°C), and informed climate action plans by identifying impacts, risks, and mitigation/adaptation strategies.
5. **How do human activities disrupt natural greenhouse gas cycles?**
   Human activities disrupt the natural carbon and nitrogen cycles by **increasing the atmospheric concentrations** of CO_2, CH_4, and N_2O through fossil fuel burning, deforestation, agriculture, and industrial emissions. This excess GHG accumulation exceeds natural sinks' capacity, intensifying the greenhouse effect and leading to global warming and other climate disruptions.
6. **What role does the U.S. Environmental Protection Agency (EPA) play in climate research?**
   The **U.S. EPA** monitors greenhouse gas concentrations (e.g., through the Climate Change Indicators project), regulates emissions under the Clean Air Act, and contributes to scientific assessments. It also plays a role in **public awareness, policy development, and emission reduction initiatives**, supporting national climate goals.
7. **How does atmospheric data inform climate projections?**
   Atmospheric data—including CO_2 levels, temperature trends, and aerosol concentrations—provide the **observational foundation** for validating and improving climate models. These data help establish historical baselines, identify trends, and enhance the accuracy of future projections by enabling comparisons between observed and simulated outcomes.
8. **What are the limitations of current climate models?**
   Limitations include:
   - Incomplete understanding of **regional precipitation patterns**, deep convection, and cloud-aerosol interactions.
   - **Biases** in simulating extreme events and tropical ocean variability.
   - **Uncertainty** in feedback mechanisms (e.g., ice sheet behavior and methane release). Despite improvements, challenges remain in achieving **fine-scale accuracy** and simulating long-term Earth system responses.
9. **How do environmental feedback loops affect climate predictions?**
   Feedback loops—such as those involving water vapor, albedo changes from ice melt, or carbon cycle dynamics—can **amplify or dampen** climate responses. For example, warming-induced permafrost thaw can release methane, enhancing warming. These complex interactions increase **uncertainty** in long-term predictions but are essential for refining projections and mitigation strategies.
10. **What are some technical innovations that aid in climate research?**
    Innovations include:
    - **High-resolution satellite observations** and reanalysis datasets.
    - **Earth system models** (e.g., CMIP6 and CESM2).
    - **Machine learning techniques** for extreme event attribution.
    - **Biogeochemical modeling** of carbon and aerosol feedbacks.
    - **Climate data integration platforms** supporting cross-disciplinary research and policy planning.

CHAPTER 5: REVIEW QUESTIONS AND ANSWER KEYS

1. **What defines Scope 1 emissions, and why are they essential in GHG accounting?**
   Scope 1 emissions are direct greenhouse gas emissions from sources controlled by an organization—such as emissions from fuel combustion in boilers, vehicles, and fugitive leaks. They are essential because they reflect emissions from operations that the company has direct control over, allowing for immediate action and responsibility in emission reduction.
2. **How do Scope 2 emissions differ from Scope 1, and what role does purchased electricity play?**
   Scope 2 emissions are indirect emissions from the generation of purchased energy (like electricity, steam, or heating) consumed by the company. Unlike Scope 1, these occur at the facility that produces the energy, not within the company itself. Purchased electricity is the major component of Scope 2 emissions, making accurate tracking essential for reflecting energy consumption impacts.
3. **What makes Scope 3 emissions challenging to measure, and why are they important?**
   Scope 3 emissions include **all other indirect emissions** in a company's value chain, both upstream (e.g., purchased goods and transport) and downstream (e.g., product use and end-of-life). They are difficult to measure due to their scope, data collection challenges, and lack of control over external sources. However, they are crucial because they often represent the largest share of a company's carbon footprint.
4. **How can companies accurately calculate and report their Scope 3 emissions?**
   Companies can calculate Scope 3 emissions by:
   - Identifying relevant categories (15 defined by the GHG Protocol).
   - Setting boundaries and collecting both primary and secondary data.
   - Using activity data and emission factors to estimate emissions.
   - Applying allocation methods when multiple products or services are involved.
   - Ensuring transparency, using assurance, and reporting category-wise totals.
5. **What role does the National Grid play in managing emissions across all three scopes?**
   The **National Grid** primarily influences **Scope 2** emissions through the carbon intensity of the electricity it supplies. Its energy mix (e.g., renewables vs. fossil fuels) affects emission factors for users. It indirectly affects Scope 3 (upstream/downstream distribution) and supports Scope 1 reductions by integrating clean energy and encouraging grid efficiency.
6. **How can companies reduce their Scope 1 emissions from stationary and mobile combustion sources?**
   Reduction strategies include:
   - **Upgrading equipment** to energy-efficient models.
   - **Switching to low-carbon fuels** or electrification.
   - **Implementing maintenance programs** to prevent fuel inefficiency.
   - **Using monitoring systems** (e.g., CEMS) and **data analysis** to identify hotspots and improve combustion efficiency.
7. **What are fugitive emissions, and how can they be minimized in industrial operations?**
   Fugitive emissions are unintentional releases of gases, often from leaks in valves, equipment, or refrigerants. They can be minimized by:
   - Using **hermetic equipment** and **low-leakage designs**.
   - Implementing **regular inspections**, **leak detection**, and **repair programs**.
   - Replacing high-risk components like rising-stem valves with ball valves.
8. **Why is it essential for companies to track emissions across their supply chain?**
   Tracking **supply chain (Scope 3)** emissions allows companies to:
   - Identify **emission hotspots** and reduction opportunities.
   - Improve **resource efficiency**.

- Meet **regulatory and customer expectations**.
- Build **transparency and brand trust** through comprehensive climate action.

9. **How does emissions data transparency affect stakeholder trust and environmental accountability?**
   Transparency in GHG data:
   - **Builds trust** with investors, customers, and regulators.
   - **Demonstrates credibility** in climate commitments.
   - Supports **decision-making** and risk assessment.
   - Encourages industry-wide **accountability and innovation** in emission reduction.
10. **What technologies are available to help organizations monitor and reduce emissions in all scopes?**
   Technologies include:
   - **Continuous Emissions Monitoring Systems (CEMS)** for real-time tracking.
   - **Carbon accounting software** and cloud-based data platforms.
   - **Advanced metering and sensors**.
   - **Life cycle assessment tools**.
   - **Machine learning** and data analytics for emission forecasting and decision support.

CHAPTER 6: REVIEW QUESTIONS AND ANSWER KEYS

1. **What are the main objectives of the U.S. Environmental Protection Agency (EPA) in addressing climate change?**
   The EPA aims to:
   - Reduce greenhouse gas emissions through regulations and voluntary programs.
   - Promote renewable energy and energy efficiency.
   - Support climate resilience across sectors like transportation, energy, agriculture, and water.
   - Foster collaboration with states, communities, and businesses.
   - Promote environmental justice by protecting vulnerable communities from climate impacts.
2. **How does the ENERGY STAR program contribute to emissions reduction in homes and businesses?**
   ENERGY STAR identifies and promotes energy-efficient products and buildings. It helps:
   - Reduce electricity usage.
   - Lower utility bills.
   - Avoid millions of tons of greenhouse gas emissions.
   - Certify buildings with superior energy performance (score ≥ 75). Since 1992, it has saved $500+ billion in energy costs and reduced 4 billion metric tons of GHGs.
3. **What is the purpose of the EPA's Climate Change Adaptation Resource Center (ARC-X)?**
   ARC-X is an interactive tool that:
   - Helps local governments prepare for climate change.
   - Provides tailored adaptation strategies and resources.
   - Supports decision-makers in integrating climate resilience into planning and operations.
4. **How does LEED certification encourage energy-efficient building practices?**
   LEED (Leadership in Energy and Environmental Design):
   - Promotes sustainable building design and construction.
   - Sets performance benchmarks for energy, water, materials, and health.
   - Encourages the use of renewable energy and resource conservation.

- Recognizes buildings that reduce GHGs, conserve resources, and improve occupant health.

5. **What are the roles of voluntary programs like the GPP in climate action?**
   The **Green Power Partnership (GPP)**:
   - Encourages organizations to purchase renewable electricity (solar, wind, etc.).
   - Provides recognition, tools, and guidance.
   - Helps develop the voluntary green power market.
   - Drives consumer and institutional support for renewable energy projects.
6. **How does the EPA collaborate with states, tribes, and local governments on climate initiatives?**
   The EPA:
   - Offers technical assistance, funding, and tools (e.g., State and Local Climate and Energy Program).
   - Supports initiatives like the Heat Island Reduction Program and the Advance Program.
   - Engages in partnerships to promote clean energy, climate resilience, and air quality at the local level.
7. **What are methane reduction programs, and how do they target high-emission sectors?**
   EPA programs are given as follows:
   - **AgSTAR** – captures methane from livestock waste via anaerobic digestion.
   - **CMOP** – reduces methane from coal mines.
   - **GMI** – supports international methane recovery projects.
   - **LMOP** – captures landfill gas for energy.
   - **Natural Gas STAR** – targets methane reduction in the oil and gas industry.
   These initiatives reduce powerful GHGs and turn waste gases into energy sources.
8. **How does the EPA's SmartWay Transport Partnership help reduce emissions in transportation?**
   SmartWay:
   - Helps freight companies track and reduce emissions.
   - Promotes efficient logistics and cleaner vehicles.
   - Encourages public-private collaboration.
   - Reduces fuel use, operational costs, and GHG emissions in freight transport.
9. **What role does the RAD Program play in reducing waste-related emissions?**
   The **Responsible Appliance Disposal (RAD) Program**:
   - Ensures proper disposal of old refrigerators and air conditioners.
   - Prevents release of high-GWP refrigerants and insulating foam gases.
   - Encourages recycling and reduces landfill methane emissions.
10. **How does the IRA influence the EPA's climate change goals?**
    The **Inflation Reduction Act (IRA)**:
    - Provides \$2 billion for the **Community Change Grants** for disadvantaged communities.
    - Provides funding for pollution reduction, climate resilience, and infrastructure.
    - Supports EPA-led programs to advance environmental justice and accelerate climate action at the community level.

CHAPTER 7: REVIEW QUESTIONS AND ANSWER KEYS

1. **How are carbon emissions measured through various metrics?**
   Carbon emissions are measured using:
   - **Total emissions** (aggregate CO_2 output of a country/entity),
   - **Per capita emissions** (average emissions per person),

- **Lifecycle emissions** (emissions across a product/service's life from production to disposal).

  These metrics help assess environmental impact, equity, and opportunities for reduction.

2. **How do data sources for emission estimations help track progress and implement effective climate policies?**

   Data sources like **EDGAR**, **UNFCCC inventories**, and **World Bank databases** enable:
   - Trend analysis,
   - Policy evaluation,
   - Transparency and accountability.
   - They help distinguish between production-based and consumption-based emissions, revealing hidden carbon footprints in global trade.
3. **Who are the top greenhouse gas emitters by total carbon emissions and by per capita?**
   - **Top total emitters (2022)**: China, USA, India, EU27, Russia, Brazil.
   - **Top per capita emitters**: USA (~17.9 tCO_2e), Russia (~17.98), much higher than China or India.

     Per capita figures highlight inequality despite high total emissions from populous developing countries.
4. **What are the main drivers of national per capita emissions in the United States?**

   Key drivers include:
   - **High income and consumption**, especially among top quintile households;
   - **Energy use in transportation, housing, and food**, with wealthier households emitting up to 2.6× more than lower-income ones.
5. **What challenges do developing nations with low total emissions but high growth rates face?**

   Challenges include:
   - Balancing economic growth with emission controls,
   - High costs of clean energy transitions, and
   - Political and social resistance to reforms, making it hard to shift to low-carbon development without support.
6. **What is the global gridded model of carbon footprints based on the EU's population model?**

   The **GGMCF** model maps carbon footprints spatially (250m resolution) across 189 countries using the EU's GHS-SMOD data.

   It shows that a small number of affluent cities produce a large share of national emissions—useful for targeted urban mitigation.
7. **How can Net Zero by 2050 be achieved, including U.S. strategies?**

   Strategies include:
   - **Energy efficiency**, **renewable energy**, and **electrification**;
   - **Carbon capture and storage**, and **afforestation**; and
   - Adoption of **hydrogen**, **low-carbon fuels**, and **nature-based solutions**.
   - U.S. plans combine these with region-specific policies and investment in innovation.
8. **How do energy sectors contribute to GHG increases in India and Brazil, and what measures are they taking?**
   - **India**: Heavy reliance on coal, especially for electricity. Steps include renewables, energy efficiency, and cleaner agriculture.
   - **Brazil**: Emissions from transport, industry, and livestock. Measures include expanding renewables, biofuels, and emissions tracking.
9. **Discuss the European Union's metal industries' contribution to global warming and their reduction efforts.**

   EU metal industries (iron, steel, cement, aluminum) are major GHG contributors, especially in Germany, Poland, etc.

The EU responds with:

- **Emission Trading System (EU ETS)**, and
- **Decarbonization of industrial sectors**, aiming to meet the 2030 targets.

10. **What are the observed changes in climate aspects illustrated in Figure 7.11?**
    Key U.S. climate changes (2002–2021 vs. 1901–1960) include:
    - **Warming** across all regions,
    - **More warm nights** except in the Northern Great Plains,
    - **Increased precipitation** in most areas,
    - **Heavier rainfall events**, and
    - **Rising relative sea levels**, showing mounting risks from climate change across the U.S.

CHAPTER 8: REVIEW QUESTIONS AND ANSWER KEYS

1. **What are zoning and land-use regulations, and how do they impact carbon emissions?**
   Zoning and land-use regulations control the use of land in urban areas. By promoting mixed-use development and transit-oriented design, they reduce reliance on personal vehicles and lower greenhouse gas emissions.
2. **How do building codes influence energy efficiency in urban development?**
   Building codes mandate energy efficiency standards for structures. This improves insulation, lighting, heating, and cooling, reducing energy consumption and emissions.
3. **What is the role of waste management regulations in reducing emissions?**
   Such regulations encourage recycling, composting, and responsible landfill practices, thus minimizing methane and other emissions from decomposing waste.
4. **How do local city initiatives contribute to climate change mitigation?**
   Cities can encourage renewable energy use, bike-sharing programs, green public transport, and tree-planting initiatives that reduce emissions and improve sustainability.
5. **How does the UNFCCC support international climate policy?**
   The UNFCCC provides a global platform for countries to negotiate treaties like the Paris Agreement and set national emissions reduction targets.
6. **What are the challenges of enforcing the Paris Agreement on a global scale?**
   Challenges include ensuring that all countries report emissions transparently, aligning national interests with climate goals, and providing sufficient financing to developing countries.
7. **How do non-state actors, like financial institutions, influence climate policy?**
   They channel investments into green sectors, advocate for sustainable policies, and adopt climate risk assessments that drive public and private sector action.
8. **What is the significance of the GHG Protocol in emissions reporting?**
   It standardizes emissions accounting for companies and organizations to help quantify, track, and reduce emissions.
9. **How does the EU's carbon neutrality goal affect global climate action?**
   The EU's aggressive targets encourage innovation and provide a model for other regions to follow.
10. **What is the role of the International Renewable Energy Agency (IRENA) in promoting clean energy?**
    IRENA supports countries through policy advice, capacity building, and facilitating financing for renewable energy deployment worldwide.

CHAPTER 9: REVIEW QUESTIONS AND ANSWER KEYS

1. **What are the most promising technologies for carbon capture and storage?**
   Direct air capture (DAC); carbon capture, utilization, and storage (CCUS); and cryogenic carbon capture show significant promise.

2. **How can AI contribute to emission reductions?**
   AI can contribute to emission reductions by analyzing emissions data to optimize energy use, predict consumption patterns, and enable smart grid management.
3. **What role does green hydrogen play in a low-carbon economy?**
   It serves as a clean fuel alternative, especially in hard-to-decarbonize sectors like steel, shipping, and heavy transport.
4. **How does big data improve emissions tracking and reporting?**
   Big data integrates diverse datasets to enhance accuracy, transparency, and timeliness of emissions reporting.
5. **What are some innovations in energy-efficient building systems?**
   Innovations in energy-efficient building systems are smart thermostats, automated lighting, high-efficiency insulation, and on-site renewable energy solutions.
6. **How does waste valorization contribute to circular economy goals?**
   Waste valorization contributes to circular economy goals by transforming waste into useful products or energy, and reducing landfill emissions and raw material consumption.
7. **What are the pros and cons of direct air capture technology?**
   Pros: Removes CO_2 directly from air; scalable.
   Cons: High energy demand and cost.
8. **How does cryogenic carbon capture work, and where is it applied?**
   It cools exhaust gases to freeze CO_2 for removal, often in industrial processes like cement or steel production.
9. **What role does digital twin technology play in carbon reduction?**
   It simulates assets digitally to optimize energy use and reduce emissions across manufacturing and building sectors.
10. **How does bio-based fuel production differ from traditional fossil fuels?**
    Biofuels use renewable feedstocks (plants, algae, waste) and emit less CO_2 over their lifecycle than fossil fuels.

CHAPTER 10: REVIEW QUESTIONS AND ANSWER KEYS

1. **What are conventional energy resources?**
   They include traditional sources like coal, oil, natural gas, and nuclear power.
2. **Discuss the adverse environmental impacts of fossil fuels.**
   Burning fossil fuels releases greenhouse gases and air pollutants and leads to ocean acidification and ecosystem degradation.
3. **Will nuclear energy help reduce GHG emissions?**
   Yes. It provides stable, low-carbon power, but concerns over safety and waste remain.
4. **What are alternative approaches to conventional energy?**
   Alternative approaches include renewable energy, smart grids, energy efficiency programs, demand-side management, and energy storage.
5. **How does energy efficiency reduce emissions?**
   Energy efficiency reduces emissions by lowering energy consumption and thus reducing the need for fossil-fuel power generation.
6. **How do smart grids offer an efficient delivery network?**
   Smart grids optimize energy flow using real-time data and AI, improving reliability and reducing waste.
7. **What are the benefits of demand-side management (DSM)?**
   DSM smooths energy demand peaks, reduces emissions, and lowers electricity costs.
8. **What are the challenges and opportunities toward sustainability?**
   Challenges: financing, outdated infrastructure, policy gaps.
   Opportunities: job creation, tech innovation, and resilience to climate impacts.

9. **What solutions help attain sustainable energy in the UNECE region?**
   Policy reform, investments in renewables and grid upgrades, energy efficiency standards, and regional cooperation.
10. **Compare the energy performance of different countries.**
    Industrialized nations show declining intensity and rapid renewable adoption; developing countries face capacity and financing constraints.

CHAPTER 11: REVIEW QUESTIONS AND ANSWER KEYS

1. **What constitutes the global carbon cycle?**
   The continuous exchange of carbon between land, ocean, biosphere, and atmosphere constitutes the global carbon cycle.
2. **What is carbon neutrality?**
   It is balancing emissions with removal or offsets so that the net CO_2 entering the atmosphere is zero.
3. **Is investing in decarbonization profitable?**
   Yes. Decarbonization can create jobs and stimulate innovation, especially in renewable energy and sustainable infrastructure.
4. **What are the drivers and barriers of decarbonization in PE sectors?**
   Drivers: ESG goals, cost savings, and innovation.
   Barriers: Capital intensity, technology uncertainty, and policy risk.
5. **Discuss the economics of decarbonization.**
   Initial investments may be high, but long-term returns include energy savings, risk reduction, and new revenue streams.
6. **Explain the Net Zero 2050 scenario.**
   It outlines policy, tech, and finance pathways to achieve net-zero emissions globally by 2050.
7. **Discuss the economic impacts of the low-carbon industry transition.**
   They include new green jobs, displaced workers needing reskilling, and cost savings over the long term.
8. **Discuss job shifts in green industries.**
   Shift from fossil fuel extraction to renewable energy operations, energy efficiency projects, and sustainable agriculture.
9. **What are the growth opportunities in the net-zero transition?**
   They include sectors like solar, wind, EVs, green hydrogen, sustainable aviation fuel, and circular economy industries.
10. **What are the challenges and opportunities in decarbonizing?**
    Challenges: financing, infrastructure, and skill gaps.
    Opportunities: innovation, green investment, better public health, and resilience.

CHAPTER 12: REVIEW QUESTIONS AND ANSWER KEYS

1. **How do global carbon emissions vary by country and sector?**
   Global carbon emissions differ widely between countries due to economic activity, energy use, and industrialization. Developed countries generally have higher emissions from sectors like power generation, transportation, and manufacturing. Developing countries may have lower per capita emissions but face growing emissions from urbanization and energy expansion.
2. **Which countries are the largest emitters of GHGs and why?**
   The largest GHG emitters are China, the United States, and India. These countries have large populations and economies heavily reliant on fossil fuels for electricity, industry, and

transportation. Industrial activity, energy demand, and reliance on coal and oil contribute significantly.

3. **How can emissions be reduced in high-emitting cities worldwide?**
   Emissions in high-emitting cities can be reduced through:
   - Investment in public transportation and electric vehicles,
   - Retrofitting buildings for energy efficiency,
   - Renewable energy adoption,
   - Urban planning that minimizes sprawl, and
   - Policies promoting green spaces and reducing traffic congestion.
4. **What role does the Carbon Disclosure Project (CDP) play in promoting environmental transparency?**
   While not detailed directly in the chapter, CDP (inferred from the text's focus on data-sharing platforms like the GCA Portal) plays a similar role by collecting environmental data from companies and cities. It encourages transparency and accountability by making emissions and climate-related data public, pushing organizations to act on their disclosures.
5. **How can data-sharing efforts support global carbon reduction initiatives?**
   Data sharing allows better tracking of emissions, facilitates cooperation between countries and stakeholders, supports evidence-based policymaking, and ensures transparency. Open access to data promotes accountability and encourages innovation and collaboration across disciplines.
6. **What are some effective strategies for decarbonization at a national level?**
   Effective national decarbonization strategies include:
   - Setting and updating Nationally Determined Contributions (NDCs),
   - Implementing carbon taxes or cap-and-trade systems,
   - Investing in renewable energy and clean technologies,
   - Electrifying transport and industrial sectors,
   - Enhancing energy efficiency, and
   - Encouraging behavior change through incentives.
7. **How does the COP28 conference set goals for future climate actions globally?**
   COP28 emphasized transitioning away from fossil fuels, increasing renewable energy, and integrating health, food, and water into the climate agenda. The first Global Stocktake assessed progress and called for tripling renewable energy and doubling energy efficiency by 2030. It also initiated frameworks for adaptation and climate-health integration.
8. **How do developed and developing countries differ in their approaches to carbon reduction?**
   Developed countries focus more on emissions reductions due to historical responsibilities, while developing countries prioritize equity, development, and adaptation. The "common but differentiated responsibilities" principle guides this divide. Financing, technology transfer, and support are critical to bridging these differing approaches.
9. **What are the implications of carbon neutrality pledges made by various nations?**
   These pledges aim to achieve net-zero emissions by mid-century, aligning with the Paris Agreement goals. While they indicate ambition, the effectiveness depends on implementation. They can drive policy reform, investment in green technology, and shift market dynamics, but without concrete action, such pledges risk becoming symbolic.
10. **How does urbanization affect global carbon emissions, and what solutions exist?**
    Urbanization increases emissions through greater energy use, transportation demand, and construction. Solutions include smart city planning, energy-efficient buildings, sustainable public transport, green infrastructure, and decentralized renewable energy systems to mitigate the carbon footprint of growing urban areas.

CHAPTER 13: REVIEW QUESTIONS AND ANSWER KEYS

1. **What are the benefits of compliance in climate action?**
   Compliance reduces greenhouse gas emissions, cuts operational costs, enhances resilience to shocks, avoids penalties, and improves public health by ensuring cleaner air and water. It also drives innovation and opens new green market opportunities.
2. **Discuss the economic incentives for compliance and climate action.**
   Economic incentives include cost savings from energy efficiency, tax breaks for green practices, emissions trading schemes (e.g. cap-and-trade), and nature performance bonds that reduce debt in exchange for sustainability commitments.
3. **Discuss "nature performance bonds," how they are issued and help reduce the climate crisis.**
   These are debt instruments where repayment terms depend on meeting sustainability goals like reforestation or emissions cuts. They offer a way for developing nations to manage debt while investing in climate action, encouraging long-term sustainability.
4. **Discuss reforming subsidies and metrics from fossil fuels to renewable energies and the benefits.**
   Redirecting the $5.2 trillion in fossil fuel subsidies toward renewables (e.g. solar and wind) would promote green growth, reduce emissions, improve energy security, and boost local jobs. Adopting well-being metrics alongside GDP would also support social equity and resilience.
5. **What are "market-based instruments," and why are they pivotal for reducing GHG emissions?**
   Market-based instruments like emissions trading, carbon pricing, and tax incentives put a price on pollution, encouraging companies to innovate and reduce emissions at lower cost. These tools leverage the market to achieve broad emissions reductions efficiently.
6. **Discuss the role of EHS in monitoring and measurement for regulatory compliance.**
   EHS ensures accurate data collection on emissions and resource use. Monitoring and audits help identify non-compliance and risks, fostering continuous improvement and reducing pollution impacts.
7. **How does real-time environmental monitoring help track emissions?**
   Continuous monitoring with IoT sensors provides up-to-date data on emissions and energy usage, allowing companies to quickly identify deviations and take corrective action to stay compliant and reduce costs.
8. **Discuss the steps involved in environmental audits.**
   Steps include the following:
   - Pre-audit planning
   - Data collection and review
   - Site inspection
   - Identifying gaps/non-conformance
   - Reporting findings
   - Implementing corrective actions
   - Follow-up audits to ensure compliance
9. **Discuss the steps involved in EIAs.**
   Steps include the following:
   - Screening
   - Scoping
   - Baseline data collection
   - Impact assessment
   - Mitigation measures
   - Public consultation

- Reporting
- Decision-making and monitoring

10. **Briefly discuss technological innovations in compliance.**
Key innovations include blockchain for transparent emissions tracking, IoT for real-time data, AI and machine learning for predictive analytics and maintenance, and advanced control systems for process optimization.

CHAPTER 14: REVIEW QUESTIONS AND ANSWER KEYS

1. **What are the lessons learned from Case Study 14.1?**
Implementing NYC's LL 97 climate law drove large buildings to reduce emissions via energy efficiency, equipment upgrades, renewable energy, and IoT-based monitoring, showing the effectiveness of strict regulations in driving measurable emissions cuts.
2. **What are the lessons learned from Case Study 14.2?**
The cement industry can achieve substantial emissions reductions by using alternative fuels, improving process efficiency, and investing in carbon capture and waste heat recovery despite high upfront costs and supply chain hurdles.
3. **What are the lessons learned from Case Study 14.3?**
Chemical companies succeeded in lowering emissions through process optimization, energy recovery, alternative feedstocks, and green certifications. Continued R&D and strong policy support help overcome technical and economic barriers.
4. **What conclusions can be drawn across all sectors?**
Across all sectors, robust data collection, proactive investments in low-carbon solutions, policy consistency, and cross-sector collaboration are key drivers of sustainable decarbonization and long-term competitive advantage.

Index

For Product Safety Concerns and Information please contact our
EU representative GPSR@taylorandfrancis.com Taylor & Francis
Verlag GmbH, Kaufingerstraße 24, 80331 München, Germany